Lie Groups: Quantization (Volume 2)

Lie Groups: Quantization (Volume 2)

Thomas Fleming

STATES
ACADEMIC PRESS
www.statesacademicpress.com

Published by States Academic Press
109 South 5th Street,
Brooklyn, NY 11249, USA
www.statesacademicpress.com

Lie Groups: Quantization (Volume 2)
Thomas Fleming

International Standard Book Number: 978-1-63989-329-4 (Hardback)

Cataloging-in-Publication Data

Lie groups : quantization. Volume 2 / Thomas Fleming.
p. cm.
Includes bibliographical references and index.
ISBN 978-1-63989-329-4
1. Lie groups. 2. Geometric quantization. 3. Symmetric spaces.
4. Topological groups. 5. Mathematics. I. Fleming, Thomas.

QA387 .L54 2022
512.55--dc23

Contents

Preface

This book has been an outcome of determined endeavour from a group of educationists in the field. The primary objective was to involve a broad spectrum of professionals from diverse cultural background involved in the field for developing new researches. The book not only targets students but also scholars pursuing higher research for further enhancement of the theoretical and practical applications of the subject.

A group is a collection of symmetries of any object, and each group is the symmetries of some object. Lie groups are groups whose elements are organized continuously and smoothly, making them differentiable manifolds. This is in contrast to discrete groups, where the elements are separated. A Lie group is a continuous group whose elements are described by several real parameters. As such, they provide a natural model for the concept of continuous symmetry, such as rotational symmetry in three dimensions. The real motivation for introducing Lie groups was to model the continuous symmetries of differential equations. They are extensively used in various parts of contemporary mathematics and physics. Lie groups also play a huge role in modern geometry on many different levels. This book outlines the processes and applications of Lie groups in detail. It covers some existent theories and innovative concepts revolving around this field. With state-of-the-art inputs by acclaimed experts of this field, this book targets students and professionals.

It was an honour to edit such a profound book and also a challenging task to compile and examine all the relevant data for accuracy and originality. I wish to acknowledge the efforts of the contributors for submitting such brilliant and diverse chapters in the field and for endlessly working for the completion of the book. Last, but not the least; I thank my family for being a constant source of support in all my research endeavours.

Thomas Fleming

5

Graded Lie Groups

In this chapter we develop the theory of pseudo-differential operators on graded Lie groups. Our approach relies on using positive Rockland operators, their fractional powers and their associated Sobolev spaces studied in Chapter 4. As we have pointed out in the introduction, the graded Lie groups then become the natural setting for such analysis in the context of general nilpotent Lie groups.

The introduced symbol classes $S^m_{\rho,\delta}$ and the corresponding operator classes

$$\Psi^m_{\rho,\delta} = \mathrm{Op}\, S^m_{\rho,\delta},$$

for (ρ,δ) with $1 \geq \rho \geq \delta \geq 0$ and $\delta \neq 1$, have an operator calculus, in the sense that the set $\bigcup_{m\in\mathbb{R}} \Psi^m_{\rho,\delta}$ forms an algebra of operators, stable under taking the adjoint, and acting on the Sobolev spaces in such a way that the loss of derivatives is controlled by the order of the operator. Moreover, the operators that are elliptic or hypoelliptic within these classes allow for a parametrix construction whose symbol can be obtained from the symbol of the original operator.

During the construction of the pseudo-differential calculus $\cup_{m\in\mathbb{R}}\Psi^m_{\rho,\delta}$ on graded Lie groups in this chapter, there are several difficulties one has to overcome and which do not appear in the case of compact Lie groups as described in Chapter 2. The immediate one is the need to find a natural framework for discussing the symbols to which we will be associating the operators (quantization) and we will do so in Section 5.1. In Section 5.2 we define symbol classes leading to algebras of symbols and operators and discuss their properties. The symbol classes that we introduce are based on a positive Rockland operator on the group and contain all the left-invariant differential operators. As with Sobolev spaces, the symbol classes can be shown to be actually independent of the choice of a positive Rockland operator used in their definition. In Section 5.3 we show that the multipliers of Rockland operators are in the introduced symbol classes. We investigate the behaviour of the kernels of operators corresponding to these symbols in Section 5.4, both at 0 and at infinity and show, in particular, that they are Calderón-Zygmund (in the sense of Coifman and Weiss, see Sections 3.2.3 and A.4). The symbolic calculus is established in Section 5.5. In Section 5.7 we show that the operators satisfy an analogue of the Calderón-Vaillancourt theorem. The construction of parametrices for elliptic and hypoelliptic operators in the calculus is carried out in Section 5.8.

Conventions

Throughout Chapter 5, G is always a graded Lie group, endowed with a family of dilations with integer weights. Its homogeneous dimension is denoted by Q. Also throughout, $\mathcal{R}$ will be a homogeneous positive Rockland operator of homogeneous degree ν. If G is a stratified Lie group, we can choose $\mathcal{R} = -\mathcal{L}$ with $\mathcal{L}$ a sub-Laplacian, or another homogeneous positive Rockland operator. Since it is a left-invariant differential operator, we denote by $\pi(\mathcal{R})$ the operator described in Definition 1.7.4. Both $\mathcal{R}$ and $\pi(\mathcal{R})$ and their properties have been extensively discussed in Chapter 4, especially Section 4.1.

Finally, when we write

$$\sup_{\pi \in \widehat{G}}$$

we always understand it as the essential supremum with respect to the Plancherel measure on $\widehat{G}$.

5.1 Symbols and quantization

The global quantization naturally occurs on any unimodular Lie (or locally compact) group of type 1 thanks to the Plancherel formula, see Subsection 1.8.2 for the Plancherel formula. The quantization was first noticed by Michael Taylor in [Tay86, Section I.3]. The case of locally compact type 1 groups was studied recently in [MR15]. The case of the compact Lie groups was described in Section 2.2.1. Here we describe the particular case of graded nilpotent Lie groups, with an emphasis on the technical meaning of the objects involved. A very brief outline of the constructions of this chapter appeared in [FR14a].

Formally, for a family of operators $\sigma(x, \pi)$ on $\mathcal{H}_\pi$ parametrised by $x \in G$ and $\pi \in \widehat{G}$, we associate the operator $T = \mathrm{Op}(\sigma)$ given by

$$T\phi(x) := \int_{\widehat{G}} \mathrm{Tr}\left(\pi(x)\sigma(x, \pi)\widehat{\phi}(\pi)\right) d\mu(\pi). \tag{5.1}$$

Again formally, the Fourier inversion formula implies that if $\sigma(x, \pi)$ does not depend on x and is the group Fourier transform of some function κ, i.e. if $\sigma(x, \pi) = \widehat{\kappa}(\pi)$, then $\mathrm{Op}(\sigma)$ is the convolution operator with right-convolution kernel κ, i.e.

$\mathrm{Op}(\sigma)\phi = \phi * \kappa$. We would like this to be true not only for (say) integrable functions κ but also for quite a large class of distributions, in order

$$\text{to quantize } X^\alpha = \mathrm{Op}(\sigma) \text{ by } \sigma(x,\pi) = \pi(X)^\alpha,$$

with $\pi(X)$ as in Definition 1.7.4.

The first problem is to make sense of the objects above. The dependence of σ on x is not problematic for the interpretation in the formula (5.1), but we have identified a unitary irreducible representation π with its equivalence class and the families of operators may be measurable in $\pi \in \mathrm{Rep}\, G$ but not defined for all $\pi \in \widehat{G}$. More worryingly, we would like to consider collections of operators which are unbounded, for instance such as $\pi(X)^\alpha$, $\pi \in \widehat{G}$. For these reasons, it may be difficult to give a meaning to the formula (5.1) in general.

Thus, our first task is to define a large class of collections of operators $\sigma(x,\pi)$, $x \in G$, $\pi \in \widehat{G}$, for which we can make sense of the quantization procedure. We will use the realisations

$$\mathcal{K}(G), \quad L^\infty(\widehat{G}), \quad \text{and} \quad \mathscr{L}_L(L^2(G))$$

of the von Neumann algebra of the group G described in Section 1.8.2. We will also use their generalisations

$$\mathcal{K}_{a,b}(G), \; L^\infty_{a,b}(\widehat{G}), \quad \text{and} \quad \mathscr{L}_L(L^2_a(G), L^2_b(G))$$

which we define in Section 5.1.2. In order to do so we use a special feature of our setting, namely the existence of positive Rockland operators and the corresponding L^2-Sobolev spaces.

5.1.1 Fourier transform on Sobolev spaces

In Section 4.3, we have discussed in detail the fractional powers of a positive Rockland operator $\mathcal{R}$ and of the operator $\mathrm{I}+\mathcal{R}$. In the sequel, we will also need to understand powers of the operators $\pi_1(\mathrm{I}+\mathcal{R})$, $\pi_1 \in \mathrm{Rep}\, G$. We now address this, and use it to extend the group Fourier transform to the Sobolev spaces $L^2_a(G)$.

From now on we will keep the same notation for the operators $\mathcal{R}$ and $\pi_1(\mathcal{R})$ (where $\pi_1 \in \mathrm{Rep}\,(G)$) and their respective self-adjoint extensions, see Proposition 4.1.15. We note that by Proposition 4.2.6 the operator $\pi_1(\mathcal{R})$ is also positive. We can consider the powers of $\mathrm{I}+\mathcal{R}$ and $\pi_1(\mathrm{I}+\mathcal{R}) = \mathrm{I}+\pi_1(\mathcal{R})$ as defined by the functional calculus

$$(\mathrm{I}+\mathcal{R})^{\frac{a}{\nu}} = \int_0^\infty (1+\lambda)^{\frac{a}{\nu}} dE(\lambda), \quad \pi_1(\mathrm{I}+\mathcal{R})^{\frac{a}{\nu}} = \int_0^\infty (1+\lambda)^{\frac{a}{\nu}} dE_{\pi_1}(\lambda),$$

where E and E_{π_1} are the spectral measures of $\mathcal{R}$ and $\pi_1(\mathcal{R})$, respectively, and ν is the homogeneous degree of $\mathcal{R}$, see Corollary 4.1.16.

Remark 5.1.1. If a/ν is a positive integer, there is no conflict of notation between

- the powers of $\pi_1(\mathrm{I}+\mathcal{R})$ as the infinitesimal representation of π_1 (see Definition 1.7.4) at $\mathrm{I}+\mathcal{R}\in\mathfrak{U}(\mathfrak{g})$
- and the operator $\pi_1(\mathrm{I}+\mathcal{R})^{\frac{a}{\nu}}$ defined by functional calculus.

Indeed, if $a=\nu$, the two coincide. If $a=\ell\nu$, $\ell\in\mathbb{N}$, then the operator $\pi_1(\mathrm{I}+\mathcal{R})^{\frac{a}{\nu}}$ defined by functional calculus coincides with the ℓ-th power of $\pi_1(\mathrm{I}+\mathcal{R})$. The case $a=0$ is trivial.

We can describe more concretely the operators $\pi_1(\mathrm{I}+\mathcal{R})^{\frac{a}{\nu}}$, $\pi_1\in\operatorname{Rep}G$.

Lemma 5.1.2. *Let $\mathcal{R}$ be a positive Rockland operator of homogeneous degree ν. As in Corollary 4.3.11, we denote by $\mathcal{B}_a$ the right-convolution kernels of its Bessel potentials $(\mathrm{I}+\mathcal{R})^{-\frac{a}{\nu}}$, $\operatorname{Re}a>0$.*

If $a\in\mathbb{C}$ with $\operatorname{Re}a<0$, then $\mathcal{B}_{-a}$ is an integrable function and

$$\forall\pi_1\in\operatorname{Rep}G\qquad \pi_1(\mathrm{I}+\mathcal{R})^{\frac{a}{\nu}}=\widehat{\mathcal{B}}_{-a}(\pi_1).$$

For any $a\in\mathbb{C}$ and any $\pi_1\in\operatorname{Rep}G$, the operator $\pi_1(\mathrm{I}+\mathcal{R})^{\frac{a}{\nu}}$ maps $\mathcal{H}_{\pi_1}^\infty$ onto $\mathcal{H}_{\pi_1}^\infty$ bijectively. Furthermore, the inverse of $\pi_1(\mathrm{I}+\mathcal{R})^{\frac{a}{\nu}}$ is $\pi_1(\mathrm{I}+\mathcal{R})^{-\frac{a}{\nu}}$ as operators acting on $\mathcal{H}_{\pi_1}^\infty$.

Proof. Let $a\in\mathbb{C}$, $\operatorname{Re}a<0$. Then the Bessel potential $(\mathrm{I}+\mathcal{R})^{\frac{a}{\nu}}$ coincides with the bounded operator with right-convolution kernel $\mathcal{B}_{-a}\in L^1(G)$, see Corollary 4.3.11. Therefore, $(\mathrm{I}+\mathcal{R})^{\frac{a}{\nu}}\in\mathscr{L}_L(L^2(G))$ and

$$\mathcal{F}_G\{(\mathrm{I}+\mathcal{R})^{\frac{a}{\nu}}f\}=\mathcal{F}_G\{f*\mathcal{B}_{-a}\}=\widehat{\mathcal{B}}_{-a}\widehat{f},\quad f\in L^2(G).$$

Now we apply Corollary 4.1.16 with the bounded multiplier given by $\phi(\lambda)=(1+\lambda)^{\frac{a}{\nu}}$, $\lambda\geq 0$. By Equality (4.5) in Corollary 4.1.16, we obtain

$$\mathcal{F}_G\{(\mathrm{I}+\mathcal{R})^{\frac{a}{\nu}}f\}=\pi(\mathrm{I}+\mathcal{R})^{\frac{a}{\nu}}\widehat{f},\quad f\in L^2(G).$$

The injectivity of the group Fourier transform on $\mathcal{K}(G)$ yields that $\widehat{\mathcal{B}}_{-a}(\pi)=\pi(\mathrm{I}+\mathcal{R})^{\frac{a}{\nu}}$ for any $\pi\in\widehat{G}$, and the first part of the statement is proved.

Let $a\in\mathbb{C}$. We apply Corollary 4.1.16 with the multiplier given by $\phi(\lambda)=(1+\lambda)^{\frac{a}{\nu}}$, $\lambda\geq 0$. Although this multiplier is unbounded, simple modifications of the proof show that Equality (4.5) in Corollary 4.1.16 still holds for f in the domain of the operator. Recall that the domain of $(\mathrm{I}+\mathcal{R})^{\frac{a}{\nu}}$ contains $\mathcal{S}(G)$ by Corollary 4.3.16 and moreover $(\mathrm{I}+\mathcal{R})^{\frac{a}{\nu}}\mathcal{S}(G)=\mathcal{S}(G)$. Consequently, if $\pi_1\in\operatorname{Rep}G$, we have

$$\pi_1\{(\mathrm{I}+\mathcal{R})^{\frac{a}{\nu}}f\}v=\pi_1(\mathrm{I}+\mathcal{R})^{\frac{a}{\nu}}\pi_1(f)v,\quad f\in\mathcal{S}(G),\ v\in\mathcal{H}_{\pi_1},$$

with $\pi_1(\mathrm{I}+\mathcal{R})^{\frac{a}{\nu}}$ defined spectrally. Recall that $\pi_1(f)v\in\mathcal{H}_{\pi_1}^\infty$ when $f\in\mathcal{S}(G)$ by Proposition 1.7.6 (iv), hence here $\pi_1\{(\mathrm{I}+\mathcal{R})^{\frac{a}{\nu}}f\}v\in\mathcal{H}_\pi^\infty$ as well. By Lemma 1.8.19,

$\pi_1(\mathrm{I}+\mathcal{R})^{\frac{a}{\nu}}$ maps $\mathcal{H}_{\pi_1}^\infty$ to $\mathcal{H}_{\pi_1}^\infty$. The spectral calculus implies that as operators acting on $\mathcal{H}_{\pi_1}^\infty$, we have

$$\pi_1(\mathrm{I}+\mathcal{R})^{\frac{a}{\nu}}\pi_1(\mathrm{I}+\mathcal{R})^{-\frac{a}{\nu}} = \mathrm{I}_{\mathcal{H}_{\pi_1}^\infty} \quad \text{and} \quad \pi_1(\mathrm{I}+\mathcal{R})^{-\frac{a}{\nu}}\pi_1(\mathrm{I}+\mathcal{R})^{\frac{a}{\nu}} = \mathrm{I}_{\mathcal{H}_{\pi_1}^\infty}.$$

Consequently, the inverse of $\pi_1(\mathrm{I}+\mathcal{R})^{\frac{a}{\nu}}$ is $\pi_1(\mathrm{I}+\mathcal{R})^{-\frac{a}{\nu}}$ as operators defined on $\mathcal{H}_{\pi_1}^\infty$ and $\pi_1(\mathrm{I}+\mathcal{R})^{\frac{a}{\nu}}\mathcal{H}_{\pi_1}^\infty = \mathcal{H}_{\pi_1}^\infty$. □

Lemma 5.1.2 and Remark 4.1.17 now imply easily

Corollary 5.1.3. *Let $\mathcal{R}$ be a positive Rockland operator of homogeneous degree ν. For any $a \in \mathbb{C}$, $\{\pi(\mathrm{I}+\mathcal{R})^{\frac{a}{\nu}} : \mathcal{H}_\pi^\infty \to \mathcal{H}_\pi^\infty, \pi \in \widehat{G}\}$ is a measurable $\widehat{G}$-field of operators acting on smooth vectors (in the sense of Definition 1.8.14).*

Lemma 5.1.2 together with the Plancherel formula (see Section 1.8.2) and Corollary 4.3.11 also imply

Corollary 5.1.4. *Let $\mathcal{R}$ be a positive Rockland operator of homogeneous degree ν. For any $a \in \mathbb{R}$, we have*

$$a > Q/2 \qquad \Longrightarrow \qquad \{\pi(\mathrm{I}+\mathcal{R})^{-\frac{a}{\nu}}, \pi \in \widehat{G}\} \in L^2(\widehat{G}),$$

and also, for $a > Q/2$,

$$\|\pi(\mathrm{I}+\mathcal{R})^{-\frac{a}{\nu}}\|_{L^2(\widehat{G})} = \|\widehat{\mathcal{B}}_a(\pi)\|_{L^2(\widehat{G})} = \|\mathcal{B}_a\|_{L^2(G)} < \infty.$$

Note that an analogue of Corollary 5.1.4 for compact Lie groups may be obtained by noticing that (2.15) yields

$$m > n/2 \qquad \Longrightarrow \qquad \sum_{\pi\in\widehat{G}} d_\pi \|\pi(\mathrm{I}-\mathcal{L}_G)^{-\frac{m}{2}}\|_{\mathrm{HS}}^2 = \sum_{\pi\in\widehat{G}} d_\pi^2 \langle\pi\rangle^{-2m} < \infty.$$

The following statement describes an important property of the field $\{\pi(\mathrm{I}+\mathcal{R})^{\frac{a}{\nu}}, \pi \in \widehat{G}\}$, in relation with the right Sobolev spaces (see Section 4.4.8 for right Sobolev spaces):

Proposition 5.1.5. *Let $\mathcal{R}$ be a positive Rockland operator on G of homogeneous degree ν. Let also $a \in \mathbb{R}$.*

If $f \in \tilde{L}_a^2(G)$, then $(\mathrm{I}+\tilde{\mathcal{R}})^{\frac{a}{\nu}} f \in L^2(G)$ and there exists a field of operators $\{\sigma_\pi : \mathcal{H}_\pi^\infty \to \mathcal{H}_\pi,\ \pi \in \widehat{G}\}$ such that

$$\{\sigma_\pi \pi(\mathrm{I}+\mathcal{R})^{\frac{a}{\nu}} : \mathcal{H}_\pi^\infty \to \mathcal{H}_\pi,\ \pi \in \widehat{G}\} \in L^2(\widehat{G}), \tag{5.2}$$

and for almost all $\pi \in \widehat{G}$,

$$\mathcal{F}_G\{(\mathrm{I}+\tilde{\mathcal{R}})^{\frac{a}{\nu}} f\}(\pi) = \sigma_\pi \pi(\mathrm{I}+\mathcal{R})^{\frac{a}{\nu}}. \tag{5.3}$$

Conversely, if $\{\sigma_\pi : \mathcal{H}_\pi^\infty \to \mathcal{H}_\pi,\ \pi \in \widehat{G}\}$ satisfies (5.2) then there exists a unique function $f \in \tilde{L}_a^2(G)$ satisfying (5.3).

In Proposition 5.1.5, $\sigma_\pi\pi(\mathrm{I}+\mathcal{R})^{\frac{a}{\nu}}$ is not obtained as the composition of (possibly) unbounded operators as in Definition A.3.2. Instead, for $\sigma_\pi\pi(\mathrm{I}+\mathcal{R})^{\frac{a}{\nu}}$, it is viewed as the composition of a field of operators defined on smooth vectors with a field of operators acting on smooth vectors, see Section 1.8.3.

In Proposition 5.1.5, we use the right Sobolev spaces associated with the positive Rockland operator $\mathcal{R}$. These spaces are in fact independent of the choice of a positive Rockland operator used in their definition, see Sections 4.4.5 and 4.4.8. Consequently, if (5.2) holds for one positive Rockland operator then (5.2) and (5.3) hold for any positive Rockland operator and the Sobolev norm of $f \in L^2(G)$, using one particular positive Rockland operator $\mathcal{R}$, is equal to the $L^2(\widehat{G})$-norm of (5.2).

Proof of Proposition 5.1.5. If $f \in \tilde{L}^2_a(G)$, then by Theorem 4.4.3 (3) (see also Section 4.4.8), we have that $f_a := (\mathrm{I}+\tilde{\mathcal{R}})^{\frac{a}{\nu}} f$ is in $L^2(G)$ and its Fourier transform is a field of bounded operators (in fact in the Hilbert-Schmidt class). By Lemma 5.1.2, $\pi(\mathrm{I}+\mathcal{R})^{-\frac{a}{\nu}}$ maps $\mathcal{H}_\pi^\infty$ onto itself. Hence we can define

$$\sigma_\pi := \pi(f_a)\pi(\mathrm{I}+\mathcal{R})^{-\frac{a}{\nu}},$$

as an operator defined on $\mathcal{H}_\pi^\infty$. One readily checks that the operators σ_π, $\pi \in \widehat{G}$, satisfy (5.2) and (5.3).

For the converse, if $\{\sigma_\pi : \mathcal{H}_\pi^\infty \to \mathcal{H}_\pi \; : \; \pi \in \widehat{G}\}$ satisfies (5.2) then we define the function

$$L^2(G) \ni f_a := \mathcal{F}_G^{-1}\{\sigma_\pi\pi(\mathrm{I}+\mathcal{R})^{\frac{a}{\nu}}\},$$

which is square integrable by the Plancherel theorem (see Theorem 1.8.11), and the function

$$f := (\mathrm{I}+\tilde{\mathcal{R}})^{-\frac{a}{\nu}} f_a,$$

which will be in $\tilde{L}^2_a(G)$ by Theorem 4.4.3 (3). One readily checks that the function f satisfies the properties described in the statement. □

We now aim at stating and proving a property similar to Proposition 5.1.5 for the left Sobolev spaces. It will use the composition of a field with $\pi(\mathrm{I}+\mathcal{R})^{\frac{a}{\nu}}$ on the left and this is problematic when we consider any general field $\sigma = \{\sigma_\pi : \mathcal{H}_\pi^\infty \to \mathcal{H}_\pi\}$ without utilising the composition of unbounded operators as in Definition A.3.2. To overcome this problem, we introduce the following notion:

Definition 5.1.6. Let $\pi_1 \in \operatorname{Rep} G$ and $a \in \mathbb{R}$. We denote by $\mathcal{H}_{\pi_1}^a$ the Hilbert space obtained by completion of $\mathcal{H}_{\pi_1}^\infty$ for the norm

$$\|\cdot\|_{\mathcal{H}_{\pi_1}^a} : v \longmapsto \|\pi_1(\mathrm{I}+\mathcal{R})^{\frac{a}{\nu}} v\|_{\mathcal{H}_{\pi_1}} := \|v\|_{\mathcal{H}_{\pi_1}^a},$$

where $\mathcal{R}$ is a positive Rockland operator on G of homogeneous degree ν.

We may call them the $\mathcal{H}_{\pi_1}$-Sobolev spaces. Note that in the case of the Schrödinger representation for the Heisenberg group, they coincide with Shubin-Sobolev spaces, see Section 6.4.3. More generally, if we realise an element $\pi \in \widehat{G}$ as a representation π_1 acting on some $L^2(\mathbb{R}^m)$ via the orbit methods, see Section 1.8.1, then we view the corresponding Sobolev spaces as tempered distributions: $\mathcal{H}^a_{\pi_1} \subset \mathcal{S}'(\mathbb{R}^m)$.

The following lemma is a routine exercise.

Lemma 5.1.7. *Let $\pi_1 \in \operatorname{Rep} G$ and $a \in \mathbb{R}$.*

1. *If $a = 0$, then $\mathcal{H}^a_{\pi_1} = \mathcal{H}_{\pi_1}$. If $a > 0$, we realise $\mathcal{H}^a_{\pi_1}$ as a subspace of $\mathcal{H}_{\pi_1}$ and it is the domain of the operator $\pi_1(\mathrm{I} + \mathcal{R})^{\frac{a}{\nu}}$. If $a < 0$, we realise $\mathcal{H}^a_{\pi_1}$ as a Hilbert space containing $\mathcal{H}_{\pi_1}$ and the operator $\pi_1(\mathrm{I} + \mathcal{R})^{\frac{a}{\nu}}$ extends uniquely to a bounded operator $\mathcal{H}^a_{\pi_1} \to \mathcal{H}_{\pi_1}$.*

2. *For any $a \in \mathbb{R}$, realising $\mathcal{H}^a_{\pi_1}$ as in Part 1, this space is independent of the positive Rockland operator $\mathcal{R}$ and two positive Rockland operators yield equivalent norms.*

3. *We have the continuous inclusions*
$$a < b \implies \mathcal{H}^b_{\pi_1} \subset \mathcal{H}^a_{\pi_1}.$$
For any $a, b \in \mathbb{R}$, the operator $\pi_1(\mathrm{I} + \mathcal{R})^{\frac{a}{\nu}}$ maps $\mathcal{H}^b_{\pi_1}$ to $\mathcal{H}^{b-a}_{\pi_1}$ injectively and continuously. In this way, $\mathcal{H}^a_{\pi_1}$ and $\mathcal{H}^{-a}_{\pi_1}$ are in duality via
$$\langle u, v\rangle_{\mathcal{H}^a_{\pi_1} \times \mathcal{H}^{-a}_{\pi_1}} := (\pi_1(\mathrm{I} + \mathcal{R})^{\frac{a}{\nu}} u, \pi_1(\mathrm{I} + \bar{\mathcal{R}})^{-\frac{a}{\nu}} \bar{v})_{\mathcal{H}_{\pi_1}}.$$
This duality extends the $\mathcal{H}_{\pi_1}$ duality in the sense that
$$\forall u \in \mathcal{H}^a_{\pi_1} \cap \mathcal{H}_{\pi_1},\ v \in \mathcal{H}^{-a}_{\pi_1} \cap \mathcal{H}_{\pi_1} \qquad \langle u, v\rangle_{\mathcal{H}^a_{\pi_1} \times \mathcal{H}^{-a}_{\pi_1}} = (u, \bar{v})_{\mathcal{H}_{\pi_1}}.$$

4. *If π_2 is another strongly continuous representation such that $\pi_1 \sim_T \pi_2$, that is, T is a unitary operator satisfying $T\pi_1 = \pi T_2$, then T maps $\mathcal{H}^\infty_{\pi_1}$ to $\mathcal{H}^\infty_{\pi_2}$ bijectively by Lemma 1.8.12 and extends uniquely to an isometric operator $\mathcal{H}^a_{\pi_1} \to \mathcal{H}^a_{\pi_2}$.*

Lemma 5.1.7, especially Part 4, shows that $\widehat{G}$-fields with domain or range on these Sobolev spaces make sense:

Definition 5.1.8. Let $a \in \mathbb{R}$. A $\widehat{G}$-field of operators $\sigma = \{\sigma_\pi : \mathcal{H}^\infty_\pi \to \mathcal{H}_\pi, \pi \in \widehat{G}\}$ defined on smooth vectors is *defined on the Sobolev spaces* $\mathcal{H}^a_\pi$ when for each $\pi_1 \in \operatorname{Rep} G$, the operator σ_{π_1} is bounded on $\mathcal{H}^a_{\pi_1}$ in the sense that
$$\exists C \quad \forall v \in \mathcal{H}^\infty_{\pi_1} \quad \|\sigma_{\pi_1} v\|_{\mathcal{H}_{\pi_1}} \leq C \|v\|_{\mathcal{H}^a_{\pi_1}}.$$

Thus, by density of $\mathcal{H}_{\pi_1}^{\infty}$ in $\mathcal{H}_{\pi_1}^{a}$, σ_{π_1} extends uniquely to a bounded operator defined on $\mathcal{H}_{\pi_1}^{a}$ for which we keep the same notation $\sigma_{\pi_1} : \mathcal{H}_{\pi_1}^{a} \to \mathcal{H}_{\pi_1}$.

Example 5.1.9. For any positive Rockland operator of degree ν, the field $\{\pi(\mathrm{I} + \mathcal{R})^{\frac{a}{\nu}}, \pi \in \widehat{G}\}$, is defined on the Sobolev spaces $\mathcal{H}_{\pi}^{a}$. This is an easy consequence of Lemma 5.1.7, especially Part 3.

We will allow ourselves the shorthand notation

$$\sigma = \{\sigma_{\pi} : \mathcal{H}_{\pi}^{a} \to \mathcal{H}_{\pi}, \pi \in \widehat{G}\},$$

to indicate that the $\widehat{G}$-field of operators is defined on the Sobolev spaces $\mathcal{H}_{\pi}^{a}$.

Instead of Definition 5.1.8, we could also have defined $\widehat{G}$-fields of operators defined on $\mathcal{H}_{\pi}^{a}$-Sobolev spaces in a way similar to Definition 1.8.13 (where $\widehat{G}$-fields of operators defined on smooth vectors were defined). Naturally, these two viewpoints are equivalent since $\mathcal{H}_{\pi_1}^{\infty}$ is dense in $\mathcal{H}_{\pi_1}^{a}$.

However, in order to define $\widehat{G}$-fields of operators with range in the $\mathcal{H}_{\pi}^{a}$-Sobolev spaces, we have to adopt the latter viewpoint in the sense that we modify Definitions 1.8.13 and 1.8.14 (in this way, we make no further assumptions on the fields or on the Sobolev spaces):

Definition 5.1.10. Let $a \in \mathbb{R}$.

- A *$\widehat{G}$-field of operators defined on smooth vectors* with *range in the Sobolev spaces $\mathcal{H}_{\pi}^{a}$* is a family of classes of operators $\{\sigma_{\pi}, \pi \in \widehat{G}\}$ where

  $$\sigma_{\pi} := \{\sigma_{\pi_1} : \mathcal{H}_{\pi_1}^{\infty} \to \mathcal{H}_{\pi_1}^{a}, \pi_1 \in \pi\}$$

  for each $\pi \in \widehat{G}$ viewed as a subset of $\mathrm{Rep}\, G$, satisfying for any two elements σ_{π_1} and σ_{π_2} in σ_{π}:

  $$\pi_1 \sim_T \pi_2 \Longrightarrow \sigma_{\pi_2} T = T \sigma_{\pi_1} \text{ on } \mathcal{H}_{\pi}^{\infty}.$$

  (Here we have kept the same notation for the intertwining operator T and its unique extension between Sobolev spaces $\mathcal{H}_{\pi_1}^{a} \to \mathcal{H}_{\pi_2}^{a}$, see Lemma 5.1.7 Part 4.)

- It is measurable when for one (and then any) choice of realisation $\pi_1 \in \pi$ and any vector $v_{\pi_1} \in \mathcal{H}_{\pi_1}^{a}$, as π runs over $\widehat{G}$, the resulting field $\{\sigma_{\pi} v_{\pi}, \pi \in \widehat{G}\}$ is μ-measurable whenever $\int_{\widehat{G}} \|v_{\pi}\|_{\mathcal{H}_{\pi}^{a}}^{2} d\mu(\pi) < \infty$. (Here we assume that all the $\mathcal{H}_{\pi}^{a}$-norms are realised via a fixed positive Rockland operator.)

Unless otherwise stated, a $\widehat{G}$-field of operators defined on smooth vectors with range in the Sobolev spaces $\mathcal{H}_{\pi}^{a}$ is always assumed measurable. We will allow ourselves the shorthand notation

$$\sigma = \{\sigma_{\pi} : \mathcal{H}_{\pi}^{\infty} \to \mathcal{H}_{\pi}^{a}, \pi \in \widehat{G}\}$$

to indicate that the $\widehat{G}$-field of operators has range in the Sobolev space $\mathcal{H}_\pi^a$.

Naturally, if a $\widehat{G}$-field of operators is defined on smooth vectors $\sigma = \{\sigma_\pi : \mathcal{H}_\pi^\infty \to \mathcal{H}_\pi, \pi \in \widehat{G}\}$ with the usual range $\mathcal{H}_\pi = \mathcal{H}_\pi^0$, then it has *range in the Sobolev spaces* $\mathcal{H}_\pi^a$ when for each $\pi_1 \in \operatorname{Rep} G$ and any $v \in \mathcal{H}_{\pi_1}^\infty$, we have $\sigma_{\pi_1} v \in \mathcal{H}_{\pi_1}^a$.

Moreover, the following property of composition is easy to check: if σ_1 has range in $\mathcal{H}_\pi^a$ and σ_2 is defined on $\mathcal{H}_\pi^a$,

$$\text{i.e.}\quad \sigma_1 = \{\sigma_{1,\pi} : \mathcal{H}_\pi^\infty \to \mathcal{H}_\pi^a, \pi \in \widehat{G}\} \quad \text{and} \quad \sigma_2 = \{\sigma_{2,\pi} : \mathcal{H}_\pi^a \to \mathcal{H}_\pi, \pi \in \widehat{G}\},$$

then the following field

$$\sigma_2\sigma_1 := \{\sigma_{2,\pi}\sigma_{1,\pi} : \mathcal{H}_\pi^\infty \to \mathcal{H}_\pi, \pi \in \widehat{G}\}$$

makes sense as a $\widehat{G}$-field of operators defined on smooth vectors. This coincides or extends the definition of composition of fields (the first one acting on smooth vectors) given in Section 1.8.3.

We can apply this property of composition to $\sigma = \{\sigma_\pi : \mathcal{H}_\pi^\infty \to \mathcal{H}_\pi^a, \pi \in \widehat{G}\}$ and $\{\pi(\mathrm{I}+\mathcal{R})^{\frac{a}{\nu}}, \pi \in \widehat{G}\}$, see Example 5.1.9 for the latter, to obtain the $\widehat{G}$-field defined on smooth vectors by

$$\pi(\mathrm{I}+\mathcal{R})^{\frac{a}{\nu}}\sigma = \{\pi(\mathrm{I}+\mathcal{R})^{\frac{a}{\nu}}\sigma_\pi : \mathcal{H}_\pi^\infty \to \mathcal{H}_\pi, \pi \in \widehat{G}\}. \tag{5.4}$$

We can now state the proposition which will enable us to define the group Fourier transform of a function in a left or right Sobolev space.

Proposition 5.1.11. *Let* $a \in \mathbb{R}$.

(L) *If* $f \in L_a^2(G)$, *then* $(\mathrm{I}+\mathcal{R})^{\frac{a}{\nu}} f \in L^2(G)$ *and there exists a field of operators* $\{\sigma_\pi : \mathcal{H}_\pi^\infty \to \mathcal{H}_\pi^a\,,\ \pi \in \widehat{G}\}$ *such that*

$$\{\pi(\mathrm{I}+\mathcal{R})^{\frac{a}{\nu}}\sigma_\pi : \mathcal{H}_\pi^\infty \to \mathcal{H}_\pi\,,\ \pi \in \widehat{G}\} \in L^2(\widehat{G}), \tag{5.5}$$

$$\mathcal{F}_G\{(\mathrm{I}+\mathcal{R})^{\frac{a}{\nu}} f\}(\pi) = \pi(\mathrm{I}+\mathcal{R})^{\frac{a}{\nu}}\sigma_\pi, \quad \text{for almost all } \pi \in \widehat{G}, \tag{5.6}$$

where $\mathcal{R}$ *is a positive Rockland operator on* G *of homogeneous degree* ν.

Conversely, if $\{\sigma_\pi : \mathcal{H}_\pi^\infty \to \mathcal{H}_\pi^a\,,\ \pi \in \widehat{G}\}$ *satisfies (5.5) for one positive Rockland operator* $\mathcal{R}$, *then there exists a unique function* $f \in L_a^2(G)$ *satisfying (5.6).*

(R) *If* $f \in \tilde{L}_a^2(G)$, *then the (unique) field* σ *obtained in Proposition 5.1.5 can be extended uniquely into a field* $\{\sigma_\pi : \mathcal{H}_\pi^a \to \mathcal{H}_\pi\,,\ \pi \in \widehat{G}\}$ *defined on* $\mathcal{H}_\pi^a$.

Properties (L) *and* (R) *are independent of the choice of* $\mathcal{R}$.

In Proposition 5.1.11, $\pi(\mathrm{I}+\mathcal{R})^{\frac{a}{\nu}}\sigma_\pi$ is not obtained as the composition of (possibly) unbounded operators as in Definition A.3.2 but is understood via (5.4).

In Proposition 5.1.11, we use the left and right Sobolev spaces associated with the positive Rockland operator $\mathcal{R}$. These spaces are in fact independent of the choice of a positive Rockland operator used in their definition, see Sections 4.4.5 and 4.4.8. Consequently, if (5.5) hold for one positive Rockland operator then (5.5) and (5.6) hold for any positive Rockland operator and the Sobolev norm of $f \in L^2(G)$, using one particular positive Rockland operator $\mathcal{R}$, is equal to the $L^2(\widehat{G})$-norm of (5.5).

Proof of Proposition 5.1.11. Property (L). If $f \in L^2_a(G)$, then by Theorem 4.4.3 (3), we have that $f_a := (\mathrm{I}+\mathcal{R})^{\frac{a}{\nu}} f$ is in $L^2(G)$ and its Fourier transform is a field of bounded operators (in fact in the Hilbert-Schmidt class). By (5.4) we can define $\sigma = \{\sigma_\pi : \mathcal{H}_\pi^\infty \to \mathcal{H}_\pi^a\}$ via $\sigma_\pi := \pi(\mathrm{I}+\mathcal{R})^{-\frac{a}{\nu}}\pi(f_a)$. One readily checks that the field σ satisfies (5.2) and (5.3).

For the converse, if $\{\sigma_\pi : \mathcal{H}_\pi^\infty \to \mathcal{H}_\pi^a \, : \pi \in \widehat{G}\}$ satisfies (5.2) then we define the function

$$L^2(G) \ni f_a := \mathcal{F}_G^{-1}\{\pi(\mathrm{I}+\mathcal{R})^{\frac{a}{\nu}}\sigma_\pi\},$$

which is square integrable by the Plancherel theorem (see Theorem 1.8.11), and the function

$$f := (\mathrm{I}+\mathcal{R})^{-\frac{a}{\nu}} f_a,$$

which will be in $L^2_a(G)$ by Theorem 4.4.3 (3). One readily checks that the function f satisfies the properties described in the statement. This shows the property (L).

Property (R) follows easily from (5.2). □

From the proof above, one can check easily that if $f \in L^2_a(G)$ or $\tilde{L}^2_a(G)$ is also in any of the spaces where the group Fourier transform has already been defined, namely, $L^2(G)$ or $\mathcal{K}(G)$, then $\sigma = \{\sigma_\pi : \mathcal{H}_\pi^\infty \to \mathcal{H}_\pi, \pi \in \widehat{G}\}$ will coincide with the group Fourier transform of f. Hence we can extend the definition of the group Fourier transform to Sobolev spaces:

Definition 5.1.12. Let $a \in \mathbb{R}$. The group Fourier transform of $f \in L^2_a(G)$ or $f \in \tilde{L}^2_a(G)$ is the field σ of operators defined on smooth vectors given in Proposition 5.1.11.

This leads us to define the following spaces of fields of operators:

Definition 5.1.13. (L) Let $L^2_a(\widehat{G})$ denote the space of fields of operators σ with range in $\mathcal{H}_\pi^a$ and satisfying (5.5), that is,

$$\sigma = \{\sigma_\pi : \mathcal{H}_\pi^\infty \to \mathcal{H}_\pi^a \, , \, \pi \in \widehat{G}\},$$
$$\{\pi(\mathrm{I}+\mathcal{R})^{\frac{a}{\nu}}\sigma_\pi : \mathcal{H}_\pi^\infty \to \mathcal{H}_\pi \, , \, \pi \in \widehat{G}\} \in L^2(\widehat{G}),$$

for one (and then any) positive Rockland operator of homogeneous degree ν. We also set

$$\|\sigma\|_{L^2_a(\widehat{G})} := \|\pi(\mathrm{I}+\mathcal{R})^{\frac{a}{\nu}}\sigma_\pi\|_{L^2(\widehat{G})}. \tag{5.7}$$

(R) Let $\tilde{L}^2_a(\widehat{G})$ denote the space of fields of operators σ defined on $\mathcal{H}^a_\pi$ and satisfying (5.2), that is,

$$\sigma = \{\sigma_\pi : \mathcal{H}^a_\pi \to \mathcal{H}_\pi\,,\, \pi \in \widehat{G}\},$$
$$\{\sigma_\pi \pi(\mathrm{I}+\mathcal{R})^{\frac{a}{\nu}} : \mathcal{H}^\infty_\pi \to \mathcal{H}_\pi\,,\, \pi \in \widehat{G}\} \in L^2(\widehat{G}),$$

for one (and then any) positive Rockland operator of homogeneous degree ν. We also set

$$\|\sigma\|_{\tilde{L}^2_a(\widehat{G})} := \|\sigma_\pi \pi(\mathrm{I}+\mathcal{R})^{\frac{a}{\nu}}\|_{L^2(\widehat{G})}.$$

It is a routine exercise, using Proposition 5.1.11 and the properties of the Sobolev spaces (see Section 4.4), to show that

Proposition 5.1.14. *Let $a \in \mathbb{R}$. If $\mathcal{R}$ is a positive Rockland operator of homogeneous degree ν, the map $\|\cdot\|_{L^2_a(\widehat{G})}$ given by (5.7) is a norm on the vector space $L^2_a(\widehat{G})$. Endowed with this norm, $L^2_a(\widehat{G})$ is a Banach space which is independent of $\mathcal{R}$. Two norms corresponding to any two choices of Rockland operators via (5.7) are equivalent.*

The Fourier transform $\mathcal{F}_G$ is an isomorphism between Banach spaces acting from $L^2_a(G)$ onto $L^2_a(\widehat{G})$. It coincides with the usual Fourier transform on $L^2(G)$ for $a=0$.

Let $\sigma = \{\sigma_\pi, \pi \in \widehat{G}\}$ be in $L^2_a(\widehat{G})$. Then

$$\{\pi(X)^\alpha \sigma_\pi, \pi \in \widehat{G}\}$$

is in $L^2_{a-[\alpha]}(\widehat{G})$ for any $\alpha \in \mathbb{N}_0^n$, and

$$\{\pi(\mathrm{I}+\mathcal{R})^{s/\nu} \sigma_\pi, \pi \in \widehat{G}\}$$

is in $L^2_{a-s}(\widehat{G})$ for any $s \in \mathbb{R}$. Furthermore, if $f = \mathcal{F}_G^{-1}\sigma \in L^2_a(G)$ then

$$\mathcal{F}_G(X^\alpha f)(\pi) = \pi(X)^\alpha \widehat{f}(\pi) \quad \text{and} \quad \mathcal{F}_G((\mathrm{I}+\mathcal{R})^{s/\nu} f)(\pi) = \pi(\mathrm{I}+\mathcal{R})^{s/\nu} \widehat{f}(\pi).$$

We have similar results for the right Sobolev spaces. Furthermore the adjoint map $\sigma \mapsto \sigma^*$ maps $L^2_a(\widehat{G}) \to \tilde{L}^2_a(\widehat{G})$ and $\tilde{L}^2_a(\widehat{G}) \to L^2_a(\widehat{G})$ isomorphically as Banach spaces.

Recall that the tempered distributions $X^\alpha f$ and $(\mathrm{I}+\mathcal{R})^{s/\nu} f$ used in the statement just above are respectively defined via

$$\langle X^\alpha f, \phi\rangle = \langle f, \{X^\alpha\}^t \phi\rangle, \quad \phi \in \mathcal{S}(G), \tag{5.8}$$

and

$$\langle (\mathrm{I}+\mathcal{R})^{s/\nu} f, \phi\rangle = \langle f, (\mathrm{I}+\bar{\mathcal{R}})^{s/\nu} \phi\rangle, \quad \phi \in \mathcal{S}(G). \tag{5.9}$$

For (5.9), see Definition 4.3.17. For (5.8), this is the composition of the formula obtained for one vector field (with polynomial coefficients) by integration by parts. See also (1.10) for the definition of $\{X^\alpha\}^t$.

In Corollary 1.8.3, we stated the inversion formula valid for any Schwartz function on any connected simply connected Lie group. Here we weaken the hypothesis using the Sobolev spaces in the context of a graded Lie group G:

Proposition 5.1.15 (Fourier inversion formula). *Let f be in the left Sobolev space $L^2_s(G)$ or in the right Sobolev space $\tilde{L}^2_s(G)$ with $s > Q/2$. Then for almost every $\pi \in \operatorname{Rep} G$, the operator $\widehat{f}(\pi)$ is trace class with*

$$\int_{\widehat{G}} \operatorname{Tr}|\widehat{f}(\pi)| d\mu(\pi) < \infty. \tag{5.10}$$

Furthermore, f is continuous on G, and for every $x \in G$ we have

$$f(x) = \int_{\widehat{G}} \operatorname{Tr}\left(\pi(x)\widehat{f}(\pi)\right) d\mu(\pi) = \int_{\widehat{G}} \operatorname{Tr}\left(\widehat{f}(\pi)\pi(x)\right) d\mu(\pi). \tag{5.11}$$

In the statement above, as $s > Q/2 > 0$, the field $\widehat{f}$ is in $L^2(\widehat{G})$, it is then a field of bounded operators (even in Hilbert-Schmidt classes) and so can be composed on the left and the right with $\pi(x)$. The (possibly infinite) traces

$$\operatorname{Tr}\left|\pi_1(x)\widehat{f}(\pi_1)\right|, \quad \operatorname{Tr}\left|\widehat{f}(\pi_1)\pi_1(x)\right| \quad \text{and} \quad \operatorname{Tr}\left|\widehat{f}(\pi_1)\right|$$

are equal for $\pi_1 \in \operatorname{Rep} G$ as π_1 is unitary. They are constant on the class of $\pi_1 \in \operatorname{Rep} G$ in $\widehat{G}$ and are, therefore, treated as depending on $\pi \in \widehat{G}$. They are finite for μ-almost all $\pi \in \widehat{G}$ in view of (5.10).

Note that (5.10) implies not only that the two expressions

$$\int_{\widehat{G}} \operatorname{Tr}\left(\pi(x)\widehat{f}(\pi)\right) d\mu(\pi) \quad \text{and} \quad \int_{\widehat{G}} \operatorname{Tr}\left(\widehat{f}(\pi)\pi(x)\right) d\mu(\pi)$$

make sense but that they are also equal by the properties of the trace since $\pi(x)$ is bounded.

Proof of Proposition 5.1.15. Let $\mathcal{R}$ be a positive Rockland operator of homogeneous degree ν. Let $f \in L^2_s(G)$ with $s > Q/2$. We set

$$f_s := (\mathrm{I} + \mathcal{R})^{\frac{s}{\nu}} f \in L^2(G).$$

The properties of the trace imply

$$\operatorname{Tr}|\widehat{f}(\pi)| = \operatorname{Tr}\left|\pi(\mathrm{I} + \mathcal{R})^{-\frac{s}{\nu}} \widehat{f_s}(\pi)\right| \le \|\pi(\mathrm{I} + \mathcal{R})^{-\frac{s}{\nu}}\|_{\mathtt{HS}} \|\widehat{f_s}(\pi)\|_{\mathtt{HS}}.$$

Integrating against the Plancherel measure, we obtain by the Cauchy-Schwartz inequality

$$\int_{\widehat{G}} \mathrm{Tr}|\widehat{f}(\pi)| d\mu(\pi) \le \|\pi(\mathrm{I}+\mathcal{R})^{-\frac{s}{\nu}}\|_{L^2(\widehat{G})} \|\widehat{f}_s\|_{L^2(\widehat{G})}.$$

By Corollary 5.1.4, $C_s := \|\pi(\mathrm{I}+\mathcal{R})^{-\frac{s}{\nu}}\|_{L^2(\widehat{G})}$ is a positive finite constant. Since $\|\widehat{f}_s(\pi)\|_{L^2(\widehat{G})}$ is equal to $\|f\|_{L^2_s(G)}$ which is finite, we have obtained (5.10).

Let $\phi \in \mathcal{S}(G)$. By the Plancherel formula, especially (1.30), we have

$$\begin{aligned}
(f,\phi)_{L^2(G)} &= (f_s, (\mathrm{I}+\mathcal{R})^{-\frac{s}{\nu}}\phi)_{L^2(G)} \\
&= \int_{\widehat{G}} \mathrm{Tr}\left(\mathcal{F}_G\{f_s\}(\pi)\ \left(\mathcal{F}_G\{(\mathrm{I}+\mathcal{R})^{-\frac{s}{\nu}}\phi\}(\pi)\right)^*\right) d\mu(\pi) \\
&= \int_{\widehat{G}} \mathrm{Tr}\left(\pi(\mathrm{I}+\mathcal{R})^{\frac{s}{\nu}}\widehat{f}(\pi)\ \widehat{\phi}(\pi)^*\pi(\mathrm{I}+\mathcal{R})^{-\frac{s}{\nu}}\right) d\mu(\pi) \\
&= \int_{\widehat{G}} \mathrm{Tr}\left(\widehat{f}(\pi)\ \widehat{\phi}(\pi)^*\right) d\mu(\pi).
\end{aligned}$$

Note that the two functions f_s and $(\mathrm{I}+\mathcal{R})^{\frac{s}{\nu}}\phi$ are both square integrable so all the traces above are finite.

We now fix a non-negative function $\chi \in \mathcal{D}(G)$ with compact support containing 0 and satisfying $\int_G \chi = 1$. We apply what precedes to $\phi := \chi_\epsilon$ given by

$$\chi_\epsilon(y) := \epsilon^{-Q}\chi(\epsilon^{-1}y), \quad \epsilon > 0,\ y \in G,$$

and obtain

$$(f,\chi_\epsilon)_{L^2(G)} = \int_{\widehat{G}} \mathrm{Tr}\left(\widehat{f}(\pi)\ \widehat{\chi}_\epsilon(\pi)^*\right) d\mu(\pi). \tag{5.12}$$

Let us show that the right hand-side of (5.12) converges to

$$\int_{\widehat{G}} \mathrm{Tr}\left(\widehat{f}(\pi)\ \widehat{\chi}_\epsilon(\pi)^*\right) d\mu(\pi) \longrightarrow_{\epsilon\to 0} \int_{\widehat{G}} \mathrm{Tr}\left(\widehat{f}(\pi)\right) d\mu(\pi). \tag{5.13}$$

Note that the right hand-side of (5.13) is finite by (5.10).

The integrand on the left-hand side is bounded by

$$\left|\mathrm{Tr}\left(\widehat{f}(\pi)\ \widehat{\chi}_\epsilon(\pi)^*\right)\right| \le \|\widehat{\chi}_\epsilon(\pi)\|_{\mathscr{L}(\mathcal{H}_\pi)} \mathrm{Tr}|\widehat{f}(\pi)|,$$

and

$$\|\widehat{\chi}_\epsilon(\pi)\|_{\mathscr{L}(\mathcal{H}_\pi)} \le \|\chi_\epsilon\|_{L^1(G)} = \|\chi\|_{L^1(G)}.$$

Hence

$$\left|\mathrm{Tr}\left(\widehat{f}(\pi)\ \widehat{\chi}_\epsilon(\pi)^*\right)\right| \le \|\chi\|_{L^1(G)} \mathrm{Tr}|\widehat{f}(\pi)|,$$

and the right-hand side is μ-integrable by (5.10).

Let us show the convergence for every $\pi \in \widehat{G}$

$$\mathrm{Tr}\left(\widehat{f}(\pi)\ \widehat{\chi}_\epsilon(\pi)^*\right) \longrightarrow_{\epsilon\to 0} \mathrm{Tr}\left(\widehat{f}(\pi)\right). \tag{5.14}$$

In order to do this, we want to estimate the difference

$$\begin{aligned}\left|\mathrm{Tr}\left(\widehat{f}(\pi)\ \widehat{\chi}_\epsilon(\pi)^*\right) - \mathrm{Tr}\left(\widehat{f}(\pi)\right)\right| &= \left|\mathrm{Tr}\left(\widehat{f}(\pi)\ (\widehat{\chi}_\epsilon(\pi)^* - \mathrm{I})\right)\right| \\ &\leq \|\widehat{\chi}_\epsilon(\pi)^* - \mathrm{I}\|_{L^\infty(\widehat{G})} \mathrm{Tr}\left|\widehat{f}(\pi)\right|.\end{aligned}$$

Since

$$\widehat{\chi}_\epsilon(\pi)^* = \int_G \chi_\epsilon(y)\pi(y)dy = \int_G \epsilon^{-Q}\chi(\epsilon^{-1}y)\pi(y)dy = \int_G \chi(z)\pi(\epsilon z)dz,$$

and as $\int_G \chi = 1$, we have

$$\begin{aligned}\|\widehat{\chi}_\epsilon(\pi)^* - \mathrm{I}\|_{\mathscr{L}(\mathcal{H}_\pi)} &= \|\int_G \chi(z)\left(\pi(\epsilon z) - \mathrm{I}\right)dz\|_{\mathscr{L}(\mathcal{H}_\pi)} \\ &\leq \int_G |\chi(z)|\ \|\pi(\epsilon z) - \mathrm{I}\|_{\mathscr{L}(\mathcal{H}_\pi)}dz \\ &\leq \sup_{z\in\mathrm{supp}\chi} \|\pi(\epsilon z) - \mathrm{I}\|_{\mathscr{L}(\mathcal{H}_\pi)} \int_G |\chi(z)|dz.\end{aligned}$$

As π is strongly continuous and suppχ compact, we know that

$$\sup_{z\in\mathrm{supp}\chi} \|\pi(\epsilon z) - \mathrm{I}\|_{\mathscr{L}(\mathcal{H}_\pi)} \longrightarrow_{\epsilon\to 0} 0.$$

This implies the convergence in (5.14) for each $\pi \in \widehat{G}$.

We can now apply Lebesgue's dominated convergence theorem to obtain the convergence in (5.13).

By the Sobolev embeddings (see Theorem 4.4.25), f is continuous on G and it is a simple exercise to show that the left hand-side of (5.12) converges to

$$(f, \chi_\epsilon)_{L^2(G)} \longrightarrow_{\epsilon\to 0} f(0).$$

Hence we have obtained the inversion formula given in (5.11) at $x = 0$. Replacing f by its left translation $f(x\,\cdot)$ which is still in $L^2_s(G)$ with the same Sobolev norm, it is then easy to obtain (5.11) for every $x \in G$.

For the case of $f \in \tilde{L}^2_s(G)$ with $s > Q/2$, we set $f_s := (\mathrm{I} + \tilde{\mathcal{R}})^{\frac{s}{\nu}} f \in L^2(G)$ and we obtain similar properties as above, ending by using right translations to obtain (5.11). □

5.1.2 The spaces $\mathcal{K}_{a,b}(G)$, $\mathscr{L}_L(L^2_a(G), L^2_b(G))$, and $L^\infty_{a,b}(\widehat{G})$

In this section we describe the spaces $\mathcal{K}_{a,b}(G)$, $\mathscr{L}_L(L^2_a(G), L^2_b(G))$ and $L^\infty_{a,b}(\widehat{G})$, extending the notion of the group von Neumann algebras discussed in Section 1.8.2, to the setting of Sobolev spaces.

Definition 5.1.16 (Spaces $\mathscr{L}_L(L^2_a(G), L^2_b(G))$ and $\mathcal{K}_{a,b}(G)$)**.** Let $a, b \in \mathbb{R}$. We denote by

$$\mathscr{L}_L(L^2_a(G), L^2_b(G))$$

the subspace of operators $T \in \mathscr{L}(L^2_a(G), L^2_b(G))$ which are left-invariant.

We denote by

$$\mathcal{K}_{a,b}(G)$$

the subspace of tempered distributions $f \in \mathcal{S}'(G)$ such that the operator $\mathcal{S}(G) \ni \phi \mapsto \phi * f$ extends to a bounded operator from $L^2_a(G)$ to $L^2_b(G)$.

If a positive Rockland operator $\mathcal{R}$ of homogeneous degree ν is fixed, then the $\mathcal{K}_{a,b}(G)$-norm is defined for any $f \in \mathcal{K}_{a,b}(G)$, as the operator norm of $\phi \mapsto \phi * f$ viewed as an operator from $L^2_a(G)$ to $L^2_b(G)$, i.e.

$$\|f\|_{\mathcal{K}_{a,b}} := \|\phi \mapsto \phi * f\|_{\mathscr{L}(L^2_a(G), L^2_b(G))}. \tag{5.15}$$

Here we have considered the Sobolev norms $\phi \mapsto \|(\mathrm{I} + \mathcal{R})^{\frac{c}{\nu}}\phi\|_2$ for $c = a, b$ for $L^2_a(G)$ and $L^2_b(G)$, respectively.

The vector space $\mathscr{L}_L(L^2_a, L^2_b)$ is a Banach subspace of $\mathscr{L}(L^2_a, L^2_b)$. Since the Sobolev spaces $L^2_a(G)$ are independent of the choice of a positive Rockland operator $\mathcal{R}$ (see Section 4.4.5), so are $\mathscr{L}_L(L^2_a(G), L^2_b(G))$ and also $\mathcal{K}_{a,b}(G)$. However, the norms on these spaces do depend on a choice of a positive Rockland operator $\mathcal{R}$.

We may often write $\mathcal{K}_{a,b}$ instead of $\mathcal{K}_{a,b}(G)$ to ease the notation when no confusion is possible.

We have the immediate properties:

Proposition 5.1.17. *1. If $a = b = 0$ then*

$$\mathcal{K}_{0,0} = \mathcal{K} \quad \textit{and} \quad \mathscr{L}_L(L^2_a, L^2_b) = \mathscr{L}_L(L^2).$$

The norms $\|\cdot\|_{\mathcal{K}_{0,0}}$ and $\|\cdot\|_{\mathcal{K}}$ (defined in (5.15) and in (1.37) respectively) coincide. For any $f \in \mathcal{K}$ we have

$$\|f^*\|_{\mathcal{K}} = \|f\|_{\mathcal{K}} \quad \textit{where} \quad f^*(x) = \bar{f}(x^{-1}),$$

and

$$\forall r > 0 \qquad \|f \circ D_r\|_{\mathcal{K}} = r^{-Q}\|f\|_{\mathcal{K}}.$$

2. *Fixing a positive Rockland operator $\mathcal{R}$, the mapping $f \mapsto \|f\|_{\mathcal{K}_{a,b}}$ defines a norm on the vector space $\mathcal{K}_{a,b}$ which becomes a Banach space. Any two positive Rockland operators produce equivalent norms on $\mathcal{K}_{a,b}$.*

3. *Let $a, b \in \mathbb{R}$. We have the continuous inclusion*

$$\mathcal{K}_{a,b}(G) \subset \mathcal{S}'(G).$$

Moreover if T_f denotes the convolution operator $\phi \mapsto \phi * f$ for $f \in \mathcal{S}'(G)$, then the following are equivalent:

$$\begin{aligned} f \in \mathcal{K}_{a,b} \quad &\Longleftrightarrow \quad T_f \in \mathscr{L}_L(L^2_a(G), L^2_b(G)) \\ &\Longleftrightarrow \quad (\mathrm{I}+\mathcal{R})^{\frac{b}{\nu}} T_f (\mathrm{I}+\mathcal{R})^{-\frac{a}{\nu}} \in \mathscr{L}_L(L^2(G)) \\ &\Longleftrightarrow \quad (\mathrm{I}+\mathcal{R})^{\frac{b}{\nu}} (\mathrm{I}+\tilde{\mathcal{R}})^{-\frac{a}{\nu}} f \in \mathcal{K}(G), \end{aligned}$$

where $\mathcal{R}$ is any positive Rockland operator of homogeneous degree ν.

4. *For any $c_1, c_2 \geq 0$ we have the inclusions*

$$\mathscr{L}_L(L^2_a, L^2_b) \subset \mathscr{L}_L(L^2_{a+c_1}, L^2_{b-c_2})$$

and

$$\mathcal{K}_{a,b} \subset \mathcal{K}_{a+c_1, b-c_2}.$$

5. *If $f \in \mathcal{K}_{a,b}$ then $X^\alpha f \in \mathcal{K}_{a,b-[\alpha]}$ for any $\alpha \in \mathbb{N}_0^n$ and $(\mathrm{I}+\mathcal{R})^{s/\nu} f \in \mathcal{K}_{a,b-s}$ for any $s \in \mathbb{R}$. Furthermore, X^α and $(\mathrm{I}+\mathcal{R})^{s/\nu}$ are bounded on $\mathcal{K}_{a,b}$:*

$$\|X^\alpha f\|_{\mathcal{K}_{a,b-[\alpha]}} \leq C_{a,b,[\alpha]} \|f\|_{\mathcal{K}_{a,b}}$$

and

$$\|(\mathrm{I}+\mathcal{R})^{s/\nu} f\|_{\mathcal{K}_{a,b-s}} \leq C'_{a,b,s} \|f\|_{\mathcal{K}_{a,b}}$$

for some positive finite constants $C_{a,b,[\alpha]}$ and $C'_{a,b,s}$ independent of f.

If $-a$ and b are in $\nu\mathbb{N}_0$, a norm equivalent to the $\mathcal{K}_{a,b}$-norm is

$$f \longmapsto \sum_{[\alpha] \leq -a,\ [\beta] \leq b} \|\tilde{X}^\alpha X^\beta f\|_{\mathcal{K}},$$

and if $a' \in [a, 0]$ and $b' \in [0, b]$ then

$$\|f\|_{\mathcal{K}_{a',b'}} \leq C_{a,b,a',b,\mathcal{R}} \sum_{[\alpha] \leq -a,\ [\beta] \leq b} \|\tilde{X}^\alpha X^\beta f\|_{\mathcal{K}}.$$

The definitions of the tempered distributions $X^\alpha f$ and $(\mathrm{I}+\mathcal{R})^{s/\nu} f$ were recalled in (5.8) and (5.9) respectively. For the proper definition of the operators $(\mathrm{I}+\mathcal{R})^{\frac{b}{\nu}}$, $(\mathrm{I}+\tilde{\mathcal{R}})^{-\frac{a}{\nu}}$, see Definitions 4.3.17 and 4.4.31.

Proof of Proposition 5.1.17. Part (1) follows from the properties of the von Neumann-algebras $\mathcal{K}(G)$ and $\mathscr{L}_L(L^2(G))$ as well as from the following two easy observations:

$$\forall\psi \in L^2(G) \qquad \|\psi \circ D_r\|_2 = r^{-\frac{Q}{2}} \|\psi\|_2,$$

and for any $f \in \mathcal{K}$, $\phi \in \mathcal{S}(G)$ and $r > 0$,

$$\phi * (f \circ D_r)(x) = r^{-Q} \left(\left(\phi \circ D_{\frac{1}{r}} \right) * f \right)(rx).$$

Part (2) is easy to check. Part (3) follows from the Schwartz kernel theorem, see Corollary 3.2.1. Parts (4) and (5), follow easily from the properties of the Sobolev spaces and Part (3). □

We now show that we can make sense of convolution of distributions in some $\mathcal{K}_{a,b}(G)$-spaces. The following lemma is almost immediate to check.

Lemma 5.1.18. *Let $f \in \mathcal{K}_{a,b}(G)$ and $g \in \mathcal{K}_{b,c}(G)$ for $a, b, c \in \mathbb{R}$, and let $T_f : \phi \mapsto \phi * f$ and $T_g : \phi \mapsto \phi * g$ be the associated operators. Then the operator $T_g T_f$ is continuous from $L^2_a(G)$ to $L^2_c(G)$ and its right-convolution kernel (as a continuous linear operator from $\mathcal{S}(G)$ to $\mathcal{S}'(G)$) is denoted by $h \in \mathcal{K}_{a,c}(G)$.*

If (f_n) and (g_n) are sequences of Schwartz functions converging to f in $\mathcal{K}_{a,b}(G)$ and g in $\mathcal{K}_{b,c}(G)$, respectively, then h is the limit of $f_n * g_n$ in $\mathcal{K}_{a,c}(G)$.

Consequently, with the notation of the lemma above, h coincides with the convolution of f with g whenever the convolution of f with g makes any technical sense, for instance, if the tempered distributions f and g (which are already assumed to be in $\mathcal{K}_{a,b}(G)$ and $\mathcal{K}_{b,c}(G)$ respectively) satisfy

- f and g are locally integrable functions with $|f| * |g| \in L^1(G)$,
- or at least one of the distributions f or g has compact support,
- or at least one of the distributions f or g is Schwartz.

Hence we may extend the notation and define:

Definition 5.1.19. If $f \in \mathcal{K}_{a,b}(G)$ and $g \in \mathcal{K}_{b,c}(G)$ for $a, b, c \in \mathbb{R}$, and $T_f : \phi \mapsto \phi * f$, $T_g : \phi \mapsto \phi * g$ are the associated operators, we denote by $f * g$ the distribution in $\mathcal{K}_{a,c}(G)$ which is the right convolution kernel of $T_g T_f$.

We obtain easily the following properties:

Corollary 5.1.20. *Let $f \in \mathcal{K}_{a,b}(G)$ and $g \in \mathcal{K}_{b,c}(G)$ for $a, b, c \in \mathbb{R}$. Then we have the following property of associativity for any $\phi \in \mathcal{S}(G)$*

$$\phi * (f * g) = (\phi * f) * g,$$

and more generally for any $h \in \mathcal{K}_{c,d}(G)$ (where $d \in \mathbb{R}$)

$$f * (g * h) = (f * g) * h,$$

as convolutions of an element of $\mathcal{K}_{a,b}(G)$ with an element of $\mathcal{K}_{b,d}(G)$ for the left-hand side, and of an element of $\mathcal{K}_{a,c}(G)$ with an element of $\mathcal{K}_{c,d}(G)$ for the right-hand side.

The rest of this section is devoted to the definition of the group Fourier transform of a distribution in $\mathcal{K}_{a,b}(G)$. We start by defining what will turn out to be the image of the group Fourier transform on $\mathcal{K}_{a,b}(G)$. We recall that $L^\infty(\widehat{G})$ is the space of measurable fields of operators on $\widehat{G}$ which are uniformly bounded, see Definition 1.8.8.

Definition 5.1.21. Let $a, b \in \mathbb{R}$. We denote by $L^\infty_{a,b}(\widehat{G})$ the space of fields of operators $\sigma = \{\sigma_\pi : \mathcal{H}_\pi^\infty \to \mathcal{H}_\pi^b\,,\ \pi \in \widehat{G}\}$ satisfying

$$\exists C > 0 \quad \forall \phi \in \mathcal{S}(G) \qquad \|\sigma\widehat{\phi}\|_{L^2_b(\widehat{G})} \le C\|\phi\|_{L^2_a(G)}. \tag{5.16}$$

Here we assume that a positive Rockland operator has been fixed to define the norms on $L^2_b(\widehat{G})$ and $L^2_a(G)$.

For such a field σ, $\|\sigma\|_{L^\infty_{a,b}(\widehat{G})}$ denotes the infimum of the constant $C > 0$ satisfying (5.16).

We may sometimes abuse the notation and write $\|\sigma_\pi\|_{L^\infty_{a,b}(\widehat{G})}$ when no confusion is possible.

Note that as $\phi \in \mathcal{S}(G)$, its group Fourier transform acts on smooth vectors, see Example 1.8.18. Hence the composition $\sigma\widehat{\phi}$ above makes sense, see Section 1.8.3.

Naturally, the space $L^\infty_{a,b}(\widehat{G})$ introduced in Definition 5.1.21 is independent of the choice of a Rockland operator used to define the norms on $L^2_b(\widehat{G})$ and $L^2_a(G)$:

Lemma 5.1.22. *If $\{\sigma_\pi : \mathcal{H}_\pi^\infty \to \mathcal{H}_\pi^b, \pi \in \widehat{G}\}$ satisfies the condition in Definition 5.1.21 for one positive Rockland operator, then it satisfies the same property for any positive Rockland operator. Moreover, if $\mathcal{R}_1$ and $\mathcal{R}_2$ are two positive Rockland operators, and if $\|\sigma\|_{L^\infty_{a,b,\mathcal{R}_1}(\widehat{G})}$ and $\|\sigma\|_{L^\infty_{a,b,\mathcal{R}_2}(\widehat{G})}$ denote the corresponding infima, then there exists $C > 0$ independent of σ such that*

$$C^{-1}\|\sigma\|_{L^\infty_{a,b,\mathcal{R}_2}(\widehat{G})} \le \|\sigma\|_{L^\infty_{a,b,\mathcal{R}_1}(\widehat{G})} \le C\|\sigma\|_{L^\infty_{a,b,\mathcal{R}_2}(\widehat{G})}.$$

Proof. This follows easily from the independence of the Sobolev spaces on G and $\widehat{G}$ of the positive Rockland operators, see Section 4.4.5 and Proposition 5.1.14. □

If the field acts on smooth vectors, we can simplify Definition 5.1.21:

Lemma 5.1.23. *Let $\sigma = \{\sigma_\pi : \mathcal{H}_\pi^\infty \to \mathcal{H}_\pi^\infty\,,\ \pi \in \widehat{G}\}$ be a field acting on smooth vectors. Then $\sigma \in L^\infty_{a,b}(\widehat{G})$ if and only if*

$$\{\pi(\mathrm{I}+\mathcal{R})^{\frac{b}{\nu}}\sigma_\pi\,\pi(\mathrm{I}+\mathcal{R})^{-\frac{a}{\nu}} : \mathcal{H}_\pi^\infty \to \mathcal{H}_\pi^\infty, \pi \in \widehat{G}\} \in L^\infty(\widehat{G}), \tag{5.17}$$

where $\mathcal{R}$ is a positive Rockland operator of degree ν, and in this case,

$$\|\sigma\|_{L^\infty_{a,b}(\widehat{G})} = \|\pi(\mathrm{I}+\mathcal{R})^{\frac{b}{\nu}}\sigma_\pi\, \pi(\mathrm{I}+\mathcal{R})^{-\frac{a}{\nu}}\|_{L^\infty(\widehat{G})}.$$

Proof. This follows easily from the density of $\mathcal{S}(G)$ in $L^2_b(G)$. □

Note that the composition in (5.17) makes sense as all the fields involved act on smooth vectors. In Corollary 5.1.30, we will see a sufficient condition (which will be useful later) for a field to be acting on smooth vectors.

We can now characterise the elements of $\mathcal{K}_{a,b}(G)$ in terms of $L^\infty_{a,b}(\widehat{G})$:

Proposition 5.1.24. *Let $a, b \in \mathbb{R}$.*

(i) If $\sigma \in L^\infty_{a,b}(\widehat{G})$, then the operator $T_\sigma : \mathcal{S}(G) \to \mathcal{S}'(G)$ defined via

$$\widehat{T_\sigma\phi}(\pi) := \sigma_\pi \widehat{\phi}(\pi), \quad \phi \in \mathcal{S}(G),\ \pi \in \widehat{G}, \tag{5.18}$$

extends uniquely to an operator in $\mathscr{L}(L^2_a, L^2_b)$. Moreover,

$$\|T_\sigma\|_{\mathscr{L}(L^2_a, L^2_b)} = \|\sigma\|_{L^\infty_{a,b}(\widehat{G})}, \tag{5.19}$$

where the Sobolev norms are defined using a chosen positive Rockland operator $\mathcal{R}$ with homogeneous degree ν. The right convolution kernel $f \in \mathcal{S}'(G)$ of T_σ is in $\mathcal{K}_{a,b}(G)$.

(ii) Conversely, if $f \in \mathcal{K}_{a,b}(G)$ then there exists a unique $\sigma \in L^\infty_{a,b}(\widehat{G})$ such that

$$\widehat{\phi * f}(\pi) = \sigma_\pi \widehat{\phi}(\pi), \quad \phi \in \mathcal{S}(G),\ \pi \in \widehat{G}. \tag{5.20}$$

Furthermore, if f is also in any of the spaces where the group Fourier transform has already been defined, namely any Sobolev space $L^2_a(G)$ or $\mathcal{K}(G)$, then $\sigma = \{\sigma_\pi, \pi \in \widehat{G}\}$ will coincide with the group Fourier transform of f.

Proof. The properties of T_σ in Part (i) follow from the Plancherel theorem (Theorem 1.8.11) and the density of $\mathcal{S}(G)$ in $L^2(G)$. The right convolution kernel $f \in \mathcal{S}'(G)$ of T_σ is in $\mathcal{K}_{a,b}(G)$ by Proposition 5.1.17.

Conversely, let $f \in \mathcal{K}_{a,b}(G)$. By assumption the operator $T_f : \mathcal{S}(G) \ni \phi \mapsto \phi * f$ admits a bounded extension from $L^2_a(G)$ to $L^2_b(G)$. Thus the operator $(\mathrm{I}+\mathcal{R})^{\frac{b}{\nu}} T_f (\mathrm{I}+\mathcal{R})^{-\frac{a}{\nu}}$ is bounded on $L^2(G)$ and we denote by $f_{a,b} \in \mathcal{K}(G)$ its right convolution kernel. For any $\phi \in \mathcal{S}(G)$, we have $\phi_a := (\mathrm{I}+\mathcal{R})^{\frac{a}{\nu}}\phi \in \mathcal{S}(G)$ by Corollary 4.3.16 thus $\phi_a * f_{a,b} \in L^2(G)$ and we have

$$T_f\phi \in L^2_b(G) \quad \text{with} \quad T_f\phi = (\mathrm{I}+\mathcal{R})^{-\frac{b}{\nu}}(\phi_a * f_{a,b}).$$

Consequently $\mathcal{F}_G(T_f\phi) \in L^2_b(\widehat{G})$ and

$$\mathcal{F}_G(T_f\phi) = \pi(\mathrm{I}+\mathcal{R})^{-\frac{b}{\nu}} \widehat{f}_{a,b}\widehat{\phi}_a = \pi(\mathrm{I}+\mathcal{R})^{-\frac{b}{\nu}} \widehat{f}_{a,b}\pi(\mathrm{I}+\mathcal{R})^{\frac{a}{\nu}}\widehat{\phi}.$$

One checks easily that $\{\sigma_\pi : \mathcal{H}_\pi^\infty \to \mathcal{H}_\pi^b, \pi \in \widehat{G}\}$ defined via

$$\sigma_\pi := \pi(\mathrm{I}+\mathcal{R})^{-\frac{b}{\nu}} \widehat{f}_{a,b}(\pi)\, \pi(\mathrm{I}+\mathcal{R})^{\frac{a}{\nu}}$$

is in $L^\infty_{a,b}(\widehat{G})$ and satisfies (5.20). The rest of the proof of Part (ii) follows easily from the computations above and the uniqueness of the group Fourier transforms already defined. □

Thanks to Proposition 5.1.24, we can extend the definition of the group Fourier transform to $\mathcal{K}_{a,b}(G)$:

Definition 5.1.25 (The group Fourier transform on $\mathcal{K}_{a,b}(G)$)**.** The group Fourier transform of $f \in \mathcal{K}_{a,b}(G)$ is the field of operators $\{\sigma_\pi : \mathcal{H}_\pi^\infty \to \mathcal{H}_\pi^b, \pi \in \widehat{G}\}$ in $L^\infty_{a,b}(\widehat{G})$ associated to f by Proposition 5.1.24, and we write

$$\widehat{f}(\pi) := \pi(f) := \sigma_\pi, \quad \pi \in \widehat{G}.$$

As the next example implies, any left-invariant vector field is in some $\mathcal{K}_{a,b}(G)$ and their Fourier transform can be defined via Definition 5.1.25. As is shown in the proof below, this coincides with the infinitesimal representation of the corresponding element of $\mathfrak{U}(\mathfrak{g})$ defined in Section 1.7.

Example 5.1.26. Let $\alpha \in \mathbb{N}_0^n$. The operator X^α is in $\mathscr{L}(L^2_{[\alpha]}(G), L^2(G))$ and more generally in $\mathscr{L}(L^2_{[\alpha]+s}(G), L^2_s(G))$ for any $s \in \mathbb{R}$. Its right convolution kernel is the distribution $X^\alpha\delta_0$ defined via (see (5.8))

$$\langle X^\alpha\delta_0, \phi\rangle = \langle \delta_0, \{X^\alpha\}^t\phi\rangle = \{X^\alpha\}^t\phi(0),$$

which is in $\mathcal{K}_{[\alpha],0}$, and more generally in $\mathcal{K}_{s+[\alpha],s}$ for any $s \in \mathbb{R}$. Its group Fourier transform is

$$\mathcal{F}_G(X^\alpha\delta_0)(\pi) = \pi(X^\alpha) = \pi(X)^\alpha$$

and coincides with the infinitesimal representation on $\mathfrak{U}(\mathfrak{g})$. It is in $L^\infty_{s+[\alpha],s}(\widehat{G})$ for any $s \in \mathbb{R}$.

Proof. By Theorem 4.4.16, X^α maps $L^2_{[\alpha]}(G)$ continuously to $L^2(G)$ and, more generally, $L^2_{s+[\alpha]}(G)$ continuously to $L^2_s(G)$.

By Proposition 1.7.6, we have for any $\phi \in \mathcal{S}(G)$

$$\mathcal{F}_G(X^\alpha\phi)(\pi) = \pi(X^\alpha)\widehat{\phi}(\pi) = \pi(X)^\alpha\widehat{\phi}(\pi).$$

This shows that $\mathcal{F}_G(X^\alpha\delta_0)$ coincides with $\{\pi(X^\alpha), \pi \in \widehat{G}\}$. □

As our next example shows, when multipliers in a positive Rockland operator are in $\mathscr{L}_L(L^2_s(G), L^2_{s-b}(G))$, the group Fourier transform of their right convolution kernels can also be given via the functional calculus of the Rockland operators:

Example 5.1.27. Let $\mathcal{R}$ be a positive Rockland operator of homogeneous degree ν. Let m be a measurable function on $[0,\infty)$ satisfying

$$\exists C>0 \quad \forall \lambda \geq 0 \quad |m(\lambda)| \leq C(1+\lambda)^{\frac{b}{\nu}}.$$

Then the operator $m(\mathcal{R})$ defined by the functional calculus of $\mathcal{R}$ extends uniquely to an operator in $\mathscr{L}_L(L^2_{s+b}(G), L^2_s(G))$ for any $s \in \mathbb{R}$. Its right convolution kernel $m(\mathcal{R})\delta_0$ is in $\mathcal{K}_{s+b,s}$ for any $s \in \mathbb{R}$. Its group Fourier transform is

$$\mathcal{F}_G(m(\mathcal{R})\delta_0)(\pi) = m(\pi(\mathcal{R}))$$

defined by the functional calculus of $\pi(\mathcal{R})$. It is in $L^\infty_{s+b,s}(\widehat{G})$ for any $s \in \mathbb{R}$. For a fixed $s \in \mathbb{R}$, we have

$$\begin{aligned} \|m(\mathcal{R})\|_{\mathscr{L}_L(L^2_{s+b}(G),L^2_s(G))} = \|m(\mathcal{R})\delta_0\|_{\mathcal{K}_{s+b,s}} &= \|m(\pi(\mathcal{R}))\|_{L^\infty_{s+b,s}(\widehat{G})} \\ &\leq \sup_{\lambda>0}(1+\lambda)^{-\frac{b}{\nu}}|m(\lambda)|, \end{aligned}$$

if we realise the Sobolev norms with $\mathcal{R}$.

We refer to Section 4.1.3 and Corollary 4.1.16 for the properties of the functional calculus of $\mathcal{R}_2$ and $\pi(\mathcal{R})$.

Proof. The function m_1 given by

$$m_1(\lambda) := m(\lambda)(1+\lambda)^{-\frac{b}{\nu}}, \quad \lambda \geq 0,$$

is measurable and bounded on $[0,\infty)$. The operator $m_1(\mathcal{R})$ defined by the functional calculus of $\mathcal{R}$ is therefore bounded on $L^2(G)$ with

$$\|m_1(\mathcal{R})\|_{\mathscr{L}(L^2(G))} \leq \sup_{\lambda\geq 0}|m_1(\lambda)|.$$

Again from the properties of the functional calculus of $\mathcal{R}$, we also have

$$m(\mathcal{R}) \supset m_1(\mathcal{R})(\mathrm{I}+\mathcal{R})^{\frac{b}{\nu}},$$

in the sense of operators. Since $\mathrm{Dom}(\mathrm{I}+\mathcal{R})^{b/\nu} \supset \mathcal{S}(G)$ (see Corollary 4.3.16), this shows that the domain of $m(\mathcal{R})$ contains $\mathcal{S}(G)$ and that

$$m_1(\mathcal{R}) = m(\mathcal{R})(\mathrm{I}+\mathcal{R})^{-\frac{b}{\nu}} \quad \text{on } \mathcal{S}(G).$$

The properties of the functional calculus of $\mathcal{R}$ yield for any $s \in \mathbb{R}$,

$$\begin{aligned} \|m_1(\mathcal{R})\|_{\mathscr{L}(L^2(G))} &= \|m_1(\mathcal{R})\|_{\mathscr{L}(L^2_s(G))} \\ &= \|m(\mathcal{R})(\mathrm{I}+\mathcal{R})^{-\frac{b}{\nu}}\|_{\mathscr{L}(L^2_s(G))} \\ &= \|m(\mathcal{R})\|_{\mathscr{L}(L^2_{s+b}(G),L^2_s(G))}. \end{aligned}$$

By Corollary 4.1.16, the kernel of $m_1(\mathcal{R})$ is the tempered distribution $m_1(\mathcal{R})\delta_0$ with Fourier transform $\{m_1(\pi(\mathcal{R})), \pi \in \widehat{G}\}$. Adapting the proof of Corollary 4.1.16, we see that

$$m_1(\pi(\mathcal{R})) = m(\pi(\mathcal{R}))(\mathrm{I} + \pi(\mathcal{R}))^{-\frac{b}{\nu}} \quad \text{on } \mathcal{H}_\pi^\infty, \quad \pi \in \widehat{G}.$$

It is now straightforward to check that the kernel of the operator $m(\mathcal{R})$ is in $\mathcal{K}_{s+b,s}$ and its Fourier transform is $\{m(\pi(\mathcal{R})), \pi \in \widehat{G}\}$. □

Naturally, any Schwartz function is in any $\mathcal{K}_{a,b}$ and one can readily estimate the associated norm:

Example 5.1.28. If $\phi \in \mathcal{S}(G)$, then for any $a, b \in \mathbb{R}$, the operator $T_\phi : \psi \mapsto \psi * \phi$ is in $\mathscr{L}(L_a^2(G), L_b^2(G))$, $\phi \in \mathcal{K}_{a,b}$ and $\widehat{\phi} \in L_{a,b}^\infty$. If we fix a positive Rockland operator $\mathcal{R}$ of homogeneous degree ν, then we have

$$\|T_\phi\|_{\mathscr{L}(L_a^2(G), L_b^2(G))} = \|\phi\|_{\mathcal{K}_{a,b}} = \|\widehat{\phi}\|_{L_{a,b}^\infty} \le \|(\mathrm{I} + \mathcal{R})^{\frac{b}{\nu}}(\mathrm{I} + \tilde{\mathcal{R}})^{-\frac{a}{\nu}}\phi\|_{L^1(G)} < \infty,$$

where the norms on $\mathscr{L}(L_a^2(G), L_b^2(G))$, $\mathcal{K}_{a,b}$ and $L_{a,b}^\infty$ are defined with $\mathcal{R}$.

With Definition 5.1.25, we can reformulate Proposition 5.1.24 and parts of Proposition 5.1.17 and Corollary 5.1.20 as the following proposition.

Proposition 5.1.29. *1. Let $a, b \in \mathbb{R}$. The Fourier transform $\mathcal{F}_G$ maps $\mathcal{K}_{a,b}(G)$ onto $L_{a,b}^\infty(\widehat{G})$. Furthermore, $\mathcal{F}_G : \mathcal{K}_{a,b}(G) \to L_{a,b}^\infty(\widehat{G})$ is an isomorphism between Banach spaces. In particular, for $f \in \mathcal{K}_{a,b}(G)$,*

$$\|f\|_{\mathcal{K}_{a,b}} = \|\widehat{f}\|_{L_{a,b}^\infty(\widehat{G})}.$$

It coincides with the Fourier transform on $\mathcal{K}(G)$ for $a = b = 0$.

2. If $\sigma_1 \in L_{a_1,b_1}^\infty(\widehat{G})$ and $\sigma_2 \in L_{a_2,b_2}^\infty(\widehat{G})$ with $b_2 = a_1$, then their product $\sigma_1\sigma_2$ makes sense as the element of $L_{a_2,b_1}^\infty(\widehat{G})$ given by the Fourier transform of $(\mathcal{F}_G^{-1}\sigma_2) * (\mathcal{F}_G^{-1}\sigma_1)$.

In other words, if $f_1 \in \mathcal{K}_{a_1,b_1}(\widehat{G})$ and $f_2 \in \mathcal{K}_{a_2,b_2}(\widehat{G})$ with $b_2 = a_1$, then the Fourier transform of $f_2 * f_1 \in \mathcal{K}_{a_2,b_1}(\widehat{G})$ is

$$\mathcal{F}_G(f_2 * f_1) = \mathcal{F}_G(f_1)\mathcal{F}_G(f_2).$$

3. Let $\sigma = \{\sigma_\pi : \mathcal{H}_\pi^\infty \to \mathcal{H}_\pi, \pi \in \widehat{G}\} \in L_{a,b}^\infty(\widehat{G})$. Then we have for any $\alpha \in \mathbb{N}_0^n$,

$$\{\pi(X)^\alpha\sigma_\pi : \mathcal{H}_\pi^\infty \to \mathcal{H}_\pi, \pi \in \widehat{G}\} \in L_{a,b-[\alpha]}^\infty(\widehat{G}), \tag{5.21}$$

and for any $s \in \mathbb{R}$,

$$\{\pi(\mathrm{I} + \mathcal{R})^{s/\nu}\sigma_\pi : \mathcal{H}_\pi^\infty \to \mathcal{H}_\pi, \pi \in \widehat{G}\} \in L_{a,b-s}^2(\widehat{G}). \tag{5.22}$$

Furthermore, if $f = \mathcal{F}_G^{-1}\sigma \in \mathcal{K}_{a,b}(G)$ *then*

$$\mathcal{F}_G(X^\alpha f)(\pi) = \pi(X)^\alpha \widehat{f}(\pi) \quad \text{and} \quad \mathcal{F}_G((\mathrm{I}+\mathcal{R})^{s/\nu} f)(\pi) = \pi(\mathrm{I}+\mathcal{R})^{s/\nu} \widehat{f}(\pi).$$

The fields of operators in (5.21) and (5.22) are understood as compositions of fields of operators in $L^\infty_{a_2,b_2}$ and $L^\infty_{a_1,b_1}$ with $b_2 = a_1$, see Part 2 and Examples 5.1.26 and 5.1.27.

With the help of Proposition 5.1.29, we can now give a usefull sufficient condition for a field to act on smooth vectors and reformulate Corollary 4.4.10 into

Corollary 5.1.30. *Let* $a, b \in \mathbb{R}$ *and let* $\{\gamma_\ell, \ell \in \mathbb{Z}\}$ *be a sequence of real numbers which tends to* $\pm\infty$ *as* $\ell \to \pm\infty$. *Let* $\sigma \in L^\infty_{a+\gamma_\ell, b+\gamma_\ell}(\widehat{G})$ *for every* $\ell \in \mathbb{Z}$. *Then* σ *is a field of operators acting on smooth vectors:*

$$\sigma = \{\sigma_\pi : \mathcal{H}_\pi^\infty \to \mathcal{H}_\pi^\infty, \pi \in \widehat{G}\}.$$

Furthermore $\sigma \in L^\infty_{a+\gamma,b+\gamma}(\widehat{G})$ *for every* $\gamma \in \mathbb{R}$ *and for any* $c \geq 0$, *we have*

$$\sup_{|\gamma|\leq c} \|\sigma\|_{L^\infty_{a+\gamma,b+\gamma}(\widehat{G})} \leq C_c \max\left(\|\sigma\|_{L^\infty_{a+\gamma_\ell,b+\gamma_\ell}(\widehat{G})}, \|\sigma\|_{L^\infty_{a+\gamma_{-\ell},b+\gamma_{-\ell}}(\widehat{G})} \right),$$

where $\ell \in \mathbb{N}_0$ *is the smallest integer such that* $\gamma_\ell \geq c$ *and* $-\gamma_{-\ell} \geq c$.

Proof. By Proposition 5.1.29, $\pi(X)^\alpha \sigma \in L^\infty_{a+\gamma_\ell, b+\gamma_\ell - [\alpha]}$ for every $\alpha \in \mathbb{N}_0^n$ and every $\ell \in \mathbb{Z}$. Thus choosing $\gamma_\ell \geq [\alpha] - b$, we have $\pi(X)^\alpha \sigma \widehat{\phi} \in L^2(\widehat{G})$ for every $\phi \in \mathcal{S}(G)$. Realising $\pi \in \widehat{G}$ as a representation of G and fixing $v \in \mathcal{H}_\pi^\infty$, this implies that the mapping $x \mapsto \pi(x)\sigma_\pi \widehat{\phi}(\pi) v$ is smooth. Hence $\sigma_\pi \widehat{\phi}(\pi) v$ is smooth and $\sigma\widehat{\phi}$ acts on smooth vectors. As this holds for every $\phi \in \mathcal{D}(G)$, so does σ by Lemma 1.8.19. We conclude with Corollary 4.4.10. □

We end this section with one more technical property:

Lemma 5.1.31. *Let* $\sigma \in L^\infty_{a,b}(\widehat{G})$ *where* $a, b \in \mathbb{R}$. *Let* $\phi \in \mathcal{S}(G)$. *Then we have* $\sigma\widehat{\phi} \in \tilde{L}^2_s(\widehat{G})$ *for any* $s \in \mathbb{R}$ *and*

$$\int_{\widehat{G}} \mathrm{Tr} \left| \sigma_\pi \widehat{\phi}(\pi) \right| d\mu(\pi) < \infty. \tag{5.23}$$

Setting $f := \mathcal{F}_G^{-1}\sigma \in \mathcal{K}_{a,b}$, *the function* $\phi * f$ *is smooth and we have for any* $x \in G$ *the equality*

$$\phi * f(x) = \int_{\widehat{G}} \mathrm{Tr}\left(\pi(x)\sigma_\pi \widehat{\phi}(\pi) \right) d\mu(\pi).$$

Remark 5.1.32. The composition $\sigma\widehat{\phi}$ makes sense since σ is defined on smooth vectors and $\widehat{\phi}$ acts on smooth vectors. The composition $\pi(x)\sigma_\pi\pi(\phi)$ makes sense since $\pi(x)$ is bounded and $\sigma\widehat{\phi}$ is bounded (even in Hilbert Schmidt classes) since it is stated first that $\sigma\widehat{\phi} \in \tilde{L}^2_s(\widehat{G})$ for any s, hence in particular in $L^2(\widehat{G})$.

Proof. Let T_σ be the operator with right convolution kernel $f := \mathcal{F}_G^{-1}\sigma$. Then $T_\sigma \in \mathscr{L}(L^2_a(G), L^2_b(G))$ and $T_\sigma^* T_\sigma$ extends to an operator in $\mathscr{L}(L^2_a(G))$. For any $\phi \in \mathcal{S}(G)$, the definition of the adjoint and the duality between Sobolev spaces yield

$$\begin{aligned} \|T_\sigma\phi\|^2_{L^2(G)} &= \langle T_\sigma^* T_\sigma \phi, \bar{\phi}\rangle_{L^2_a(G)\times L^2_{-a}(G)} \\ &\leq \|T_\sigma^* T_\sigma\|_{\mathscr{L}(L^2_a(G))} \|\phi\|_{L^2_a(G)} \|\phi\|_{L^2_{-a}(G)}. \end{aligned}$$

This last expression is finite since $T_\sigma^* T_\sigma \in \mathscr{L}(L^2_a(G))$ and $\mathcal{S}(G) \subset L^2_{s'}(G)$ for any $s' \in \mathbb{R}$. Thus $T_\sigma\phi \in L^2(G)$ and its Fourier transform is $\sigma\widehat{\phi} \in L^2(\widehat{G})$. For any $s \in \mathbb{R}$, we may replace ϕ with $\phi_s = (\mathrm{I} + \mathcal{R})^{s/\nu}\phi \in \mathcal{S}(G)$ and $\sigma\widehat{\phi}_s \in L^2(\widehat{G})$ yields $\sigma\widehat{\phi} \in L^2_s(\widehat{G})$.

Applying Proposition 5.1.15 to $\sigma\widehat{\phi} \in \tilde{L}^2_s(\widehat{G})$ for some $s > Q/2$, we obtain (5.23). Note that $f := \mathcal{F}_G^{-1}\sigma$ is a tempered distribution so $\phi * f$ is smooth (see Lemma 3.1.55). The group Fourier transform of $\phi * f$ is $\sigma\widehat{\phi}$ by Proposition 5.1.29 Part 2 and Example 5.1.28. We now conclude with the inversion formula given in Proposition 5.1.15. □

5.1.3 Symbols and associated kernels

In this section we aim at establishing a one-to-one correspondence between a collection σ of operators parametrised by $G \times \widehat{G}$ and a function κ; this function will turn out to be the kernel of the operator naturally associated to σ. For the abstract setting behind measurable fields of operators and some of their properties we refer to Section B.1.6, especially to Proposition B.1.17, as well as Section 1.8.3.

Definition 5.1.33 (Symbols). A *symbol* is a field of operators $\{\sigma(x,\pi) : \mathcal{H}_\pi^\infty \to \mathcal{H}_\pi, \pi \in \widehat{G}\}$ depending on $x \in G$, satisfying for each $x \in G$

$$\exists a, b \in \mathbb{R} \qquad \sigma(x,\cdot) := \{\sigma(x,\pi) : \mathcal{H}_\pi^\infty \to \mathcal{H}_\pi, \pi \in \widehat{G}\} \in L^\infty_{a,b}(\widehat{G}).$$

Here we use the usual identifications of a strongly continuous irreducible unitary representation from $\operatorname{Rep} G$ with its equivalence class in $\widehat{G}$, and of a field of operators acting on the smooth vectors parametrised by $\widehat{G}$ with its equivalence class with respect to the Plancherel measure μ.

We will usually assume that the symbols are uniformly regular in x:

Definition 5.1.34 (Continuous and smooth symbols).

- A symbol $\{\sigma(x,\pi) : \mathcal{H}_\pi^\infty \to \mathcal{H}_\pi, \pi \in \widehat{G}\}$ is said to be *continuous* in $x \in G$ whenever there exists $a, b \in \mathbb{R}$ such that

$$\forall x \in G \quad \sigma(x,\cdot) := \{\sigma(x,\pi) : \mathcal{H}_\pi^\infty \to \mathcal{H}_\pi, \pi \in \widehat{G}\} \in L_{a,b}^\infty(\widehat{G}),$$

  and the map $x \mapsto \sigma(x,\cdot)$ is continuous from $G \sim \mathbb{R}^n$ to the Banach space $L_{a,b}^\infty(\widehat{G})$.

- A symbol $\sigma = \{\sigma(x,\pi) : \mathcal{H}_\pi^\infty \to \mathcal{H}_\pi, \pi \in \widehat{G}\}$ is said to be *smooth* in $x \in G$ whenever it is a field of operators depending smoothly in $x \in G$ (see Remark 1.8.16) and, for every $\beta \in \mathbb{N}_0^n$, the field $\{\partial_x^\beta \sigma(x,\pi) : \mathcal{H}_\pi^\infty \to \mathcal{H}_\pi, \pi \in \widehat{G}\}$ is continuous.

Important note: In the sequel, whenever we talk about symbols (on graded Lie groups), we always mean the symbols which are smooth in $x \in G$ in the sense of Definition 5.1.34 unless stated otherwise.

For a symbol as in Definition 5.1.34, we will usually write

$$\sigma = \{\sigma(x,\pi), (x,\pi) \in G \times \widehat{G}\},$$

but we may sometimes abuse the notation and refer to the symbol simply as $\sigma(x,\pi)$.

Lemma 5.1.35. *If $\sigma = \{\sigma(x,\pi), (x,\pi) \in G \times \widehat{G}\}$ is a symbol, then*

$$\kappa_x := \mathcal{F}_G^{-1}\{\sigma(x,\cdot)\}$$

is a tempered distribution and the map

$$G \ni x \longmapsto \kappa_x \in \mathcal{S}'(G)$$

is smooth.

In other words,

$$\kappa \in C^\infty(G, \mathcal{S}'(G)).$$

Here $C^\infty(G, \mathcal{S}'(G))$ denotes the set of smooth functions from G to $\mathcal{S}'(G)$.

Proof. As σ is a smooth symbol, for every $\beta \in \mathbb{N}_0^n$, there exists $a_\beta, b_\beta \in \mathbb{R}$ such that $G \ni x \mapsto \partial_x^\beta \sigma(x,\cdot) \in L_{a_\beta,b_\beta}^\infty(\widehat{G})$ is continuous. By Proposition 5.1.29, composing this with $\mathcal{F}_G^{-1}$ implies that $G \ni x \mapsto \partial_x^\beta \kappa_x \in \mathcal{K}_{a_\beta,b_\beta}$ is continuous. Since the inclusion $\mathcal{K}_{a_\beta,b_\beta} \subset \mathcal{S}'(G)$ is continuous, this implies that each map $G \ni x \mapsto \partial_x^\beta \kappa_x \in \mathcal{S}'(G)$ is continuous. Hence $G \ni x \mapsto \kappa_x \in \mathcal{S}'(G)$ is smooth. □

Definition 5.1.36 (Associated kernels). If σ is a symbol, then the tempered distribution

$$\kappa_x := \mathcal{F}_G^{-1}\{\sigma(x,\cdot)\} \in \mathcal{S}'(G)$$

is called its *associated kernel*, sometimes its right convolution kernel, or just a *kernel*. We may also call the smooth map $G \ni x \mapsto \kappa_x \in \mathcal{S}'(G)$ or the map $(x,y) \mapsto \kappa_x(y) = \kappa(x,y)$ the *kernel* associated with σ.

The smoothness of the map $x \mapsto \sigma(x,\cdot)$ implies easily:

Lemma 5.1.37. *If $\sigma = \{\sigma(x,\pi)\}$ is a symbol with kernel κ_x then for any $\beta \in \mathbb{N}_0^n$,*

$$X^\beta\sigma := \{X_x^\beta\sigma(x,\pi)\}, \quad \tilde{X}^\beta\sigma := \{\tilde{X}_x^\beta\sigma(x,\pi)\}, \text{ and } \partial_x^\beta\sigma := \{\partial_x^\beta\sigma(x,\pi)\},$$

are symbols with respective kernels

$$X_x^\beta\kappa_x, \quad \tilde{X}_x^\beta\kappa_x, \quad \text{and} \quad \partial_x^\beta\kappa_x.$$

Examples of symbols are the symbols in the classes $S^m_{\rho,\delta}(G)$ defined later on. Here are more specific examples of symbols which do not depend on $x \in G$.

Example 5.1.38. If $f \in \mathcal{K}_{a,b}(G)$, then $\widehat{f} = \{\widehat{f}(\pi) : \mathcal{H}_\pi^\infty \to \mathcal{H}_\pi, \pi \in \widehat{G}\}$ is a symbol with kernel f.

The following are particular instances of this case:

- $\widehat{\delta}_0 = \mathrm{I} = \{\mathrm{I} : \mathcal{H}_\pi^\infty \to \mathcal{H}_\pi^\infty, \pi \in \widehat{G}\}$ is a symbol and its kernel is the Dirac measure δ_0.
- For any $\alpha \in \mathbb{N}_0^n$, $\{\pi(X)^\alpha : \mathcal{H}_\pi^\infty \to \mathcal{H}_\pi^\infty, \pi \in \widehat{G}\}$ is a symbol with kernel $X^\alpha\delta_0$, see Example 5.1.26. It acts on smooth vectors, see Example 1.8.17, or alternatively Example 5.1.26 together with Corollary 5.1.30.
- If $\mathcal{R}$ is a positive Rockland operator of homogeneous degree ν and if m is a measurable function on $[0,\infty)$ satisfying

$$\exists C > 0 \quad \forall \lambda \geq 0 \quad |m(\lambda)| \leq C(1+\lambda)^{b/\nu},$$

then $\{m(\pi(\mathcal{R})) : \mathcal{H}_\pi^\infty \to \mathcal{H}_\pi, \pi \in \widehat{G}\}$ is a symbol with kernel $m(\mathcal{R})\delta_0$, see Example 5.1.27. By Corollary 5.1.30, this symbol also acts on smooth vectors

$$\{m(\pi(\mathcal{R})) : \mathcal{H}_\pi^\infty \to \mathcal{H}_\pi^\infty, \pi \in \widehat{G}\}.$$

5.1.4 Quantization formula

With the notion of symbol explained in Section 5.1.3, our quantization makes sense:

Theorem 5.1.39 (Quantization). *The quantization defined by formula (5.1) makes sense for any symbol $\sigma = \{\sigma(x,\pi)\}$. More precisely, for any $\phi \in \mathcal{S}(G)$ and $x \in G$, we have*

$$\mathrm{Op}(\sigma)\phi(x) = \int_{\widehat{G}} \mathrm{Tr}\left(\pi(x)\sigma(x,\pi)\widehat{\phi}(\pi)\right) d\mu(\pi) = \phi * \kappa_x(x), \tag{5.24}$$

where κ_x denotes the kernel of σ. The integral over $\widehat{G}$ in (5.24) is well-defined and absolutely convergent. We also have $\mathrm{Op}(\sigma)\phi \in C^\infty(G)$. Furthermore, the quantization mapping $\sigma \mapsto \mathrm{Op}(\sigma)$ is one-to-one and linear.

Proof. Lemma 5.1.31 (see also Remark 5.1.32) implies that the integral in (5.24) is well defined, absolutely convergent and is equal to $\phi * \kappa_x(x)$.

By Lemma 3.1.55, for each $x \in G$, the function $\phi * \kappa_x$ is smooth. By Lemma 5.1.35, $x \mapsto \kappa_x \in \mathcal{S}'(G)$ is smooth. Hence by composition, $x \mapsto \phi * \kappa_x(x)$ is smooth.

The quantization is clearly linear. Since the kernel is in one-to-one linear correspondence with the operator, and by Lemma 5.1.35 also with the symbol, the quantization $\sigma \mapsto \mathrm{Op}(\sigma)$ is one-to-one. □

Definition 5.1.40 (Notation). If an operator T is given by the formula (5.24) with symbol $\sigma(x,\pi)$, so that $T = \mathrm{Op}(\sigma)$, we will also write

$$\sigma = \sigma_T \quad \text{or} \quad \sigma(x,\pi) = \sigma_T(x,\pi) \quad \text{or even} \quad \sigma = \mathrm{Op}^{-1}(T).$$

This notation is justified since the quantization given by (5.24) is one-to-one by Theorem 5.1.39.

The operators associated with the symbols given in Example 5.1.38 are the ones alluded to in the introduction of this Section:

Continued Example 5.1.38: If $f \in \mathcal{K}_{a,b}(G)$, then $\mathrm{Op}(\widehat{f})$ is the convolution operator $\phi \mapsto \phi * f$ with the right convolution kernel f.

The following are particular instances of this case:

- $\mathrm{Op}(\mathrm{I}) = \mathrm{I}$ and, more generally, for any $\alpha \in \mathbb{N}_0^n$, $\mathrm{Op}(\pi(X)^\alpha) = X^\alpha$.

  These relations can also be expressed as

  $$\sigma_{\mathrm{I}}(x,\pi) = \mathrm{I}_{\mathcal{H}_\pi} \text{ and } \sigma_{X^\alpha}(x,\pi) = \pi(X)^\alpha.$$

- If $\mathcal{R}$ is a positive Rockland operator of homogeneous degree ν and if m is a measurable function on $[0,\infty)$ satisfying

  $$\exists C > 0 \quad \forall \lambda \geq 0 \quad |m(\lambda)| \leq C(1+\lambda)^{b/\nu},$$

  then $\mathrm{Op}(m(\pi(\mathcal{R}))) = m(\mathcal{R})$.

In these examples, the symbols are independent of x. However it is easy to produce x-dependent symbols out of them using the following two observations.

- If $\sigma = \{\sigma(x,\pi), (x,\pi) \in G \times \widehat{G}\}$ is a symbol and $c : G \to \mathbb{C}$ is a smooth function, then $c\sigma := \{c(x)\sigma(x,\pi), (x,\pi) \in G \times \widehat{G}\}$ is a symbol.
- If $\sigma = \{\sigma(x,\pi), (x,\pi) \in G \times \widehat{G}\}$ and $\tau = \{\tau(x,\pi), (x,\pi) \in G \times \widehat{G}\}$ are two symbols, then so is their sum $\sigma + \tau = \{\sigma(x,\pi) + \tau(x,\pi), (x,\pi) \in G \times \widehat{G}\}$.

Remark 5.1.41. 1. The observations just above together with Example 5.1.38 and its continuation above imply that any differential operator of the form

$$\sum_{[\alpha]\leq M} c_\alpha(x) X^\alpha \quad \text{with smooth coefficients } c_\alpha \tag{5.25}$$

may be quantized, in the sense that $\sum_{[\alpha]\leq M} c_\alpha(x)\pi(X)^\alpha$ is a (smooth) symbol and we have

$$\sum_{[\alpha]\leq M} c_\alpha(x) X^\alpha = \text{Op}\left(\sum_{[\alpha]\leq M} c_\alpha(x)\pi(X)^\alpha\right).$$

The differential calculus is, by definition, the space of differential operators of the form

$$\sum_{|\alpha|\leq d} b_\alpha(x)\partial_x^\alpha \quad \text{with smooth coefficients } b_\alpha,$$

or, equivalently, of the form (5.25), see (3.1.5). Hence, we have obtained that the differential calculus may be quantized. This could be viewed as 'the minimum requirement' for a notion of symbol and quantization on a manifold.

2. In order to achieve this, we had to consider and use fields of operators defined on smooth vectors in our definition of symbol. Indeed, for instance, the symbol associated to a left-invariant vector field X is $\{\pi(X)\}$ while $\pi(X)$ are defined on $\mathcal{H}_\pi^\infty$ but is not bounded on $\mathcal{H}_\pi$.

   This technicality has also the following advantage when we apply our theory in the setting of the Heisenberg group $\mathbb{H}_{n_o}$ in Chapter 6. Realising (almost all of) its dual group $\widehat{\mathbb{H}}_{n_o}$ via Schrödinger representations, the spaces of smooth vectors will coincide with the Schwartz space $\mathcal{S}(\mathbb{R}^{n_o})$. In this context, the symbols will be operators acting on $\mathcal{S}(\mathbb{R}^{n_o})$ (which are smoothly parametrised by points in $\mathbb{H}_{n_o}$).

3. With our notion of symbols and quantization, we also obtain part of the functional calculus of any Rockland operators. More precisely, if $\mathcal{R}$ is a positive Rockland operator, we obtain all the operators of the form $m(\mathcal{R})$ with $m : [0,\infty) \to \mathbb{C}$ a measurable function of (at most) polynomial growth at infinity.

4. The symbol classes that we have introduced are based on the quantization relying on writing the operators as operators with right-convolution kernels. There is an obvious parallel theory of quantization and of the corresponding symbols and their classes suited for problems based on the right-invariant operators. With natural modifications we could have considered at the same time right-invariant vector fields in Part (1) above and a quantization involving left-convolution kernels of operators, i.e. writing the same operators but now in the form $\phi \mapsto \kappa_x * \phi$. As an outcome, with natural modifications we would obtain a parallel theory with the same parallel collection of results to those presented here.

$\mathrm{Op}(\sigma)$ as a limit of nice operators

The operators we have obtained as $\mathrm{Op}(\sigma)$ for symbols σ are limits of 'nice operators' in the following sense:

Lemma 5.1.42. *If $\sigma = \{\sigma(x,\pi)\}$ is a symbol, we can construct explicitly a family of symbols $\sigma_\epsilon = \{\sigma_\epsilon(x,\pi)\}$, $\epsilon > 0$, in such a way that*

1. *the kernel $\kappa_\epsilon(x,y)$ of σ_ϵ is smooth in both x and y, and compactly supported in x,*
2. *if $\phi \in \mathcal{S}(G)$ then $\mathrm{Op}(\sigma_\epsilon)\phi \in \mathcal{D}(G)$, and*
3. *$\mathrm{Op}(\sigma_\epsilon)\phi \underset{\epsilon\to 0}{\longrightarrow} \mathrm{Op}(\sigma)\phi$ uniformly on any compact subset of G.*

Proof of Lemma 5.1.42. We fix a number p such that $p/2$ is a positive integer divisible by all the weights $\upsilon_1, \ldots, \upsilon_n$. Therefore, if $|\cdot|_p$ is the quasi-norm given by (3.21), then the mapping $x \mapsto |x|_p^p$ is a p-homogeneous polynomial. We also fix $\chi_o \in C_c^\infty(\mathbb{R})$ with $\chi_o \geq 0$, $\chi_o = 1$ on $[1/2, 2]$ and $\chi_o = 0$ outside of $[1/4, 4]$. For any $\epsilon > 0$, we write

$$\chi_\epsilon(x) := \chi_o(\epsilon |x|_p^p).$$

Clearly $\chi_\epsilon \in \mathcal{D}(G)$.

If $\pi \in \widehat{G}$, we denote by $|\pi|$ the distance between the co-adjoint orbits corresponding to π and 1.

Applying the orbit method, one can construct explicitly for each $\pi \in \widehat{G}$ a basis $(v_{\ell,\pi})_{\ell=1}^\infty$ formed by smooth vectors and such that the field of vectors $\widehat{G} \ni \pi \mapsto v_{\ell,\pi}$ is measurable. We denote by $\mathrm{proj}_{\epsilon,\pi}$ the orthogonal projection on the subspaces spanned by $v_{1,\pi}, \ldots, v_{\ell,\pi}$ where ℓ is the smallest integer such that $\ell > \epsilon^{-1}$.

We consider for any $\epsilon \in (0,1)$ the mapping

$$\sigma_\epsilon(x,\pi) := \chi_\epsilon(x) 1_{|\pi| \leq \epsilon^{-1}} \sigma(x,\pi) \circ \mathrm{proj}_{\epsilon,\pi}.$$

By Definition 5.1.36, the symbol and the kernel are related by

$$\mathcal{F}_G(\kappa_{\epsilon,x})(\pi) = \sigma_\epsilon(x,\pi).$$

By the Fourier inversion formula (1.26), the corresponding kernel is

$$\kappa_{\epsilon,x}(y) = \kappa_\epsilon(x,y) = \chi_\epsilon(x) \int_{|\pi|\leq\epsilon^{-1}} \mathrm{Tr}\left(\sigma(x,\pi)\ \mathrm{proj}_{\epsilon,\pi}\pi(y)\right) d\mu(\pi),$$

which is smooth in x and y and compactly supported in x.

The corresponding operator is $\mathrm{Op}(\sigma_\epsilon)$, given for any $\phi \in \mathcal{S}(G)$ and $x \in G$ by

$$\begin{aligned}\mathrm{Op}(\sigma_\epsilon)\phi(x) &= \int_{\widehat{G}} \mathrm{Tr}\left(\pi(x)\sigma_\epsilon(x,\pi)\widehat{\phi}(\pi)\right) d\mu(\pi) \\ &= \chi_\epsilon(x) \int_{|\pi|\leq\epsilon^{-1}} \mathrm{Tr}\left(\pi(x)\sigma(x,\pi)\ \mathrm{proj}_{\epsilon,\pi}\widehat{\phi}(\pi)\right) d\mu(\pi).\end{aligned}$$

It is also given by

$$\mathrm{Op}(\sigma_\epsilon)\phi(x) = \phi * \kappa_{\epsilon,x}(x).$$

Clearly $\mathrm{Op}(\sigma_\epsilon)\phi$ is smooth and compactly supported.

Since

$$\widehat{G} \ni \pi \mapsto \mathrm{Tr}\left|\sigma(x,\pi)\widehat{\phi}(\pi)\right|$$

is integrable against μ, using the dominated convergence theorem, we obtain easily the uniform convergence of $\mathrm{Op}(\sigma_\epsilon)\phi$ to $\mathrm{Op}(\sigma)\phi$ on any compact set. □

5.2 Symbol classes $S^m_{\rho,\delta}$ and operator classes $\Psi^m_{\rho,\delta}$

In Section 5.2, we will define and study classes of symbols $S^m_{\rho,\delta} = S^m_{\rho,\delta}(G)$. By applying the quantization procedure described in Section 5.1, we will then obtain the corresponding classes of operators

$$\Psi^m_{\rho,\delta} = \mathrm{Op}(S^m_{\rho,\delta}).$$

In Section 5.5, we will show that this collection of operators $\cup_{m\in\mathbb{R}}\Psi^m_{\rho,\delta}$ forms an algebra and satisfies the usual properties expected from a symbolic calculus.

Before defining symbol classes, we need to define difference operators.

5.2.1 Difference operators

On compact Lie groups the difference operators were defined as acting on Fourier coefficients, see Definition 2.2.6. Its adaptation to our setting leads us to (densely) defined difference operators on $\mathcal{K}_{a,b}(G)$ viewed as fields.

Definition 5.2.1. For any $q \in C^\infty(G)$, we set

$$\Delta_q \widehat{f}(\pi) := \widehat{qf}(\pi) \equiv \pi(qf),$$

for any distribution $f \in \mathcal{D}'(G)$ such that $f \in \mathcal{K}_{a,b}$ and $qf \in \mathcal{K}_{a',b'}$ for some $a, b, a', b' \in \mathbb{R}$.

Recall that if $f \in \mathcal{D}'(G)$ and $q \in C^\infty(G)$, then the distribution $qf \in \mathcal{D}'(G)$ is defined via

$$\langle qf, \phi \rangle := \langle f, q\phi \rangle, \quad \phi \in \mathcal{D}(G), \tag{5.26}$$

which makes sense since $q\phi \in \mathcal{D}(G)$. In Definition 5.2.1, we assume that the two distributions f and qf are in $\cup_{a'',b''\in\mathbb{R}} \mathcal{K}_{a'',b''}$. Note that, as all the definitions of group Fourier transform coincide, different values for the parameters a, b, a', b' in Definition 5.2.1 yield the same fields of operators $\{\widehat{f}(\pi) : \mathcal{H}_\pi^\infty \to \mathcal{H}_\pi, \pi \in \widehat{G}\}$ and $\{\widehat{qf}(\pi) : \mathcal{H}_\pi^\infty \to \mathcal{H}_\pi, \pi \in \widehat{G}\}$. This justifies our use of the notation Δ_q without reference to the parameters a, b, a', b'.

Remark 5.2.2. In general, it is not possible to define an operator Δ_q on a single π, and it has to be viewed as acting on the 'whole' fields parametrised by $\widehat{G}$. For example, already on the commutative group $(\mathbb{R}^n, +)$, the difference operators corresponding to coordinate functions will satisfy

$$\Delta^\alpha \widehat{\phi}(\xi) = \left(\frac{1}{i}\frac{\partial}{\partial \xi}\right)^\alpha \widehat{\phi}(\xi), \quad \xi \in \mathbb{R}^n,$$

with appropriately chosen functions q, thus involving derivatives in the dual variable, see Example 5.2.6. Furthermore if q is not a coordinate function but for instance a (non-zero) smooth function with compact support, the corresponding difference operator is not local.

Also, on the Heisenberg group $\mathbb{H}_{n_o}$ (see Example 1.6.4), taking $q = t$ the central variable, and π_λ the Schrödinger representations (see Section 6.3.2), then Δ_t is expressed using derivatives in λ, see Lemma 6.3.6 and Remark 6.3.7.

Let us fix a basis of $\mathfrak{g}$. For the notation of the following proposition we refer to Section 3.1.3 where the spaces of polynomials on homogeneous Lie groups have been discussed, with the set $\mathcal{W}$ defined in (3.60). We will define the difference operators associated with the polynomials appearing in the Taylor expansions:

Proposition 5.2.3. 1. *For each $\alpha \in \mathbb{N}_0^n$, there exists a unique homogeneous polynomial q_α of degree $[\alpha]$ satisfying*

$$\forall \beta \in \mathbb{N}_0^n \qquad X^\beta q_\alpha(0) = \delta_{\alpha,\beta} = \begin{cases} 1 & \text{if } \beta = \alpha, \\ 0 & \text{otherwise.} \end{cases}$$

2. *The polynomials q_α, $\alpha \in \mathbb{N}_0^n$, form a basis of $\mathcal{P}$. Furthermore, for each $M \in \mathcal{W}$, the polynomials q_α, $[\alpha] = M$, form a basis of $\mathcal{P}_{[\alpha]=M}$.*

3. *The Taylor polynomial of a suitable function f at a point $x \in G$ of homogeneous degree $M \in \mathcal{W}$ is*

$$P^{(f)}_{x,M}(y) = \sum_{[\alpha]\leq M} q_\alpha(y) X^\alpha f(x). \tag{5.27}$$

4. *For any $\alpha \in \mathbb{N}_0^n$, we have for any $x, y \in G$,*

$$q_\alpha(xy) = \sum_{[\alpha_1]+[\alpha_2]=[\alpha]} c_{\alpha_1,\alpha_2} q_{\alpha_1}(x) q_{\alpha_2}(y)$$

for some coefficients $c_{\alpha_1,\alpha_2} \in \mathbb{R}$ independent of x and y. Moreover, we have

$$c_{\alpha_1,0} = \begin{cases} 1 & \text{if } \alpha_1 = \alpha \\ 0 & \text{otherwise} \end{cases}, \quad c_{0,\alpha_2} = \begin{cases} 1 & \text{if } \alpha_2 = \alpha \\ 0 & \text{otherwise} \end{cases}.$$

Proof. For each $M \in \mathcal{W}$, by Corollary 3.1.31, there exists a unique polynomial $q_\alpha \in \mathcal{P}_{=M}$ satisfying $X^\beta q_\alpha(0) = \delta_{\alpha,\beta}$ for every $\beta \in \mathbb{N}_0^n$ with $[\beta] = M$, therefore for every $\beta \in \mathbb{N}_0^n$. This shows parts (1) and (2). Part (3) follows from the definition of a Taylor polynomial.

It remains to prove Part (4). For this it suffices to consider $q_\alpha(xy)$ as a polynomial in x and in y, using the bases $(q_{\alpha_1}(x))$ and $(q_{\alpha_2}(y))$. Therefore, $q_\alpha(xy)$ can be written as a finite linear combination of $q_{\alpha_1}(x)q_{\alpha_2}(y)$. Since

$$q_\alpha((rx)(ry)) = r^{[\alpha]} q_\alpha(xy),$$

this forces this linear combination to be over $\alpha_1, \alpha_2 \in \mathbb{N}_0^n$ satisfying $[\alpha_1] + [\alpha_2] = [\alpha]$. The conclusions about the coefficients follow by setting $y = 0$ and then $x = 0$, see also (3.14). □

In the case of $(\mathbb{R}^n, +)$ the polynomials q_α are the usual normalised monomials $(\alpha_1! \dots \alpha_n!)^{-1} x^\alpha$. But it is not usually the case on other groups:

Example 5.2.4. On the three dimensional Heisenberg group $\mathbb{H}_1$ where a point is described as $(x, y, t) \in \mathbb{R}^3$ (see Example 1.6.4), we compute directly that for degree 1 we have

$$q_{(1,0,0)} = x, \ q_{(0,1,0)} = y,$$

and for degree 2,

$$q_{(2,0,0)} = x^2, \ q_{(0,2,0)} = y^2, \ q_{(1,1,0)} = xy, \ q_{(0,0,1)} = t - \frac{1}{2}xy.$$

Definition 5.2.5. For each $\alpha \in \mathbb{N}_0^n$, the *difference operators* are

$$\Delta^\alpha := \Delta_{\tilde{q}_\alpha}, \quad \alpha \in \mathbb{N}_0^n,$$

where

$$\tilde{q}_\alpha(x) := q_\alpha(x^{-1})$$

and $q_\alpha \in \mathcal{P}_{=[\alpha]}$ is defined in Proposition 5.2.3.

The difference operators generalise the Euclidean derivatives with respect to the Fourier variable on $(\mathbb{R}^n,+)$ in the following sense:

Example 5.2.6. Let us consider the abelian group $G=(\mathbb{R}^n,+)$. We identify $\widehat{\mathbb{R}}^n$ with $\mathbb{R}^n$. If the Fourier transform of a function $\phi\in\mathcal{S}(\mathbb{R}^n)$ is given by

$$\mathcal{F}_G\phi(\xi)=(2\pi)^{-\frac{n}{2}}\int_{\mathbb{R}^n}e^{-ix\cdot\xi}\phi(x)dx,\quad \xi\in\mathbb{R}^n,$$

then

$$\Delta^\alpha\mathcal{F}_G\phi(\xi)=\int_{\mathbb{R}^n}e^{-ix\cdot\xi}(-x)^\alpha\phi(x)dx=\left(\frac{1}{i}\frac{\partial}{\partial\xi}\right)^\alpha\mathcal{F}_G\phi(\xi).$$

Thus, Δ^α coincide with the operators $D^\alpha=\left(\frac{1}{i}\frac{\partial}{\partial\xi}\right)^\alpha$ usually appearing in the Fourier analysis on $\mathbb{R}^n$.

Example 5.2.7. Δ^0 is the identity operator on each $\mathcal{K}_{a,b}(G)$.

Example 5.2.8. For $\mathrm{I}=\widehat{\delta}_o=\{\mathrm{I}:\mathcal{H}_\pi^\infty\to\mathcal{H}_\pi^\infty,\ \pi\in\widehat{G}\}$ and any $\alpha\in\mathbb{N}_0^n\backslash\{0\}$, we have $\Delta^\alpha\mathrm{I}=0$.

Proof. We know that $\mathrm{I}=\widehat{\delta}_0$ (see Example 5.1.38). The distribution $\tilde{q}_\alpha\delta_0$ is defined by

$$\langle\tilde{q}_\alpha\delta_0,\phi\rangle=\langle\delta_0,\tilde{q}_\alpha\phi\rangle,\quad \phi\in\mathcal{D}(G),$$

see (5.26). Since

$$\langle\delta_0,\tilde{q}_\alpha\phi\rangle=(\tilde{q}_\alpha\phi)(0)=\tilde{q}_\alpha(0)\ \phi(0)=0$$

we must have $q\delta_0=0$. Therefore, $\Delta^\alpha\mathrm{I}=\widehat{q\delta_0}=0$. □

More generally, we have

Lemma 5.2.9. *Let $\alpha,\beta\in\mathbb{N}_0^n$. Then the symbol $\{\pi(X)^\beta:\mathcal{H}_\pi^\infty\to\mathcal{H}_\pi^\infty,\pi\in\widehat{G}\}$ (see Example 5.1.38) satisfies*

$$\Delta^\alpha\pi(X)^\beta=0\quad \textit{if } [\alpha]>[\beta].$$

If $[\alpha]\le[\beta]$, then $\Delta^\alpha\pi(X)^\beta$ is a linear combination depending only on α,β, of the terms $\pi(X)^{\beta_2}$ with $[\beta_2]=[\beta]-[\alpha]$, that is,

$$\Delta^\alpha\pi(X)^\beta=\overline{\sum_{[\alpha]+[\beta_2]=[\beta]}}\pi(X)^{\beta_2}.$$

Proof of Lemma 5.2.9. We see that $\Delta^\alpha\pi(X)^\beta$ is the group Fourier transform of the distribution $\tilde{q}_\alpha X^\beta\delta_0$ defined via

$$\langle\tilde{q}_\alpha X^\beta\delta_0,\phi\rangle=\langle X^\beta\delta_0,\tilde{q}_\alpha\phi\rangle=\{X^\beta\}^t\{\tilde{q}_\alpha\phi\}(0)$$

for any $\phi\in\mathcal{D}(G)$, see Example 5.1.38. This is so as long as we prove that $\tilde{q}_\alpha X^\beta\delta_0$ is in some $\mathcal{K}_{a,b}$. Let us find another expression for this distribution. As $\{X^\beta\}^t$ is a $[\beta]$-homogeneous left-invariant differential operators, by the Leibniz formula for vector fields, we have

$$\{X^\beta\}^t\{\tilde{q}_\alpha\phi\} = \overline{\sum_{[\beta_1]+[\beta_2]=[\beta]}} X^{\beta_1}\tilde{q}_\alpha\ X^{\beta_2}\phi.$$

We easily see that $X^{\beta_1}\tilde{q}_\alpha \in \mathcal{P}_{=[\alpha]-[\beta_1]}$ and, therefore, by Part (2) of Proposition 5.2.3 we have

$$X^{\beta_1}\tilde{q}_\alpha = \overline{\sum_{[\alpha']=[\alpha]-[\beta_1]}} \tilde{q}_{\alpha'}.$$

Hence we have obtained

$$\{X^\beta\}^t\{\tilde{q}_\alpha\phi\} = \overline{\sum_{\substack{[\beta_1]+[\beta_2]=[\beta]\\ [\alpha']=[\alpha]-[\beta_1]}}} \tilde{q}_{\alpha'}\ X^{\beta_2}\phi,$$

and

$$\langle \tilde{q}_\alpha X^\beta\delta_0, \phi\rangle = \overline{\sum_{\substack{[\beta_1]+[\beta_2]=[\beta]\\ [\alpha']=[\alpha]-[\beta_1]}}} (\tilde{q}_{\alpha'}X^{\beta_2}\phi)(0) = \overline{\sum_{\substack{[\beta_1]+[\beta_2]=[\beta]\\ 0=[\alpha]-[\beta_1]}}} X^{\beta_2}\phi(0),$$

with the convention that the sum is zero if there are no such β_1, β_2. Thus

$$\tilde{q}_\alpha X^\beta\delta_0 = \overline{\sum_{\substack{[\beta_1]+[\beta_2]=[\beta]\\ [\alpha]=[\beta_1]}}} X^{\beta_2}\delta_0.$$

Since $X^{\beta_2}\delta_0 \in \mathcal{K}_{[\beta_2],0}$ (see Example 5.1.26), we see that $\tilde{q}_\alpha X^\beta\delta_0 \in \mathcal{K}_{[\beta],0}$. Furthermore, taking the group Fourier transform we obtain

$$\Delta^\alpha\pi(X)^\beta = \overline{\sum_{\substack{[\beta_1]+[\beta_2]=[\beta]\\ [\alpha]=[\beta_1]}}} \pi(X)^{\beta_2}.$$

This sum is zero if there are no such β_1, β_2, for instance if $[\beta] < [\alpha]$. □

Let us collect some properties of the difference operators.

Proposition 5.2.10. *(i) For any $\alpha \in \mathbb{N}_0^n$, the operator Δ^α is linear, its domain of definition contains $\mathcal{F}_G(\mathcal{S}(G))$ and $\Delta^\alpha\mathcal{F}_G(\mathcal{S}(G)) \subset \mathcal{F}_G(\mathcal{S}(G))$.*

(ii) For any $\alpha_1, \alpha_2 \in \mathbb{N}_0^n$, there exist constants $c_{\alpha_1,\alpha_2,\alpha} \in \mathbb{R}$, with $\alpha \in \mathbb{N}_0^n$ such that $[\alpha] = [\alpha_1] + [\alpha_2]$, so that for any $\phi \in \mathcal{S}(G)$, we have

$$\Delta^{\alpha_1}\left(\Delta^{\alpha_2}\widehat{\phi}\right) = \Delta^{\alpha_2}\left(\Delta^{\alpha_1}\widehat{\phi}\right) = \sum_{[\alpha]=[\alpha_1]+[\alpha_2]} c_{\alpha_1,\alpha_2,\alpha}\Delta^\alpha\widehat{\phi},$$

where the sum is taken over all $\alpha \in \mathbb{N}_0^n$ satisfying $[\alpha] = [\alpha_1] + [\alpha_2]$.

(iii) For any $\alpha \in \mathbb{N}_0^n$, there exist constants $c_{\alpha,\alpha_1,\alpha_2} \in \mathbb{R}$, $\alpha_1, \alpha_2 \in \mathbb{N}_0^n$, with $[\alpha_1] + [\alpha_2] = [\alpha]$, such that for any $\phi_1, \phi_2 \in \mathcal{S}(G)$, we have

$$\Delta^\alpha \left(\widehat{\phi_1}\ \widehat{\phi_2}\right) = \sum_{[\alpha_1]+[\alpha_2]=[\alpha]} c_{\alpha,\alpha_1,\alpha_2}\ \Delta^{\alpha_1}\widehat{\phi_1}\ \Delta^{\alpha_2}\widehat{\phi_2}, \tag{5.28}$$

where the sum is taken over all $\alpha_1, \alpha_2 \in \mathbb{N}_0^n$ satisfying $[\alpha_1] + [\alpha_2] = [\alpha]$. Moreover,

$$c_{\alpha,\alpha_1,0} = \begin{cases} 1 & \text{if } \alpha_1 = \alpha \\ 0 & \text{otherwise} \end{cases}, \quad c_{\alpha,0,\alpha_2} = \begin{cases} 1 & \text{if } \alpha_2 = \alpha \\ 0 & \text{otherwise} \end{cases}.$$

The coefficients $c_{\alpha_1,\alpha_2,\alpha}$ in (ii) and $c_{\alpha,\alpha_1,\alpha_2}$ in (iii) are different in general. We interpret Formula (5.28) as the *Leibniz formula.*

Proof. Since the Schwartz space is stable under multiplication by polynomials, $\tilde{q}_\alpha\phi$ is Schwartz for any $\phi \in \mathcal{S}(G)$, and $\Delta^\alpha\widehat{\phi}(\pi) = \pi(\tilde{q}_\alpha\phi)$. This shows (i).

For Part (ii), we see that the polynomial $q_{\alpha_1}q_{\alpha_2}$ is homogeneous of degree $[\alpha_1]+[\alpha_2]$. Since $\{q_\alpha, [\alpha] = M\}$ is a basis of $\mathcal{P}_{=M}$ by Proposition 5.2.3, there exist constants $c_{\alpha_1,\alpha_2,\alpha} \in \mathbb{R}$, $\alpha_1, \alpha_2 \in \mathbb{N}_0^n$ with $[\alpha_1] + [\alpha_2] = [\alpha]$, satisfying

$$q_{\alpha_1}q_{\alpha_2} = \sum_{[\alpha_1]+[\alpha_2]=[\alpha]} c_{\alpha_1,\alpha_2,\alpha}\, q_\alpha.$$

Therefore

$$\begin{aligned} \Delta^{\alpha_1}\left(\Delta^{\alpha_2}\widehat{\phi}(\pi)\right) = \pi(\tilde{q}_{\alpha_1}\tilde{q}_{\alpha_2}\phi) &= \sum_{[\alpha_1]+[\alpha_2]=[\alpha]} c_{\alpha_1,\alpha_2,\alpha}\pi(\tilde{q}_\alpha\phi) \\ &= \sum_{[\alpha_1]+[\alpha_2]=[\alpha]} c_{\alpha_1,\alpha_2,\alpha}\Delta^\alpha\widehat{\phi}(\pi). \end{aligned}$$

This and the equality $\tilde{q}_{\alpha_1}\tilde{q}_{\alpha_2} = \tilde{q}_{\alpha_2}\tilde{q}_{\alpha_1}$ show (ii).

Let us prove (iii). By Proposition 5.2.3 (4),

$$\begin{aligned} \tilde{q}_\alpha(x)\ (\phi_2 * \phi_1)(x) &= \int_G q_\alpha(x^{-1}y\, y^{-1})\ \phi_2(y)\ \phi_1(y^{-1}x)\ dy \\ &= \sum_{[\alpha_1]+[\alpha_2]=[\alpha]} c_{\alpha_1,\alpha_2} \int_G q_{\alpha_2}(y^{-1})\phi_2(y)\ q_{\alpha_1}(x^{-1}y)\phi_1(y^{-1}x)\ dy \\ &= \sum_{[\alpha_1]+[\alpha_2]=[\alpha]} c_{\alpha_1,\alpha_2}\ (\tilde{q}_{\alpha_2}\phi_2) * (\tilde{q}_{\alpha_1}\phi_1), \end{aligned}$$

with constants depending on $\alpha, \alpha_1, \alpha_2$. Taking the Fourier transform implies the formula (5.28), with conclusions on coefficients following from Proposition 5.2.3. □

We will see that the difference operators Δ^α defined in Definition 5.2.5 appear in the general asymptotic formulae for adjoint and product of pseudo-differential operators in our context, see Sections 5.5.3 and 5.5.2.

5.2.2 Symbol classes $S^m_{\rho,\delta}$

In this section we define the symbol classes $S^m_{\rho,\delta} = S^m_{\rho,\delta}(G)$ of symbols on a graded Lie group G and discuss their properties. We use the notation for the symbol classes similar to the familiar ones on the Euclidean space and also on compact Lie groups.

Let us give the formal definition of our symbol classes.

Definition 5.2.11. Let $m, \rho, \delta \in \mathbb{R}$ with $0 \leq \rho \leq \delta \leq 1$. Let $\mathcal{R}$ be a positive Rockland operator of homogeneous degree ν. A symbol

$$\sigma = \{\sigma(x,\pi) : \mathcal{H}^\infty_\pi \to \mathcal{H}_\pi, (x,\pi) \in G \times \widehat{G}\}$$

is called a *symbol of order m and of type (ρ,δ)* whenever, for each $\alpha, \beta \in \mathbb{N}^n_0$ and $\gamma \in \mathbb{R}$, we have

$$\sup_{x\in G} \|X^\beta_x \Delta^\alpha \sigma(x,\cdot)\|_{L^\infty_{\gamma,\rho[\alpha]-m-\delta[\beta]+\gamma}(\widehat{G})} < \infty. \tag{5.29}$$

The *symbol class* $S^m_{\rho,\delta} = S^m_{\rho,\delta}(G)$ is the set of symbols of order m and of type (ρ,δ).

By Corollary 5.1.30, the symbols $X^\beta_x \Delta^\alpha \sigma$ are fields acting on smooth vectors. By Lemma 5.1.23, we can reformulate (5.29) as

$$\sup_{x\in G,\pi\in\widehat{G}} \|\pi(\mathrm{I}+\mathcal{R})^{\frac{\rho[\alpha]-m-\delta[\beta]+\gamma}{\nu}} X^\beta_x \Delta^\alpha \sigma(x,\pi)\pi(\mathrm{I}+\mathcal{R})^{-\frac{\gamma}{\nu}}\|_{\mathscr{L}(\mathcal{H}_\pi)} < \infty. \tag{5.30}$$

Recall that, as usual, the supremum in π in (5.30) has to be understood as the essential supremum with respect to the Plancherel measure.

Clearly, the converse holds: if σ is a symbol such that $X^\beta_x \Delta^\alpha \sigma$ are fields acting on smooth vectors for which (5.30) holds, then σ is in $S^m_{\rho,\delta}$.

We note that condition (5.30) requires one to fix a positive Rockland operator $\mathcal{R}$ in order to fix the norms of $L^\infty_{a',b'}(\widehat{G})$. However, the resulting class $S^m_{\rho,\delta}$ does not depend on the choice of $\mathcal{R}$, see Lemma 5.1.22.

If a positive Rockland operator $\mathcal{R}$ of homogeneous degree ν is fixed, then we set for $\sigma \in S^m_{\rho,\delta}$ and $a, b, c \in \mathbb{N}_0$,

$$\|\sigma\|_{S^m_{\rho,\delta},a,b,c} := \sup_{\substack{|\gamma|\leq c\\ [\alpha]\leq a,\ [\beta]\leq b}} \sup_{x\in G} \|X^\beta_x \Delta^\alpha \sigma(x,\cdot)\|_{L^\infty_{\gamma,\rho[\alpha]-m-\delta[\beta]+\gamma}(\widehat{G})}.$$

This quantity is also equal to

$$\|\sigma\|_{S^m_{\rho,\delta},a,b,c} = \sup_{x\in G,\ \pi\in\widehat{G}} \|\sigma(x,\pi)\|_{S^m_{\rho,\delta},a,b,c},$$

where we define for any symbol σ, $a, b, c \in \mathbb{N}_0$, and $(x,\pi) \in G \times \widehat{G}$ (fixed)

$$\|\sigma(x,\pi)\|_{S^m_{\rho,\delta},a,b,c} := \sup_{\substack{|\gamma|\leq c\\ [\alpha]\leq a,\ [\beta]\leq b}} \|\pi(\mathrm{I}+\mathcal{R})^{\frac{\rho[\alpha]-m-\delta[\beta]+\gamma}{\nu}} X^\beta_x \Delta^\alpha \sigma(x,\pi)\pi(\mathrm{I}+\mathcal{R})^{-\frac{\gamma}{\nu}}\|_{\mathscr{L}(\mathcal{H}_\pi)}.$$

Here, as always, the supremum has to be understood as the essential supremum with respect to the Plancherel measure.

Before making some comments, let us say that the classes of symbols we have just defined have the usual structures of symbol classes.

Proposition 5.2.12. *The symbol class* $S^m_{\rho,\delta}$ *is a vector space independent of any Rockland operator* $\mathcal{R}$ *used in (5.29) to consider the* $L^\infty_{\gamma,\rho[\alpha]-m-\delta[\beta]+\gamma}(\widehat{G})$*-norms. We have the continuous inclusions*

$$m_1 \le m_2, \quad \delta_1 \le \delta_2, \quad \rho_1 \ge \rho_2 \quad \Longrightarrow \quad S^{m_1}_{\rho_1,\delta_1} \subset S^{m_2}_{\rho_2,\delta_2}. \tag{5.31}$$

We fix a positive Rockland operator $\mathcal{R}$*. For any* $m \in \mathbb{R}$, $\rho, \delta \ge 0$*, the resulting maps* $\|\cdot\|_{S^m_{\rho,\delta},a,b,c}$, $a,b,c \in \mathbb{N}_0$*, are seminorms over the vector space* $S^m_{\rho,\delta}$ *which endow* $S^m_{\rho,\delta}$ *with the structure of a Fréchet space.*

We may replace the family of seminorms $\|\cdot\|_{S^m_{\rho,\delta},a,b,c}$, $a,b,c \in \mathbb{N}_0$*, by*

$$\sigma \longmapsto \sup_{\substack{[\alpha]\le a,\\ [\beta]\le b}} \sup_{x\in G} \|X^\beta_x \Delta^\alpha \sigma(x,\cdot)\|_{L^\infty_{\gamma_\ell,\rho[\alpha]-m-\delta[\beta]+\gamma_\ell}(\widehat{G})}, \qquad a,b \in \mathbb{N}_0, \ \ell \in \mathbb{Z},$$

where the sequence $\{\gamma_\ell, \ell \in \mathbb{Z}\}$ *of real numbers satisfies* $\gamma_\ell \underset{\ell\to\pm\infty}{\longrightarrow} \pm\infty$.

Two different positive Rockland operators give equivalent families of seminorms. The topology on $S^m_{\rho,\delta}$ *is independent of the choice of the Rockland operator* $\mathcal{R}$.

Proof. Using Corollary 5.1.30 and Lemma 5.1.22, this is a routine exercise. □

Remark 5.2.13. Let us make some comments about Definition 5.2.11:

1. In the abelian case, that is, $\mathbb{R}^n$ endowed with the addition law, and $\mathcal{R} = -\mathcal{L}$ with $\mathcal{L}$ being the Laplace operator, $S^m_{\rho,\delta}$ boils down to the usual Hörmander class, in view of the difference operators corresponding to the derivatives, see Example 5.2.6.

2. In the case of compact Lie groups with $\mathcal{R}$ being the (positive) Laplacian, a similar definition leads to the one considered in (2.26) since the operator $\pi(\mathrm{I}+\mathcal{R})$ is scalar. However, here, in the case of non-abelian graded Lie groups, the operator $\mathcal{R}$ can not have a scalar Fourier transform.

3. The presence of the parameter γ is included to facilitate proving that the space of symbols $\cup_{m\in\mathbb{R}} S^m_{\rho,\delta}$, with suitable restrictions on ρ, δ, forms an algebra of operators later on. It already has enabled us to see that the symbols are fields of operators acting on smooth vectors and therefore can be composed without using the composition of unbounded operators (in Definition A.3.2).

   We will see in Theorem 5.5.20 that in fact we can remove this γ. By this we mean that a symbol σ is in $S^m_{\rho,\delta}$ if and only if the condition in (5.29) holds for any $\alpha, \beta \in \mathbb{N}_0^n$ and $\gamma = 0$. Furthermore, the seminorms $\|\cdot\|_{S^m_{\rho,\delta},a,b,0}$, $a, b \in \mathbb{N}_0$, yield the topology of $S^m_{\rho,\delta}$.

4. We could have used other families of difference operators instead of the Δ^α's to define the symbol classes $S^m_{\rho,\delta}$. For instance, we could have used any family of difference operators associated with a family $\{p_\alpha\}_{\alpha\in\mathbb{N}_0^n}$ of homogeneous polynomials on G which satisfy
   - for each $\alpha \in \mathbb{N}_0^n$, p_α is of homogeneous degree $[\alpha]$,
   - and $\{p_\alpha\}_{\alpha\in\mathbb{N}_0^n}$ is a basis of $\mathcal{P}(G)$.

   Indeed, in this case, the following properties hold.

   - Any $\tilde{q}_\alpha$ is a linear combination of p_β, $[\beta] = [\alpha]$.
   - Conversely, any p_α is a linear combination of $\tilde{q}_\beta$, $[\beta] = [\alpha]$.

   Thus,

   - any Δ^α is a linear combination of Δ_{p_β}, $[\beta] = [\alpha]$.
   - Conversely, any Δ_{p_α} is a linear combination of Δ^β, $[\beta] = [\alpha]$.

   It is then easy to see that a symbol σ is in $S^m_{\rho,\delta}$ if and only if for each $\alpha, \beta \in \mathbb{N}_0^n$ and $\gamma \in \mathbb{R}$,

$$\sup_{x\in G} \|X_x^\beta \Delta_{p_\alpha} \sigma(x,\cdot)\|_{L^\infty_{\gamma,\rho[\alpha]-m-\delta[\beta]+\gamma}(\widehat{G})} < \infty.$$

   Note that this implies that the symbol class $S^m_{\rho,\delta}$ does not depend on a particular choice of realisation of G through a basis of $\mathfrak{g}$ (of eigenvectors for the dilations) but only on the graded Lie group G and its homogeneous structure.

   For such a family Δ_{p_α}, the same proof as for Proposition 5.2.10 shows a Leibniz formula in the sense of (5.28).

   Although we could use 'easier' difference operators to define our symbol classes, for instance Δ_{x^α}, $\alpha \in \mathbb{N}_0^n$, we choose to present our analysis with the difference operators Δ^α given in Definition 5.2.5. Note that the asymptotic formulae for composition and adjoint in (5.57) and (5.60) will be expressed in terms of the difference operators Δ^α and derivatives X_x^α.

   Note that the change of difference operators explained just above is linear, whereas in the compact case, one can use many more difference operators to define the symbol classes $S^m_{\rho,\delta}$, see Section 2.2.2.

The type $(1,0)$ can be thought of as the basic class of symbols and the types (ρ,δ) as its generalisations. There are certain limitations on the parameters (ρ,δ) coming from reasons similar to the ones in the Euclidean settings. For type $(1,0)$, we set

$$S^m := S^m_{1,0},$$

and

$$\|\sigma(x,\pi)\|_{S^m_{1,0},a,b,c} = \|\sigma(x,\pi)\|_{a,b,c}, \ \|\sigma\|_{S^m_{1,0},a,b,c} = \|\sigma\|_{a,b,c}, \ \text{etc.}\ldots$$

We also define the class of smoothing symbols

Definition 5.2.14. We set

$$S^{-\infty} := \bigcap_{m\in\mathbb{R}} S^m.$$

One checks easily that

$$S^{-\infty} = \bigcap_{m\in\mathbb{R}} S^m_{\rho,\delta},$$

independently of ρ and δ as long as $0 \le \delta \le \rho \le 1$ and $\rho \neq 0$. Moreover, $S^{-\infty}$ is equipped with the topology of projective limit induced by $\cap_{m\in\mathbb{R}} S^m_{\rho,\delta}$, again independently of ρ and δ.

We will see in Corollary 5.4.10 that the symbols in $S^{-\infty}$ really deserve to be called smoothing.

5.2.3 Operator classes $\Psi^m_{\rho,\delta}$

The pseudo-differential operators of order $m \in \mathbb{R} \cup \{-\infty\}$ and type (ρ,δ) are obtained by the quantization

$$\text{Op}(\sigma)\phi(x) = \int_{\widehat{G}} \text{Tr}\left(\pi(x)\sigma(x,\pi)\widehat{\phi}(\pi)\right) d\mu(\pi),$$

justified in Theorem 5.1.39, from the symbols of the same order and type, that is,

$$\Psi^m_{\rho,\delta} := \text{Op}(S^m_{\rho,\delta}).$$

They inherit a structure of topological vector spaces from the classes of symbols,

$$\|\text{Op}(\sigma)\|_{\Psi^m_{\rho,\delta},a,b,c} := \|\sigma\|_{S^m_{\rho,\delta},a,b,c}.$$

For type $(1,0)$, we set as for the corresponding symbol classes:

$$\Psi^m := \Psi^m_{1,0}.$$

Continuity on $\mathcal{S}(G)$

By Theorem 5.1.39, any operator in the operator classes defined above maps Schwartz functions to smooth functions. Let us show that in fact it acts continuously on the Schwartz space:

Theorem 5.2.15. *Let $T \in \Psi^m_{\rho,\delta}$ where $m \in \mathbb{R}$, $1 \geq \rho \geq \delta \geq 0$. Then for any $\phi \in \mathcal{S}(G)$, $T\phi \in \mathcal{S}(G)$. Moreover the operator T act continuously on $\mathcal{S}(G)$: for any seminorm $\|\cdot\|_{\mathcal{S}(G),N}$ there exist a constant $C > 0$ and a seminorm $\|\cdot\|_{\mathcal{S}(G),N'}$ such that for every $\phi \in \mathcal{S}(G)$,*

$$\|T\phi\|_{\mathcal{S}(G),N} \leq C\|\phi\|_{\mathcal{S}(G),N'}.$$

The constant C can be chosen as $C_1\|T\|_{\Psi^m_{\rho,\delta},a,b,c}$ where C_1 is a constant of and the seminorm $\|\cdot\|_{\Psi^m_{\rho,\delta},a,b,c}$ depend on G, m, ρ,δ, and on the seminorm $\|\cdot\|_{\mathcal{S}(G),N}$.

In other words, the mapping $T \mapsto T$ from $\Psi^m_{\rho,\delta}$ to the space $\mathscr{L}(\mathcal{S}(G))$ of continuous operators on $\mathcal{S}(G)$ is continuous (it is clearly linear).

Our proof of Theorem 5.2.15 will require the following preliminary result on the right convolution kernels:

Proposition 5.2.16. *Let $\sigma = \{\sigma(x,\pi)\}$ be in $S^m_{\rho,\delta}$ with $1 \geq \rho \geq \delta \geq 0$. Let κ_x denote its associated kernel. If $m < -Q/2$ then for any $x \in G$, the distribution κ_x is square integrable and*

$$\begin{aligned}
\|\kappa_x\|_{L^2(G)} &\leq C \sup_{\pi\in\widehat{G}} \|\pi(\mathrm{I}+\mathcal{R})^{\frac{-m}{\nu}}\sigma(x,\pi)\|_{\mathscr{L}(\mathcal{H}_\pi)},\\
\|\kappa_x\|_{L^2(G)} &\leq C \sup_{\pi\in\widehat{G}} \|\sigma(x,\pi)\pi(\mathrm{I}+\mathcal{R})^{\frac{-m}{\nu}}\|_{\mathscr{L}(\mathcal{H}_\pi)},
\end{aligned}$$

with $C = C_m > 0$ a finite constant independent of σ and x.

The proof below will show that we can choose $C_m = \|\mathcal{B}_{-m}\|_{L^2(G)}$ the L^2-norm of the right-convolution kernel of the Bessel potential of the positive Rockland operator $\mathcal{R}$.

Proof of Proposition 5.2.16. We write

$$\begin{aligned}
\|\sigma(x,\pi)\|_{\mathrm{HS}} &= \|\pi(\mathrm{I}+\mathcal{R})^{\frac{m}{\nu}}\pi(\mathrm{I}+\mathcal{R})^{\frac{-m}{\nu}}\sigma(x,\pi)\|_{\mathrm{HS}}\\
&\leq \|\pi(\mathrm{I}+\mathcal{R})^{\frac{m}{\nu}}\|_{\mathrm{HS}}\|\pi(\mathrm{I}+\mathcal{R})^{\frac{-m}{\nu}}\sigma(x,\pi)\|_{\mathscr{L}(\mathcal{H}_\pi)},
\end{aligned}$$

which shows

$$\|\sigma(x,\pi)\|_{\mathrm{HS}} \leq \sup_{\pi_1\in\widehat{G}} \|\pi_1(\mathrm{I}+\mathcal{R})^{\frac{-m}{\nu}}\sigma(x,\pi_1)\|_{\mathscr{L}(\mathcal{H}_{\pi_1})}\|\pi(\mathrm{I}+\mathcal{R})^{\frac{m}{\nu}}\|_{\mathrm{HS}}.$$

Squaring and integrating against the Plancherel measure, we obtain

$$\int_{\widehat{G}} \|\sigma(x,\pi)\|^2_{\mathrm{HS}}d\mu(\pi) \leq \sup_{\pi_1\in\widehat{G}} \|\pi_1(\mathrm{I}+\mathcal{R})^{\frac{-m}{\nu}}\sigma(x,\pi_1)\|^2_{\mathscr{L}(\mathcal{H}_{\pi_1})} \int_{\widehat{G}} \|\pi(\mathrm{I}+\mathcal{R})^{\frac{m}{\nu}}\|^2_{\mathrm{HS}}d\mu(\pi).$$

By the Plancherel formula and Corollary 5.1.4, if $m < -Q/2$, we have

$$C_m^2 := \int_{\widehat{G}} \|\pi(\mathrm{I}+\mathcal{R})^{\frac{m}{\nu}}\|_{\mathtt{HS}}^2 d\mu(\pi) = \|\mathcal{B}_{-m}\|_{L^2(G)}^2 < \infty.$$

This gives the first estimate in the statement. For the second estimate, we write

$$\sigma(x,\pi) = \sigma(x,\pi)\pi(\mathrm{I}+\mathcal{R})^{\frac{-m}{\nu}}\pi(\mathrm{I}+\mathcal{R})^{\frac{m}{\nu}},$$

and adapt the ideas above. □

We can now prove Theorem 5.2.15.

Proof of Theorem 5.2.15. Let $T \in \Psi^m_{\rho,\delta}$ where $m \in \mathbb{R}$, $1 \geq \rho \geq \delta \geq 0$. Then for any $\phi \in \mathcal{S}(G)$, $T\phi$ is smooth by Theorem 5.1.39.

Let $\kappa : (x,y) \mapsto \kappa_x(y)$ be the kernel associated with T. Let $\mathcal{R}$ be a positive Rockland operator of homogeneous degree ν. The properties of $\mathcal{R}$ (see Sections 4.3 and 4.4.8) yield for any $\phi \in \mathcal{S}(G)$ and $x \in G$ that

$$\begin{aligned} T\phi(x) &= \int_G \phi(y)\kappa_x(y^{-1}x)dy \\ &= \int_G \left[(\mathrm{I}+\mathcal{R})^{-N}\{(\mathrm{I}+\mathcal{R})^N\phi\}(y)\right] \quad \kappa_x(y^{-1}x)dy \\ &= \int_G \{(\mathrm{I}+\mathcal{R})^N\phi\}(y)\ \{(\mathrm{I}+\tilde{\mathcal{R}})^{-N}\kappa_x\}(y^{-1}x)dy, \end{aligned}$$

thus, by the Cauchy-Schwartz inequality,

$$|T\phi(x)| \leq \|(\mathrm{I}+\mathcal{R})^N\phi\|_{L^2(G)}\|(\mathrm{I}+\tilde{\mathcal{R}})^{-N}\kappa_x\|_{L^2(G)}.$$

Since $\mathcal{F}_G\{(\mathrm{I}+\tilde{\mathcal{R}})^{-N}\kappa_x\}(\pi) = \sigma(x,\pi)\pi(\mathrm{I}+\mathcal{R})^{-N}$ yields a symbol in $S^{m-N\nu}_{\rho,\delta}$, by Proposition 5.2.16, we have

$$\|(\mathrm{I}+\tilde{\mathcal{R}})^{-N}\kappa_x\|_{L^2(G)} \leq C \sup_{\pi\in\widehat{G}} \|\sigma(x,\pi)\pi(\mathrm{I}+\mathcal{R})^{-N}\|_{\mathscr{L}(\mathcal{H}_\pi)},$$

whenever $m - N\nu < -Q/2$. Note that in this case,

$$\sup_{\pi\in\widehat{G}} \|\sigma(x,\pi)\pi(\mathrm{I}+\mathcal{R})^{-N}\|_{\mathscr{L}(\mathcal{H}_\pi)} \leq \|\sigma\|_{S^m_{\rho,\delta},0,0,|m|}\|\pi(\mathrm{I}+\mathcal{R})^{-N+\frac{m}{\nu}}\|_{\mathscr{L}(\mathcal{H}_\pi)},$$

and by functional calculus

$$\|\pi(\mathrm{I}+\mathcal{R})^{-N+\frac{m}{\nu}}\|_{\mathscr{L}(\mathcal{H}_\pi)} \leq \sup_{\lambda\geq 0}(1+\lambda)^{-N+\frac{m}{\nu}} \leq 1.$$

Thus if we choose $N \in \mathbb{N}_0$ such that $N > (m+\frac{Q}{2})/\nu$, then

$$|T\phi(x)| \leq C\|\sigma\|_{S^m_{\rho,\delta},0,0,|m|}\|(\mathrm{I}+\mathcal{R})^N\phi\|_{L^2(G)}.$$

This shows that $T\phi$ is bounded.

Let $\beta \in \mathbb{N}_0^n$. Using the Leibniz property of vector fields, we easily obtain

$$X^\beta T\phi(x) = \sum_{[\beta_1]+[\beta_2]=[\beta]} c_{\beta_1,\beta_2,\beta} \int_G \phi(y) X^{\beta_1}_{x_1=x} X^{\beta_2}_{x_2=y^{-1}x} \kappa_{x_1}(x_2) dy.$$

As above, we can insert powers of $\mathrm{I}+\mathcal{R}$. Noticing that the symbol

$$\mathcal{F}_G\{(\mathrm{I}+\tilde{\mathcal{R}})^{-N}_{x_1} X^{\beta_1}_{x_1=x} X^{\beta_2}\kappa_{x_1}\} = \pi(X)^{\beta_2} X_x^{\beta_1}\sigma(x,\pi)\pi(\mathrm{I}+\mathcal{R})^{-N}$$

is in $S^{m+\delta[\beta_1]+[\beta_2]-N\nu}_{\rho,\delta}$, we proceed as above to obtain

$$\begin{aligned} |X^\beta T\phi(x)| &\leq C_1 \sum_{[\beta_1]+[\beta_2]=[\beta]} \|(\mathrm{I}+\mathcal{R})^N\phi\|_{L^2(G)} \|\pi(X)^{\beta_2} X_x^{\beta_1}\sigma(x,\pi)\pi(\mathrm{I}+\mathcal{R})^{-N}\|_{L^2(\widehat{G})} \\ &\leq C_2 \|\sigma\|_{S^m_{\rho,\delta},0,[\beta],|m|+[\beta]} \|(\mathrm{I}+\mathcal{R})^N\phi\|_{L^2(G)}. \end{aligned}$$

as long as $N > (m+[\beta]+\frac{Q}{2})/\nu$.

Let $\alpha \in \mathbb{N}_0^n$. Proceeding as in the proof of Proposition 5.2.3 (4), we can write

$$(xy)^\alpha = \sum_{[\alpha_1]+[\alpha_2]=[\alpha]} c'_{\alpha,\alpha_1,\alpha_2}\, q_{\alpha_1}(x)\, q_{\alpha_2}(y).$$

Using this, we easily obtain

$$\begin{aligned} x^\alpha T\phi(x) &= \int_G (y\ y^{-1}x)^\alpha \phi(y)\kappa_x(y^{-1}x)dy \\ &= \sum_{[\alpha_1]+[\alpha_2]=[\alpha]} c'_{\alpha,\alpha_1,\alpha_2} \int_G q_{\alpha_1}(y)\phi(y) q_{\alpha_2}(y^{-1}x)\kappa_x(y^{-1}x)dy. \end{aligned}$$

Noticing that

$$\mathcal{F}_G\{(\mathrm{I}+\tilde{\mathcal{R}})^{-N}\{q_{\alpha_2}\kappa_x\} = \{\Delta^{\alpha_2}\sigma(x,\cdot)\}\ \pi(\mathrm{I}+\mathcal{R})^{-N} \in S^{m-N\nu-\rho[\alpha_2]}_{\rho,\delta},$$

we can now proceed as in the first paragraph above to obtain

$$\begin{aligned} |x^\alpha T\phi(x)| &\leq C_1 \sum_{[\alpha_1]+[\alpha_2]=[\alpha]} \|(\mathrm{I}+\mathcal{R})^N_y\{q_{\alpha_1}\phi\}\|_2 \|(\mathrm{I}+\tilde{\mathcal{R}})^{-N}\{q_{\alpha_2}\kappa_x\}\|_2 \\ &\leq C_2 \|\sigma(x,\pi)\|_{S^m_{\rho,\delta},[\alpha],0,|m|+\rho[\alpha]} \sum_{[\alpha_1]\leq[\alpha]} \|(\mathrm{I}+\mathcal{R})^N_y\{q_{\alpha_1}\phi\}\|_2 \end{aligned}$$

as long as $N > (m+Q/2)/\nu$.

We can combine the two paragraphs above to show that for any $\alpha,\beta \in \mathbb{N}_0^n$, we have

$$|x^\alpha X^\beta T\phi(x)| \leq C \|\sigma(x,\pi)\|_{S^m_{\rho,\delta},[\alpha],[\beta],|m|+[\beta]+\rho[\alpha]} \sum_{[\alpha_1]\leq[\alpha]} \|(\mathrm{I}+\mathcal{R})^N_y\{q_{\alpha_1}\phi\}\|_2,$$

as long as $N > (m + [\beta] + Q/2)/\nu$. By Lemma 3.1.56, we have

$$\sum_{[\alpha_1]\leq[\alpha]} \|(\mathrm{I}+\mathcal{R})_y^N\{q_{\alpha_1}\phi\}\|_2 \leq C'\|\phi\|_{\mathcal{S}(G),N'}$$

for some $N' \in \mathbb{N}$ depending on N and α, and $T\phi$ is a Schwartz function. Furthermore, these estimates also imply the rest of Theorem 5.2.15. □

Theorem 5.2.15 shows that composing two operators in (possibly different) $\Psi^m_{\rho,\delta}$ makes sense as the composition of operators acting on the Schwartz space. We will see that in fact, the composition of $T_1 \in \Psi^{m_1}_{\rho,\delta}$ with $T_2 \in \Psi^{m_2}_{\rho,\delta}$ is T_1T_2 in $\Psi^{m_1+m_2}_{\rho,\delta}$, see Theorem 5.5.3.

We will see that our classes of pseudo-differential operators are stable under taking the formal L^2-adjoint, see Theorem 5.5.12. This together with Theorem 5.2.15 will imply the continuity of our operators on the space $\mathcal{S}'(G)$ of tempered distributions, see Corollary 5.5.13.

Returning to our exposition, before proving that the introduced classes of symbols $\cup_{m\in\mathbb{R}} S^m_{\rho,\delta}$ and of the corresponding operators $\cup_{m\in\mathbb{R}}\Psi^m_{\rho,\delta}$ are stable under composition and taking the adjoint, let us give some examples.

5.2.4 First examples

As it should be, $\cup_{m\in\mathbb{R}}\Psi^m$ contains the left-invariant differential operators. More precisely, the following lemma implies that $\sum_{[\beta]\leq m} c_\beta X^\beta \in \Psi^m$. The coefficients c_α here are constant and it is easy to relax this condition with each function c_α being smooth and bounded together with all of its left derivatives.

Lemma 5.2.17. *For any $\beta_o \in \mathbb{N}_0^n$, the operator $X^{\beta_o} = \mathrm{Op}(\pi(X)^{\beta_o})$ is in $\Psi^{[\beta_o]}$.*

Proof. By Lemma 5.2.9, we have

$$\Delta^\alpha \pi(X)^{\beta_o} = \begin{cases} 0 & \text{if } [\alpha] > [\beta_o], \\ \sum\limits_{[\alpha]+[\beta_2]=[\beta_o]} \pi(X)^{\beta_2} & \text{if } [\alpha] \leq [\beta_o]. \end{cases}$$

Recall that, by Example 5.1.26, $\{\pi(X)^\beta, \pi \in \widehat{G}\} \in L^\infty_{\gamma+[\beta],\gamma}(\widehat{G})$ for any $\gamma \in \mathbb{R}, \beta \in \mathbb{N}_0^n$. So $\{\Delta^\alpha\pi(X)^{\beta_o}, \pi \in \widehat{G}\}$ is zero if $[\alpha] > [\beta_o]$ whereas it is in $L^\infty_{\gamma+[\beta_o]-[\alpha],\gamma}(\widehat{G})$ for any $\gamma \in \mathbb{R}$ if $[\alpha] \leq [\beta_o]$. □

Remark 5.2.18. Lemma 5.2.17 implies that $\cup_{m\in\mathbb{R}}\Psi^m$ contains the left-invariant differential calculus, that is, the space of left-invariant differential operators.

One could wonder whether it also contains the right-invariant differential calculus, since we can quantize any differential operator, see Remark 5.1.41 (1). This is false in general, see Example 5.2.19 below. Thus, if one is interested in dealing with problems based on the setting of right-invariant operators one can use the corresponding version of the theory based on the right-invariant Rockland operator, see Remark 5.1.41 (4).

Example 5.2.19. Let us consider the three dimensional Heisenberg group $\mathbb{H}_1$ and the canonical basis X, Y, T of its Lie algebra (see Example 1.6.4). Then the right invariant vector field $\tilde{X}$ can not be in $\cup_{m\in\mathbb{R}}\Psi^m$.

Proof of the statement in Example 5.2.19 . We have already seen that any operator $A \in \Psi^m$ acts continuously on the Schwartz space, cf. Theorem 5.2.15. We will see later (see Corollary 5.7.2) that it also acts on Sobolev spaces with a loss of derivative controlled by its order m. By this, we mean that, if an operator A in Ψ^m is homogeneous of degree ν_A, then we must have

$$\forall s \in \mathbb{R} \quad \exists C > 0 \quad \forall f \in \mathcal{S}(G) \qquad \|Af\|_{L^2_{s-m}} \leq C\|f\|_{L^2_s},$$

and when $s+m$ and s are non-negative, we realise the Sobolev norm as $\|f\|_{L^2_s} = \|f\|_{L^2} + \|\mathcal{R}^{\frac{s}{\nu}} f\|_{L^2}$ for some positive Rockland operator of degree ν, cf. Theorem 4.4.3 Part (2). Applying the inequality to dilated functions $f \circ D_r$ and letting $r \to \infty$ yield that $m \geq \nu_A$.

Applying this to the case of $\tilde{X}$ shows that if $\tilde{X}$ were in some Ψ^m then $m \geq 1$ and $\tilde{X}$ would map L^2_1 to L^2_{1-m} hence to L^2 continuously. We have already shown in the proof of Example 4.4.32 that this is not possible. □

An example of a smoothing operator is given via convolution with a Schwartz function:

Lemma 5.2.20. *Let $\kappa \in \mathcal{S}(G)$. We denote by $T_\kappa : \phi \mapsto \phi * \kappa$ the corresponding convolution operator. Its symbol σ_{T_κ} is independent of x and is given by*

$$\sigma_{T_\kappa}(\pi) = \widehat{\kappa}(\pi).$$

Furthermore, the mapping

$$\mathcal{S}(G) \ni \kappa \mapsto T_\kappa \in \Psi^{-\infty}$$

is continuous.

Proof. For the first part, see Example 5.1.38 and its continuation.

For any $\kappa \in \mathcal{S}(G)$, we have $\tilde{q}_\alpha \kappa \in \mathcal{S}(G)$ for any $\alpha \in \mathbb{N}_0^n$, and

$$(\mathrm{I} + \mathcal{R})^a (\mathrm{I} + \tilde{\mathcal{R}})^b \kappa \in \mathcal{S}(G)$$

for any $a, b \in \mathbb{N}$ (see also (4.34) and Proposition 4.4.30). For any $m \in \mathbb{R}$, $\gamma \in \mathbb{R}$ and $\alpha \in \mathbb{N}_0^n$, we have by (1.38)

$$\begin{aligned} \|\Delta^\alpha \widehat{\kappa}\|_{L^\infty_{\gamma,[\alpha]-m+\gamma}(\widehat{G})} &= \|\pi(\mathrm{I}+\mathcal{R})^{\frac{[\alpha]-m+\gamma}{\nu}} \Delta^\alpha \pi(\kappa) \pi(\mathrm{I}+\mathcal{R})^{-\frac{\gamma}{\nu}}\|_{L^\infty(\widehat{G})} \\ &\leq \|(\mathrm{I}+\mathcal{R})^{\frac{[\alpha]-m+\gamma}{\nu}} (\mathrm{I}+\tilde{\mathcal{R}})^{-\frac{\gamma}{\nu}} \{\tilde{q}_\alpha \kappa\}\|_{L^1(G)}. \end{aligned}$$

As $\kappa \in \mathcal{S}(G)$, this L^1-norm is finite and this shows that $\sigma_{T_\kappa} \in \Psi^{-\infty}$. More precisely, this L^1-norm is less or equal to

$$\begin{cases} \|\mathcal{B}_\gamma\|_1 \|(\mathrm{I}+\mathcal{R})^a\{\tilde{q}_\alpha \kappa\}\|_1 & \text{if } \gamma \text{ and } \frac{[\alpha]-m+\gamma}{\nu} > 0 \text{ and } a = \lceil \frac{[\alpha]-m+\gamma}{\nu} \rceil, \\ \|\mathcal{B}_{-\frac{[\alpha]-m+\gamma}{\nu}}\|_1 \|(\mathrm{I}+\tilde{\mathcal{R}})^b\{\tilde{q}_\alpha \kappa\}\|_1 & \text{if } \gamma \text{ and } \frac{[\alpha]-m+\gamma}{\nu} < 0 \text{ and } b = \lceil -\frac{\gamma}{\nu} \rceil, \end{cases}$$

where $\lceil x \rceil$ denotes the smallest integer $> x$ and $\mathcal{B}_\gamma$ is the right-convolution kernel of the Bessel potential of $\mathcal{R}$, see Corollary 4.3.11. By Proposition 4.4.27, these quantities can be estimated by Schwartz seminorms. □

More generally, the operators and symbols with kernels 'depending on x' but satisfying the following property are smoothing:

Lemma 5.2.21. *Let $\kappa : (x, y) \mapsto \kappa_x(y)$ be a smooth function on $G \times G$ such that, for each multi-index $\beta \in \mathbb{N}_0^n$ and each Schwartz seminorm $\|\cdot\|_{\mathcal{S}(G),N}$, the following quantity*

$$\sup_{x \in G} \|X_x^\beta \kappa_x\|_{\mathcal{S}(G),N} < \infty,$$

is finite.

Then the symbol σ given via $\sigma(x,\pi) = \widehat{\kappa}_x(\pi)$ is smoothing. Furthermore for any seminorm $\|\cdot\|_{S^m,a,b,c}$, there exists $C > 0$ and $\beta \in \mathbb{N}_0^n$, $N \in \mathbb{N}_0$ such that

$$\|\sigma\|_{S^m,a,b,c} \leq C \sup_{x \in G} \|X_x^\beta \kappa_x\|_{\mathcal{S}(G),N}.$$

Proof of Lemma 5.2.21. By (1.38), we have

$$\sup_{\pi \in \widehat{G}} \|\sigma(x,\pi)\|_{\mathscr{L}(\mathcal{H}_\pi)} = \sup_{\pi \in \widehat{G}} \|\widehat{\kappa}_x(\pi)\|_{\mathscr{L}(\mathcal{H}_\pi)} \leq \|\kappa_x\|_{L^1(G)}.$$

More generally, for any $\gamma_1, \gamma_2 \in \mathbb{R}$, denoting by $N_1, N_2 \in \mathbb{N}_0$ integers such that $\gamma_1 \leq N_1$ $\gamma_2 \leq N_2$, we have

$$\begin{aligned}
&\sup_{\pi \in \widehat{G}} \|\pi(\mathrm{I}+\mathcal{R})^{\gamma_1} X_x^\beta \Delta^\alpha \sigma(x,\pi)\ \pi(\mathrm{I}+\mathcal{R})^{\gamma_2}\|_{\mathscr{L}(\mathcal{H}_\pi)} \\
&\quad \leq \sup_{\pi \in \widehat{G}} \|\pi(\mathrm{I}+\mathcal{R})^{N_1} X_x^\beta \Delta^\alpha \sigma(x,\pi)\ \pi(\mathrm{I}+\mathcal{R})^{N_2}\|_{\mathscr{L}(\mathcal{H}_\pi)} \\
&\quad\quad = \sup_{\pi \in \widehat{G}} \|\mathcal{F}_G\{(\mathrm{I}+\mathcal{R})^{N_1}(\mathrm{I}+\tilde{\mathcal{R}})^{N_2} X_x^\beta q_\alpha \kappa_x\}(\pi)\|_{\mathscr{L}(\mathcal{H}_\pi)} \\
&\quad\quad \leq \|(\mathrm{I}+\mathcal{R})^{N_1}(\mathrm{I}+\tilde{\mathcal{R}})^{N_2} q_\alpha X_x^\beta \kappa_x\|_{L^1(G)}.
\end{aligned}$$

This last L^1-norm is, up to a constant, less or equal than a Schwartz seminorm of $X_x^\beta \kappa_x$, see Section 3.1.9. This implies the statement. □

In Theorem 5.4.9, we will see that the converse holds, that is, that any smoothing operator has an associated kernel as in Lemma 5.2.21.

5.2.5 First properties of symbol classes

We summarise in the next theorem some properties of the symbol classes which follow from their definition.

Theorem 5.2.22. *Let $1 \geq \rho \geq \delta \geq 0$.*

(i) Let $\sigma \in S^m_{\rho,\delta}$ have kernel κ_x and order $m \in \mathbb{R}$.

1. *For every $x \in G$ and $\gamma \in \mathbb{R}$,*

$$\tilde{q}_\alpha X^\beta \kappa_x \in \mathcal{K}_{\gamma,\rho[\alpha]-m-\delta[\beta]+\gamma}.$$

2. *If $\beta_o \in \mathbb{N}_0^n$ then the symbol $\{X_x^{\beta_o}\sigma(x,\pi),(x,\pi) \in G \times \widehat{G}\}$ is in $S^{m+\delta[\beta_o]}_{\rho,\delta}$ with kernel $X_x^{\beta_o}\kappa_x$, and*

$$\|X_x^{\beta_o}\sigma(x,\pi)\|_{S^{m+\delta[\beta_o]}_{\rho,\delta},a,b,c} \leq C_{b,\beta_o}\|\sigma(x,\pi)\|_{S^m_{\rho,\delta},a,b+[\beta_o],c}.$$

3. *If $\alpha_o \in \mathbb{N}_0^n$ then the symbol $\{\Delta^{\alpha_o}\sigma(x,\pi),(x,\pi) \in G \times \widehat{G}\}$ is in $S^{m-\rho[\alpha_o]}_{\rho,\delta}$ with kernel $\tilde{q}_{\alpha_o}\kappa_x$, and*

$$\|\Delta^{\alpha_o}\sigma(x,\pi)\|_{S^{m-\rho[\alpha_o]}_{\rho,\delta},a,b,c} \leq C_{a,\alpha_o}\|\sigma(x,\pi)\|_{S^m_{\rho,\delta},a+[\alpha_o],b,c}.$$

4. *The symbol*

$$\sigma^* := \{\sigma(x,\pi)^*,(x,\pi) \in G \times \widehat{G}\}$$

is in $S^m_{\rho,\delta}$ with kernel κ_x^* given by

$$\kappa_x^*(y) = \bar{\kappa}_x(y^{-1}),$$

and

$$\|\sigma(x,\pi)^*\|_{S^m_{\rho,\delta},a,b,c} = \sup_{\substack{|\gamma|\leq c \\ [\alpha]\leq a,\,[\beta]\leq b}} \|\pi(\mathrm{I}+\mathcal{R})^{-\frac{\gamma}{\nu}}X_x^\beta\Delta^\alpha\sigma(x,\pi)\pi(\mathrm{I}+\mathcal{R})^{\frac{\rho[\alpha]-m-\delta[\beta]+\gamma}{\nu}}\|_{\mathscr{L}(\mathcal{H}_\pi)}.$$

(ii) Let $\sigma_1 \in S^{m_1}_{\rho,\delta}$ and $\sigma_2 \in S^{m_2}_{\rho,\delta}$ have kernels κ_{1x} and κ_{2x}, respectively. Then

$$\sigma(x,\pi) := \sigma_1(x,\pi)\sigma_2(x,\pi)$$

defines the symbol σ in $S^m_{\rho,\delta}$, $m = m_1 + m_2$, with kernel $\kappa_{2x} * \kappa_{1x}$ with the convolution in the sense of Definition 5.1.19. Furthermore,

$$\|\sigma(x,\pi)\|_{S^m_{\rho,\delta},a,b,c} \leq C\|\sigma_1(x,\pi)\|_{S^{m_1}_{\rho,\delta},a,b,c+\rho a+|m_2|+\delta b}\|\sigma_2(x,\pi)\|_{S^{m_2}_{\rho,\delta},a,b,c},$$

where the constant $C = C_{a,b,c,m_1,m_2} > 0$ does not depend on σ_1, σ_2.

Note that, in Part (ii), the composition $\sigma(x,\pi) := \sigma_1(x,\pi)\sigma_2(x,\pi)$ may be understood as the composition of two fields of operators acting on smooth vectors as well as the composition of $\sigma_1(x,\cdot) \in L^\infty_{\gamma_1,\gamma_1-m_1}(\widehat{G})$ with $\sigma_2(x,\cdot) \in L^\infty_{\gamma_2,\gamma_2-m_2}(\widehat{G})$ for any choice of $\gamma_1, \gamma_2 \in \mathbb{R}$ such that $\gamma_1 - m_1 = \gamma_2$.

Proof. Properties (1), (2), (3), and (4) of (i) are straightforward to check.

Let us prove Part (ii). By Property (1) of (i), or by the definition of symbol classes,

$$\kappa_{jx} \in \mathcal{K}_{\gamma_j,-m_j+\gamma_j} \quad \text{for any} \quad \gamma_j \in \mathbb{R},\ j = 1, 2,$$

thus choosing $\gamma = \gamma_2$ and $\gamma_1 = -m_2 + \gamma_2$, we have by Corollary 5.1.20

$$\kappa_{2x} * \kappa_{1x} \in \mathcal{K}_{\gamma,-m+\gamma} \quad \text{for any} \quad \gamma \in \mathbb{R}.$$

Its group Fourier transform is

$$\pi(\kappa_{1x})\pi(\kappa_{2x}) = \sigma_1(x,\pi)\sigma_2(x,\pi) = \sigma(x,\pi).$$

Therefore, σ is a symbol with kernel $\kappa_{2x} * \kappa_{1x}$.

Let $\alpha, \beta \in \mathbb{N}_0^n$ and $\gamma \in \mathbb{R}$. From the Leibniz rules for Δ^α (see Proposition 5.2.10) and X^β, the operator

$$\pi(\mathrm{I}+\mathcal{R})^{\frac{\rho[\alpha]-m-\delta[\beta]+\gamma}{\nu}} X_x^\beta \Delta^\alpha \sigma(x,\pi)\pi(\mathrm{I}+\mathcal{R})^{-\frac{\gamma}{\nu}},$$

is a linear combination over $\beta_1, \beta_2, \alpha_1, \alpha_2 \in \mathbb{N}^n$ satisfying $[\beta_1] + [\beta_2] = [\beta]$, $[\alpha_1] + [\alpha_2] = [\alpha]$, of terms

$$\pi(\mathrm{I}+\mathcal{R})^{\frac{\rho[\alpha]-m-\delta[\beta]+\gamma}{\nu}} X_x^{\beta_1} \Delta^{\alpha_1} \sigma_1(x,\pi) X_x^{\beta_2} \Delta^{\alpha_2} \sigma_2(x,\pi)\pi(\mathrm{I}+\mathcal{R})^{-\frac{\gamma}{\nu}},$$

whose operator norm is bounded by

$$\begin{aligned}\|\pi(\mathrm{I}+\mathcal{R})^{\frac{\rho[\alpha]-m-\delta[\beta]+\gamma}{\nu}} X_x^{\beta_1} \Delta^{\alpha_1} \sigma_1(x,\pi)\pi(\mathrm{I}+\mathcal{R})^{-\frac{\rho[\alpha_2]-m_2-\delta[\beta_2]+\gamma}{\nu}}\|_{\mathscr{L}(\mathcal{H}_\pi)} \\ \|\pi(\mathrm{I}+\mathcal{R})^{\frac{\rho[\alpha_2]-m_2-\delta[\beta_2]+\gamma}{\nu}} X_x^{\beta_2} \Delta^{\alpha_2} \sigma_2(x,\pi)\pi(\mathrm{I}+\mathcal{R})^{-\frac{\gamma}{\nu}}\|_{\mathscr{L}(\mathcal{H}_\pi)}.\end{aligned}$$

This shows that the inequality between the seminorms of σ, σ_1 and σ_2 given in (ii) holds. Consequently σ is a symbol of order $m = m_1 + m_2$ and of type (ρ, δ), and (ii) is proved. □

A direct consequence of Part (ii) of Theorem 5.2.22 is that the symbols in the introduced symbol classes form an algebra:

Corollary 5.2.23. *Let $1 \geq \rho \geq \delta \geq 0$. The collection of symbols $\bigcup_{m\in\mathbb{R}} S^m_{\rho,\delta}$ forms an algebra.*

Furthermore, if $\sigma_0 \in S^{-\infty}$ and $\sigma \in S^m_{\rho,\delta}$ is of order $m \in \mathbb{R}$, then $\sigma_0\sigma$ and $\sigma\sigma_0$ are also in $S^{-\infty}$.

The fact that the symbol classes $\bigcup_{m\in\mathbb{R}} S^m_{\rho,\delta}$ form an algebra does not imply directly the same property for the operator classes $\bigcup_{m\in\mathbb{R}} \Psi^m_{\rho,\delta}$ since our quantization is not an algebra morphism, that is, $\text{Op}(\sigma_1\sigma_2)$ is not equal in general to $\text{Op}(\sigma_1)\text{Op}(\sigma_2)$. However, we will show that indeed $\bigcup_{m\in\mathbb{R}} \Psi^m_{\rho,\delta}$ is an algebra of operators, cf. Theorem 5.5.3, and we will often use the following property:

Lemma 5.2.24. *Let σ_1 and σ_2 be as in Theorem 5.2.22, (ii). We assume that σ_2 does not depend on x: $\sigma_2 = \{\sigma_2(\pi) : \pi \in \widehat{G}\}$. Then*

$$\sigma(x,\pi) := \sigma_1(x,\pi)\sigma_2(\pi)$$

defines the symbol σ in $S^m_{\rho,\delta}$, $m = m_1 + m_2$ and

$$\text{Op}(\sigma) = \text{Op}(\sigma_1)\text{Op}(\sigma_2)$$

Proof. We keep the notation of the statement. Let κ_{1x} and κ_2 be the convolution kernels of σ_1 and σ_2 respectively. Hence κ_2 is a function on G independent of x. By Theorem 5.2.22(ii), $\kappa_2 * \kappa_{1x}$ is the convolution kernel of σ, thus

$$\forall \phi \in \mathcal{S}(G) \qquad \text{Op}(\sigma)(\phi)(x) = \phi * (\kappa_2 * \kappa_{1x}).$$

As $\phi * \kappa_2 = \text{Op}(\sigma_2)\phi$, this implies easily that $\text{Op}(\sigma)$ is the composition of $\text{Op}(\sigma_1)$ with $\text{Op}(\sigma_2)$. □

The following will also be useful, for instance in the estimates for the kernels in Section 5.4.1.

Corollary 5.2.25. *Let $1 \geq \rho \geq \delta \geq 0$. Let $\sigma \in S^m_{\rho,\delta}$ have kernel κ_x. If β_1 and β_2 are in $\mathbb{N}_0^n$, then*

$$\{\pi(X)^{\beta_1}\sigma(x,\pi)\,\pi(X)^{\beta_2}, (x,\pi) \in G \times \widehat{G}\} \in S^{m+[\beta_1]+[\beta_2]}_{\rho,\delta}$$

with kernel $X_y^{\beta_1}\tilde{X}_y^{\beta_2}\kappa_x(y)$. Furthermore, for any a, b, c there exists $C = C_{a,b,c,\beta_1,\beta_2}$ independent of σ such that

$$\|\pi(X)^{\beta_1}\sigma(x,\pi)\pi(X)^{\beta_2}\|_{S^m_{\rho,\delta},a,b,c} \leq C\|\sigma\|_{S^m_{\rho,\delta},a,b,c+\rho a+[\beta_1]+[\beta_2]+\delta b}.$$

If $\beta_2 = 0$, for any a, b, c there exists $C = C_{a,b,c,\beta_1}$ independent of σ such that

$$\|\pi(X)^{\beta_1}\sigma\|_{S^m_{\rho,\delta},a,b,c} \leq C\|\sigma\|_{S^m_{\rho,\delta},a,b,c}.$$

Proof. The first part follows directly from Theorem 5.2.22 Part (ii) together with Lemma 5.2.17.

We need to show a better estimate for $\beta_2 = 0$. Let $\alpha, \beta_o \in \mathbb{N}_0^n$. By the Leibniz formula (see (5.28)), we have

$$X_x^{\beta_o}\Delta^\alpha\{\pi(X)^{\beta_1}\sigma(x,\pi)\} = \sum_{[\alpha_1]+[\alpha_2]=[\alpha]} c_{\alpha,\alpha_1,\alpha_2}\{\Delta^{\alpha_1}\pi(X)^{\beta_1}\}\,\{X_x^{\beta_o}\Delta^{\alpha_2}\sigma(x,\pi)\}.$$

Hence, denoting $m_o := m + \delta[\beta_o]$, we have

$$\begin{aligned}
&\|\pi(\mathrm{I}+\mathcal{R})^{\frac{\rho[\alpha]-m_o-[\beta_1]+\gamma}{\nu}} X_x^{\beta_o}\Delta^{\alpha}\{\pi(X)^{\beta_1}\sigma(x,\pi)\}\pi(\mathrm{I}+\mathcal{R})^{-\frac{\gamma}{\nu}}\|_{\mathscr{L}(\mathcal{H}_\pi)} \\
&\leq C \sum_{[\alpha_1]+[\alpha_2]=[\alpha]} \|\pi(\mathrm{I}+\mathcal{R})^{\frac{\rho[\alpha]-m_o-[\beta_1]+\gamma}{\nu}} \Delta^{\alpha_1}\pi(X)^{\beta_1}\pi(\mathrm{I}+\mathcal{R})^{-\frac{\rho[\alpha_2]-m_o+\gamma}{\nu}}\|_{\mathscr{L}(\mathcal{H}_\pi)} \\
&\qquad \|\pi(\mathrm{I}+\mathcal{R})^{\frac{\rho[\alpha_2]-m_o+\gamma}{\nu}} X_x^{\beta_o}\Delta^{\alpha_2}\sigma(x,\pi)\pi(\mathrm{I}+\mathcal{R})^{-\frac{\gamma}{\nu}}\|_{\mathscr{L}(\mathcal{H}_\pi)}.
\end{aligned}$$

As $\{\pi(X)^{\beta_1}\} \in S_{1,0}^{[\beta_1]}$ by Lemma 5.2.17, each quantity

$$\sup_{|\gamma|\leq c,\pi\in\widehat{G}} \|\pi(\mathrm{I}+\mathcal{R})^{\frac{\rho[\alpha]-m_o-[\beta_1]+\gamma}{\nu}} \Delta^{\alpha_1}\pi(X)^{\beta_1}\pi(\mathrm{I}+\mathcal{R})^{-\frac{\rho[\alpha_2]-m_o+\gamma}{\nu}}\|_{\mathscr{L}(\mathcal{H}_\pi)} < \infty$$

is finite for any $c > 0$ and $\alpha_1, \alpha_2 \in \mathbb{N}_0^n$ such that $[\alpha_1] + [\alpha_2] = [\alpha]$. This implies

$$\begin{aligned}
&\sup_{|\gamma|\leq c,\pi\in\widehat{G}} \|\pi(\mathrm{I}+\mathcal{R})^{\frac{\rho[\alpha]-m_o-[\beta_1]+\gamma}{\nu}} X_x^{\beta_o}\Delta^{\alpha}\{\pi(X)^{\beta_1}\sigma(x,\pi)\}\pi(\mathrm{I}+\mathcal{R})^{-\frac{\gamma}{\nu}}\|_{\mathscr{L}(\mathcal{H}_\pi)} \\
&\leq C' \sum_{[\alpha_2]\leq[\alpha]} \sup_{\substack{|\gamma|\leq c \\ \pi\in\widehat{G}}} \|\pi(\mathrm{I}+\mathcal{R})^{\frac{\rho[\alpha_2]-m_o+\gamma}{\nu}} X_x^{\beta_o}\Delta^{\alpha_2}\sigma(x,\pi)\pi(\mathrm{I}+\mathcal{R})^{-\frac{\gamma}{\nu}}\|_{\mathscr{L}(\mathcal{H}_\pi)}.
\end{aligned}$$

Taking the supremum over $[\alpha] \leq a$ and $[\beta] \leq b$ yields the stated estimate. □

5.3 Spectral multipliers in positive Rockland operators

In this section we show that multipliers in positive Rockland operators belong to the introduced symbol classes Ψ^m.

The main result is stated in Proposition 5.3.4. This will allow us to use the Littlewood-Paley decompositions associated with a positive Rockland operator, and therefore will enter most of the subsequent proofs.

5.3.1 Multipliers in one positive Rockland operator

The precise class of multiplier functions that we consider is the following:

Definition 5.3.1. Let $\mathcal{M}_m$ be the space of functions $f \in C^\infty(\mathbb{R}_+)$ such that the following quantities for all $\ell \in \mathbb{N}_0$ are finite:

$$\|f\|_{\mathcal{M}_{m,\ell}} := \sup_{\lambda>0,\,\ell'=0,\ldots,\ell} (1+\lambda)^{-m+\ell'}|\partial_\lambda^{\ell'} f(\lambda)|.$$

In other words, the class of functions f that appears in the definition above are the functions which are smooth on $\mathbb{R}_+ = (0,\infty)$ and have the symbolic behaviour at infinity of the Hörmander class $S_{1,0}^m(\mathbb{R})$ on the real line. However, we rather prefer the notation $\mathcal{M}_m$ in order not to create any confusion between these classes and the classes $S_{\rho,\delta}^m(G)$ defined on the group G.

Example 5.3.2. For any $m \in \mathbb{R}$, the function $\lambda \mapsto (1+\lambda)^m$ is in $\mathcal{M}_m$.

It is a routine exercise to check that $\mathcal{M}_m$ endowed with the family of maps $\|\cdot\|_{\mathcal{M}_{m,\ell}}$, $\ell \in \mathbb{N}_0$, is a Fréchet space. Furthermore, it satisfies the following property.

Lemma 5.3.3. *If $f_1 \in \mathcal{M}_{m_1}$ and $f_2 \in \mathcal{M}_{m_2}$ then $f_1 f_2 \in \mathcal{M}_{m_1+m_2}$ with*

$$\|f_1 f_2\|_{\mathcal{M}_{m_1+m_2,\ell}} \leq C_\ell \|f_1\|_{\mathcal{M}_{m_1,\ell}} \|f_2\|_{\mathcal{M}_{m_2,\ell}}.$$

Proof. This follows from the Leibniz formula for $|\partial^{\ell'}(f_1 f_2)|$ and from the following inequality which holds for $\lambda > 0$ and $\ell_1', \ell_2' \leq \ell$:

$$(1+\lambda)^{-m_1-m_2+\ell_1'+\ell_2'} |\partial_\lambda^{\ell_1'} f_1(\lambda)| \, |\partial_\lambda^{\ell_2'} f_2(\lambda)| \leq \|f_1\|_{\mathcal{M}_{m_1,\ell}} \|f_2\|_{\mathcal{M}_{m_2,\ell}},$$

which implies the claim. □

The main property of this section is

Proposition 5.3.4. *Let $m \in \mathbb{R}$ and let $\mathcal{R}$ be a positive Rockland operator of homogeneous degree ν. If $f \in \mathcal{M}_{\frac{m}{\nu}}$, then $f(\mathcal{R})$ is in Ψ^m and its symbol $\{f(\pi(\mathcal{R})), \pi \in \widehat{G}\}$ satisfies*

$$\forall a, b, c \in \mathbb{N}_0 \qquad \exists \ell \in \mathbb{N},\ C > 0 : \qquad \|f(\pi(\mathcal{R}))\|_{a,b,c} \leq C \|f\|_{\mathcal{M}_{\frac{m}{\nu}},\ell},$$

with ℓ and C independent of f.

Proof. First let us show that it suffices to show Proposition 5.3.4 for $m < -\nu$. If $f \in \mathcal{M}_{\frac{m}{\nu}}$ with $m \geq -\nu$, then we define

- $m_2 \geq \nu$ such that $\frac{m_2}{\nu}$ is the smallest integer strictly larger than $\frac{m}{\nu}$,
- $f_1(\lambda) := (1+\lambda)^{-\frac{m_2}{\nu}} f(\lambda)$ and $f_2(\lambda) := (1+\lambda)^{\frac{m_2}{\nu}}$.

By Example 5.3.2 and Lemma 5.3.3, we see that $f_1 \in \mathcal{M}_{\frac{m_1}{\nu}}$ with $m_1 = m - m_2$. By Lemma 5.2.17, we see that $f_2(\pi(\mathcal{R})) \in S^{m_2}$. If Proposition 5.3.4 holds for $m_1 < -\nu$, then we can apply it to f_1 and hence $f_1(\pi(\mathcal{R})) \in S^{m_1}$. Thus the product

$$f(\pi(\mathcal{R})) = f_1(\pi(\mathcal{R})) f_2(\pi(\mathcal{R}))$$

is in $S^{m_1+m_2} = S^m$.

Therefore, as claimed above, it suffices to show Proposition 5.3.4 for $m < -\nu$.

Now we show that we may assume that f is supported away from 0. Indeed, if $f \in \mathcal{M}_{\frac{m}{\nu}}$, we extend it smoothly to $\mathbb{R}$ and we write

$$f = f\chi_o + f(1-\chi_o),$$

where $\chi_o \in \mathcal{D}(\mathbb{R})$ is identically 1 on $[-1,1]$. Since $f\chi_o \in \mathcal{D}(\mathbb{R})$, by Hulanicki's theorem (cf. Corollary 4.5.2), the kernel of $(f\chi_o)(\mathcal{R})$ is Schwartz and by Lemma 5.2.20, we have $(f\chi_o)(\mathcal{R}) \in \Psi^{-\infty}$ with suitable inequalities for the seminorms. Thus we just have to prove the result for $f(1-\chi_o)$ which is supported in $[1,\infty)$ where $\lambda \asymp 1+\lambda$. The statement then follows from the following lemma. □

Showing Proposition 5.3.4 boils then down to

Lemma 5.3.5. *Let $m < -\nu$. If $f \in C^\infty(\mathbb{R})$ is supported in $[1,\infty)$ and satisfies*

$$\forall \ell \in \mathbb{N}_0 \quad \exists C_\ell \quad \forall \lambda \geq 1 \qquad |\partial_\lambda^\ell f(\lambda)| \leq C_\ell |\lambda|^{\frac{m}{\nu}-\ell},$$

then $f(\mathcal{R}) \in \Psi^m$, and for any $a, b, c \in \mathbb{N}_0$ we have

$$\|f(\mathcal{R})\|_{\Psi^m, a,b,c} \leq C \sup_{\lambda \geq 1, \ell'=0,\ldots,\ell} |\lambda|^{-\frac{m}{\nu}+\ell'} |\partial_\lambda^{\ell'} f(\lambda)|,$$

with $\ell = \ell_{m,a,b,c} \in \mathbb{N}$ and $C = C_{m,a,b,c} > 0$ independent of f.

The proof of Lemma 5.3.5 relies on the following consequence of Hulanicki's theorem (see Theorem 4.5.1).

Lemma 5.3.6. *Let $\mathcal{R}$ be a positive Rockland operator on a graded Lie group G.*

Let $m \in \mathcal{D}(\mathbb{R})$ and $\alpha_o \in \mathbb{N}_0^n$. We denote by $m(\mathcal{R})\delta_0$ the kernel of the multiplier $m(\mathcal{R})$ and we set

$$\kappa(x) := x^{\alpha_o} m(\mathcal{R})\delta_0(x).$$

The function κ is Schwartz.

For any $p \in (1,\infty)$, $N \in \mathbb{N}$ and $a \in \mathbb{R}$ with $0 \leq a \leq N\nu$, there exist $C > 0$ and $k \in \mathbb{N}$ such that for any $\phi \in \mathcal{S}(G)$,

$$\|\mathcal{R}^N(\phi * \kappa)\|_p \leq C \sup_{\substack{\lambda > 0 \\ \ell=0,\ldots,k}} (1+\lambda)^k |\partial_\lambda^\ell m(\lambda)| \; \|\mathcal{R}_p^{\frac{a}{\nu}} \phi\|_{L^p(G)}.$$

Proof of Lemma 5.3.6. By Hulanicki's Theorem 4.5.1 or Corollary 4.5.2, $\kappa \in \mathcal{S}(G)$.

It suffices to prove the result with X^α, $[\alpha] = N\nu$, instead of $\mathcal{R}^N$. By Corollary 3.1.30, we can write X^α as a finite sum of $\tilde{X}^\beta p_{\alpha,\beta}$ with $p_{\alpha,\beta}$ a homogeneous polynomial of homogeneous degree $[\beta] - [\alpha] \geq 0$. We then have

$$X^\alpha(\phi * \kappa) = \phi * X^\alpha \kappa = \sum \phi * (\tilde{X}^\beta p_{\alpha,\beta}\kappa) = \sum (X^\beta \phi) * (p_{\alpha,\beta}\kappa).$$

Therefore, by Proposition 4.4.30,

$$\begin{aligned} \|X^\alpha(\phi * \kappa)\|_p &\leq \sum \|(\mathcal{R}^{\frac{-[\beta]+a}{\nu}} X^\beta \phi) * (\tilde{\mathcal{R}}^{\frac{[\beta]-a}{\nu}} p_{\alpha,\beta}\kappa)\|_p \\ &\leq \sum \|\mathcal{R}^{\frac{-[\beta]+a}{\nu}} X^\beta \phi\|_p \|\tilde{\mathcal{R}}^{\frac{[\beta]-a}{\nu}} p_{\alpha,\beta}\kappa\|_1. \end{aligned}$$

By Theorem 4.4.16, Part 2,

$$\|\mathcal{R}^{\frac{-[\beta]+a}{\nu}} X^\beta \phi\|_p \leq C \|\mathcal{R}^{\frac{a}{\nu}} \phi\|_p.$$

And we have

$$\|\tilde{\mathcal{R}}^{\frac{[\beta]-a}{\nu}} p_{\alpha,\beta}\kappa\|_1 = \|\mathcal{R}^{\frac{[\beta]-a}{\nu}} \tilde{p}_{\alpha,\beta}\tilde{\kappa}\|_1,$$

see Section 4.4.8. By Theorem 4.3.6, since $[\beta] \geq [\alpha] = N\nu \geq a$, we obtain

$$\|\mathcal{R}^{\frac{[\beta]-a}{\nu}} \tilde{p}_{\alpha,\beta}\tilde{\kappa}\|_1 \leq C\|\tilde{p}_{\alpha,\beta}\tilde{\kappa}\|_1^{1-\frac{[\beta]-a}{\nu N}} \|\mathcal{R}^N \tilde{p}_{\alpha,\beta}\tilde{\kappa}\|_1^{\frac{[\beta]-a}{\nu N}}.$$

Note that because of (4.8), we have

$$\tilde{\kappa}(x) := (-1)^{|\alpha_o|} x^{\alpha_o} \bar{m}(\mathcal{R})\delta_0(x).$$

By Hulanicki's theorem (see Theorem 4.5.1), $\|\tilde{p}_{\alpha,\beta}\tilde{\kappa}\|_1$ and $\|\mathcal{R}^{\frac{b}{\nu}} \tilde{p}_{\alpha,\beta}\tilde{\kappa}\|_1$ are

$$\lesssim \sup_{\substack{\lambda>0 \\ \ell=0,\ldots,k}} (1+\lambda)^k |\partial_\lambda^\ell m(\lambda)|,$$

for a suitable k, therefore this is also the case for $\|\tilde{\mathcal{R}}^{\frac{[\beta]-a}{\nu}} p_{\alpha,\beta}\kappa\|_1$.

Combining all these inequalities shows the desired result. □

Proof of Lemma 5.3.5. Let f be as in the statement. We need to show for any $\alpha \in \mathbb{N}_0^n$ that the convolution operator with right convolution kernel $\tilde{q}_\alpha f(\mathcal{R})\delta_0$ maps $L^2_\gamma(G)$ boundedly to $L^2_{[\alpha]-m+\gamma}(G)$ for any $\gamma \in \mathbb{R}$. It is sufficient to prove this for γ in a sequence going to $+\infty$ and $-\infty$ (see Proposition 5.2.12) and, in fact, only for a sequence of positive γ since

$$(\tilde{q}_\alpha f(\mathcal{R})\delta_0)^* = (-1)^{|\alpha|} \tilde{q}_\alpha \bar{f}(\mathcal{R})\delta_0.$$

At the end of the proof, we will see that, because of the equivalence between the Sobolev norms, it actually suffices to prove that for a fixed γ in this sequence, the operators given by

$$\phi \longmapsto \phi * (\tilde{q}_\alpha f(\mathcal{R})\delta_0) \text{ and } \phi \longmapsto \mathcal{R}^{\frac{[\alpha]-m+\gamma}{\nu}} \left(\left\{\mathcal{R}^{-\frac{\gamma}{\nu}}\phi\right\} * (\tilde{q}_\alpha f(\mathcal{R})\delta_0)\right), \tag{5.32}$$

are bounded on $L^2(G)$. So, we first prove this by decomposing f and applying the Cotlar-Stein lemma.

We fix a dyadic decomposition: there exists a non-negative function $\eta \in \mathcal{D}(\mathbb{R})$ supported in $[1/2, 2]$ and satisfying

$$\forall \lambda \geq 1 \qquad 1 = \sum_{j\in\mathbb{N}_0} \eta_j(\lambda) \quad \text{where} \quad \eta_j(\lambda) := \eta(2^{-j}\lambda).$$

We set for $j \in \mathbb{N}_0$ and $\lambda \geq 1$,

$$\begin{aligned} f_j(\lambda) &:= \lambda^{-\frac{m}{\nu}} f(\lambda)\eta_j(\lambda), \\ f^{(j)}(\lambda) &:= f_j(2^j\lambda), \\ g_j(\lambda) &:= \lambda^{\frac{m}{\nu}} f^{(j)}(\lambda). \end{aligned}$$

One obtains easily that for any $j \in \mathbb{N}_0$ and $\ell \in \mathbb{N}_0$, we have

$$\begin{aligned} \partial^\ell f_j(\lambda) &= \overline{\sum_{\ell_1+\ell_2+\ell_3=\ell}} \lambda^{-\frac{m}{\nu}-\ell_1} \, (\partial^{\ell_2} f)(\lambda) \, 2^{-j\ell_3}(\partial^{\ell_3}\eta)(2^{-j}\lambda), \\ |\partial^\ell f_j(\lambda)| &\le C_\ell \sup_{\substack{\lambda\ge 1 \\ \ell'\le \ell}} \lambda^{-\frac{m}{\nu}+\ell'} \, |\partial_\lambda^{\ell'} f(\lambda)| \overline{\sum_{\ell_1+\ell_2+\ell_3=\ell}} \lambda^{-\ell_1}\lambda^{-\ell_2} 2^{-j\ell_3}|(\partial^{\ell_3}\eta)(2^{-j}\lambda)|, \end{aligned}$$

where $\overline{\sum}$ stands for a linear combination of its terms with some constants. As η is supported in $[1/2, 2]$ and since $\lambda \asymp 2^j$, we have

$$\lambda^{-\ell_1}\lambda^{-\ell_2}2^{-j\ell_3} \asymp 2^{-j\ell_1+\ell_2+\ell_3},$$

so that

$$\overline{\sum_{\ell_1+\ell_2+\ell_3=\ell}} \lambda^{-\ell_1}\lambda^{-\ell_2}2^{-j\ell_3}|(\partial^{\ell_3}\eta)(2^{-j}\lambda)| \le C_{\ell,\eta}2^{-j\ell}.$$

Therefore, we have obtained

$$|\partial^\ell f_j(\lambda)| \le C_\ell \sup_{\substack{\lambda\ge 1 \\ \ell'\le \ell}} \lambda^{-\frac{m}{\nu}+\ell'} \, |\partial_\lambda^{\ell'} f(\lambda)| \, 2^{-j\ell}.$$

Hence, for each $j \in \mathbb{N}_0$, $f^{(j)}$ is smooth and supported in $[1/2, 2]$, and satisfies for any $\ell \in \mathbb{N}_0$ the estimate

$$|\partial^\ell f^{(j)}(\lambda)| = |2^{j\ell}\partial^\ell f_j(\lambda)| \le C_\ell \sup_{\substack{\lambda\ge 1 \\ \ell'\le \ell}} \lambda^{-\frac{m}{\nu}+\ell'} \, |\partial_\lambda^{\ell'} f(\lambda)|.$$

Consequently, each g_j is smooth and supported in $[1/2, 2]$, and satisfies

$$\forall \ell \in \mathbb{N}_0 \qquad \sup_{\substack{\lambda\in[\frac{1}{2},2] \\ \ell'=0,\ldots,\ell}} |\partial^{\ell'} g_j(\lambda)| \le C_\ell \sup_{\substack{\lambda\ge 1 \\ \ell'\le \ell}} \lambda^{-\frac{m}{\nu}+\ell'} \, |\partial_\lambda^{\ell'} f(\lambda)|. \tag{5.33}$$

Clearly $f(\lambda)$ is the sum of the terms

$$2^{j\frac{m}{\nu}} g_j(2^{-j}\lambda) = f(\lambda)\eta_j(\lambda)$$

over $j \in \mathbb{N}_0$ and this sum is uniformly locally finite with respect to λ. Furthermore, since the functions f and g_j are continuous and bounded, the operators $f(\mathcal{R})$ and $g_j(2^{-j}\mathcal{R})$ defined by the functional calculus are bounded on $L^2(G)$ by Corollary 4.1.16. Therefore, we have in the strong operator topology of $\mathscr{L}(L^2(G))$ that

$$f(\mathcal{R}) = \sum_{j=0}^{\infty} 2^{j\frac{m}{\nu}} g_j(2^{-j}\mathcal{R}),$$

and in $\mathcal{K}(G)$ or $\mathcal{S}'(G)$ that

$$f(\mathcal{R})\delta_o = \sum_{j=0}^{\infty} 2^{j\frac{m}{\nu}} g_j(2^{-j}\mathcal{R})\delta_o.$$

We fix $\alpha \in \mathbb{N}_0^n$. For each $j \in \mathbb{N}_0$, by Hulanicki's theorem (see Corollary 4.5.2), $g_j(2^{-j}\mathcal{R})\delta_o$ is Schwartz, thus so is

$$K_j := 2^{j\frac{m}{\nu}} \tilde{q}_\alpha g_j(2^{-j}\mathcal{R})\delta_o$$

and also (see (4.8))

$$K_j^* =: K_j^*(x) = \bar{K}_j(x^{-1}) = (-1)^{|\alpha|} 2^{j\frac{m}{\nu}} \tilde{q}_\alpha \bar{g}_j(2^{-j}\mathcal{R})\delta_o(x^{-1}).$$

We claim that for any $a, b \in \mathbb{R}$ satisfying

- either $b \in \nu\mathbb{N}_0$ and $a \in [0, b)$
- or $b \geq 0$ and $a < \lfloor b/\nu \rfloor$

there exist $\ell \in \mathbb{N}$ and $C > 0$ such that for all $j \in \mathbb{N}_0$, we have

$$\|\tilde{\mathcal{R}}^{-\frac{a}{\nu}} \mathcal{R}^{\frac{b}{\nu}} K_j\|_{\mathcal{K}} \leq C(2^{\frac{j}{\nu}})^{m-[\alpha]-a+b} \sup_{\substack{\lambda \geq 1 \\ \ell' \leq \ell}} \lambda^{-\frac{m}{\nu}+\ell'} |\partial_\lambda^{\ell'} f(\lambda)|, \tag{5.34}$$

and the same is true for $\mathcal{R}^{-\frac{a}{\nu}} \tilde{\mathcal{R}}^{\frac{b}{\nu}} K_j^*$.

Let us prove this claim. By homogeneity (see (4.3)), we see that

$$g_j(2^{-j}\mathcal{R})\delta_o(x) = (2^{-\frac{j}{\nu}})^{-Q} g_j(\mathcal{R})\delta_o(2^{\frac{j}{\nu}} x),$$

thus

$$\begin{aligned} K_j(x) &= 2^{j\frac{m}{\nu}} (2^{\frac{j}{\nu}})^{-[\alpha]} \tilde{q}_\alpha(2^{\frac{j}{\nu}} x)\ (2^{-\frac{j}{\nu}})^{-Q} g_j(\mathcal{R})\delta_o(2^{\frac{j}{\nu}} x) \\ &= (2^{\frac{j}{\nu}})^{m-[\alpha]+Q} \left(\tilde{q}_\alpha g_j(\mathcal{R})\delta_o\right)(2^{\frac{j}{\nu}} x). \end{aligned}$$

More generally, by Part (7) of Theorem 4.3.6 for $\mathcal{R}$ and consequently for $\tilde{\mathcal{R}}$ (see (4.50)) we have

$$\tilde{\mathcal{R}}^{-\frac{a}{\nu}} \mathcal{R}^{\frac{b}{\nu}} K_j = (2^{\frac{j}{\nu}})^{m-[\alpha]+Q-a+b} \left(\tilde{\mathcal{R}}^{-\frac{a}{\nu}} \mathcal{R}^{\frac{b}{\nu}} \{\tilde{q}_\alpha g_j(\mathcal{R})\delta_o\} \right) \circ D_{2^{\frac{j}{\nu}}},$$

whenever it makes sense (that is, K_j is in the L^2-domain of $\mathcal{R}^{\frac{b}{\nu}}$ such that $\mathcal{R}^{\frac{b}{\nu}} K_j$ is in the L^2-domain of $\tilde{\mathcal{R}}^{-\frac{a}{\nu}}$). Consequently, by Proposition 5.1.17 (1), with norms possibly infinite, we have

$$\|\tilde{\mathcal{R}}^{-\frac{a}{\nu}} \mathcal{R}^{\frac{b}{\nu}} K_j\|_{\mathcal{K}} = (2^{\frac{j}{\nu}})^{m-[\alpha]-a+b} \left\| \tilde{\mathcal{R}}^{-\frac{a}{\nu}} \mathcal{R}^{\frac{b}{\nu}} \{\tilde{q}_\alpha g_j(\mathcal{R})\delta_o\} \right\|_{\mathcal{K}}.$$

Since $(\tilde{\mathcal{R}}^{-\frac{a}{\nu}}\mathcal{R}^{\frac{b}{\nu}}K_j)^* = \tilde{\mathcal{R}}^{\frac{b}{\nu}}\mathcal{R}^{-\frac{a}{\nu}}K_j^*$ for any a, b whenever it makes sense, or by the same argument as above, we also have

$$\|\tilde{\mathcal{R}}^{-\frac{a}{\nu}}\mathcal{R}^{\frac{b}{\nu}}K_j^*\|_{\mathcal{K}} = (2^{\frac{j}{\nu}})^{m-[\alpha]-a+b}\left\|\tilde{\mathcal{R}}^{-\frac{a}{\nu}}\mathcal{R}^{\frac{b}{\nu}}\{\tilde{q}_\alpha\bar{g}_j(\mathcal{R})\delta_o\}\right\|_{\mathcal{K}}.$$

Therefore, if $b \in \nu\mathbb{N}_0$ and $a \in [0, b)$, by Lemma 5.3.6, there exist $\ell = \ell_{a,b} \in \mathbb{N}$ such that

$$\begin{aligned}\left\|\tilde{\mathcal{R}}^{-\frac{a}{\nu}}\mathcal{R}^{\frac{b}{\nu}}\{\tilde{q}_\alpha g_j(\mathcal{R})\delta_o\}\right\|_{\mathcal{K}} &\leq C_{a,b} \sup_{\substack{\lambda>0\\ \ell'=0,\ldots,\ell}} (1+\lambda)^\ell|\partial_\lambda^{\ell'} g_j(\lambda)| \\ &\leq C_{a,b} \sup_{\substack{\lambda>0\\ \ell'=0,\ldots,\ell}} |\partial_\lambda^{\ell'} g_j(\lambda)|,\end{aligned}$$

since each g_j is supported in $[1/2, 2]$. As g_j satisfies (5.33), we have shown Claim (5.34) in the case $b \in \nu\mathbb{N}_0$ and $a \in [0, b)$.

If $a < \lfloor b/\nu \rfloor$ then we can apply the result we have just obtained to $\nu(\lfloor b/\nu \rfloor)$ and $\nu\lceil b/\nu \rceil$. Using Theorem 4.3.6 we then have for any $\phi \in \mathcal{S}(G)$, with $\theta := \lfloor \frac{b}{\nu} \rfloor \lceil \frac{b}{\nu} \rceil^{-1}$, that

$$\begin{aligned}\|\mathcal{R}^{\frac{b}{\nu}}\phi\|_2 &\leq C\|\mathcal{R}^{\lfloor\frac{b}{\nu}\rfloor}\phi\|_2^{1-\theta}\|\mathcal{R}^{\lceil\frac{b}{\nu}\rceil}\phi\|_2^\theta \\ &\leq C\left(\sup_{\substack{\lambda>0\\ \ell'=0,\ldots,\ell}} |\partial_\lambda^{\ell'} g_j(\lambda)| \; \|\mathcal{R}^{\frac{a}{\nu}}\phi\|_2\right)^{1-\theta+\theta},\end{aligned}$$

for some ℓ. This shows Claim (5.34) in the case $a < \lfloor b/\nu \rfloor$.

We set $T_j : \mathcal{S}(G) \ni \phi \mapsto \phi * K_j$. We want to apply the Cotlar-Stein lemma (Theorem A.5.2) to two families of $L^2(G)$-bounded operators: first to T_j, $j \in \mathbb{N}_0$, and then to

$$T_{j,\beta,\gamma} : \phi \longmapsto \phi * \mathcal{R}^{\frac{\beta}{\nu}}\tilde{\mathcal{R}}^{-\frac{\gamma}{\nu}}K_j, \; j \in \mathbb{N}_0.$$

where $\gamma \in \nu\mathbb{N}$ is such that $\beta := [\alpha] - m + \gamma > 0$.

Let us check the hypothesis of the Cotlar-Stein lemma for T_j. By Claim (5.34) for $a = b = 0$, there exists $\ell \in \mathbb{N}_0$ such that for any $j, k \in \mathbb{N}_0$,

$$\begin{aligned}&\max\left(\|T_j^*T_k\|_{\mathscr{L}(L^2(G))}, \|T_jT_k^*\|_{\mathscr{L}(L^2(G))}\right) \\ &\leq C\max\left(\|T_j^*\|_{\mathscr{L}(L^2(G))}\|T_k\|_{\mathscr{L}(L^2(G))}, \|T_j\|_{\mathscr{L}(L^2(G))}\|T_k^*\|_{\mathscr{L}(L^2(G))}\right) \\ &\leq C2^{\frac{j+k}{\nu}(m-[\alpha])}(\sup_{\substack{\lambda\geq 1\\ \ell'\leq\ell}} \lambda^{-\frac{m}{\nu}+\ell'} \; |\partial_\lambda^{\ell'} f(\lambda)|)^2 \\ &\leq C2^{\frac{|j-k|}{\nu}(m-[\alpha])}(\sup_{\substack{\lambda\geq 1\\ \ell'\leq\ell}} \lambda^{-\frac{m}{\nu}+\ell'} \; |\partial_\lambda^{\ell'} f(\lambda)|)^2,\end{aligned}$$

since $m-[\alpha]<0$.

Let us check the hypothesis of the Cotlar-Stein lemma for $T_{j,\beta,\gamma}$. By Proposition 4.4.30 the right convolution kernel of the operator $T^*_{j,\beta,\gamma}T_{k,\beta,\gamma}$ is given by

$$(\mathcal{R}^{\frac{\beta}{\nu}}\tilde{\mathcal{R}}^{-\frac{\gamma}{\nu}}K_k)*(\tilde{\mathcal{R}}^{\frac{\beta}{\nu}}\mathcal{R}^{-\frac{\gamma}{\nu}}K_j^*)=(\tilde{\mathcal{R}}^{-\frac{\gamma}{\nu}}\mathcal{R}^{\frac{\gamma}{\nu}}K_k)*(\tilde{\mathcal{R}}^{\frac{2\beta-\gamma}{\nu}}\mathcal{R}^{-\frac{\gamma}{\nu}}K_j^*).$$

Therefore, its operator norm is

$$\begin{aligned}\|T^*_{j,\beta,\gamma}T_{k,\beta,\gamma}\|_{\mathscr{L}(L^2(G))}&\le\|\tilde{\mathcal{R}}^{-\frac{\gamma}{\nu}}\mathcal{R}^{\frac{\gamma}{\nu}}K_k\|_{\mathcal{K}}\|\tilde{\mathcal{R}}^{\frac{2\beta-\gamma}{\nu}}\mathcal{R}^{-\frac{\gamma}{\nu}}K_j^*\|_{\mathcal{K}}.\\&\le 2^{\frac{k}{\nu}(m-[\alpha]-\gamma+\gamma)}2^{\frac{j}{\nu}(m-[\alpha]-\gamma+2\beta-\gamma)}\left(\sup_{\substack{\lambda\ge1\\ \ell'\le\ell}}\lambda^{-\frac{m}{\nu}+\ell'}\ |\partial_\lambda^{\ell'}f(\lambda)|\right)^2,\end{aligned}$$

for some ℓ, thanks to Claim (5.34) with $a=b=\gamma\in\nu\mathbb{N}$ and with $b=2\beta-\gamma=2[\alpha]-2m+\gamma$ and $a=\gamma$. So we have obtained

$$\|T^*_{j,\beta,\gamma}T_{k,\beta,\gamma}\|_{\mathscr{L}(L^2(G))}\le 2^{\frac{k-j}{\nu}(m-[\alpha])}\left(\sup_{\substack{\lambda\ge1\\ \ell'\le\ell}}\lambda^{-\frac{m}{\nu}+\ell'}\ |\partial_\lambda^{\ell'}f(\lambda)|\right)^2.$$

Since the adjoint of $T^*_{j,\beta,\gamma}T_{k,\beta,\gamma}$ is $T^*_{k,\beta,\gamma}T_{j,\beta,\gamma}$, we may replace $k-j$ above by $|k-j|$.

We proceed in a similar way for the operator norm of $T_{j,\beta,\gamma}T^*_{k,\beta,\gamma}$ whose right convolution kernel is

$$(\mathcal{R}^{\frac{\beta}{\nu}}\tilde{\mathcal{R}}^{-\frac{\gamma}{\nu}}K_k^*)*(\tilde{\mathcal{R}}^{\frac{\beta}{\nu}}\mathcal{R}^{-\frac{\gamma}{\nu}}K_j)=(\mathcal{R}^{\frac{2\beta-\gamma}{\nu}}\tilde{\mathcal{R}}^{-\frac{\gamma}{\nu}}K_k^*)*(\tilde{\mathcal{R}}^{\frac{\gamma}{\nu}}\mathcal{R}^{\frac{-\gamma}{\nu}}K_j).$$

Therefore, we obtain

$$\begin{aligned}&\max\left(\|T^*_{j,\beta,\gamma}T_{k,\beta,\gamma}\|_{\mathscr{L}(L^2(G))},\|T_{j,\beta,\gamma}T^*_{k,\beta,\gamma}\|_{\mathscr{L}(L^2(G))}\right)\\&\quad\le C2^{\frac{|k-j|}{\nu}(m-[\alpha])}(\sup_{\substack{\lambda\ge1\\ \ell'\le\ell}}\lambda^{-\frac{m}{\nu}+\ell'}\ |\partial_\lambda^{\ell'}f(\lambda)|)^2.\end{aligned}$$

By the Cotlar-Stein lemma (see Theorem A.5.2), $\sum T_j$ and $\sum_j T_{j,\beta,\gamma}$ converge in the strong operator topology of $\mathscr{L}(L^2(G))$ and the resulting operators have operator norms, up to a constant, less or equal than

$$\sup_{\lambda\ge1,\ell\le k}\ \lambda^{-\frac{m}{\nu}+\ell}\ |\partial_\lambda^\ell f(\lambda)|.$$

Clearly $\sum T_j$ and $\sum_j T_{j,\beta,\gamma}$ coincide on $\mathcal{S}(G)$ with the operators in (5.32), respectively. Using the equivalence between the two Sobolev norms (Theorem 4.4.3, Part

4), this implies

$$\begin{aligned}\|\phi * (\tilde{q}_\alpha f(\mathcal{R})\delta_0)\|_{L^2_\beta(G)} &\leq C\left(\|\phi * (\tilde{q}_\alpha f(\mathcal{R})\delta_0)\|_2 + \|\mathcal{R}^{\frac{\beta}{\nu}}\left(\phi * (\tilde{q}_\alpha f(\mathcal{R})\delta_0\right)\|_2\right)\\ &\leq C \sup_{\substack{\lambda\geq 1\\ \ell'\leq\ell}} \lambda^{-\frac{m}{\nu}+\ell'} \, |\partial_\lambda^{\ell'} f(\lambda)| \left(\|\phi\|_2 + \|\mathcal{R}^{\frac{\gamma}{\nu}}\phi\|_2\right)\\ &\leq C \sup_{\substack{\lambda\geq 1\\ \ell'\leq\ell}} \lambda^{-\frac{m}{\nu}+\ell'} \, |\partial_\lambda^{\ell'} f(\lambda)| \|\phi\|_{L^2_\gamma(G)}.\end{aligned}$$

We have obtained that the convolution operator with the right convolution kernel $\tilde{q}_\alpha f(\mathcal{R})\delta_0$ maps $L^2_\gamma(G)$ boundedly to $L^2_{m-[\alpha]+\gamma}(G)$ for any $\gamma \in \nu\mathbb{N}$ such that $m - [\alpha] + \gamma > 0$, with operator norm less or equal than

$$\sup_{\lambda\geq 1, \ell'\leq\ell} \lambda^{-\frac{m}{\nu}+\ell'} \, |\partial_\lambda^{\ell'} f(\lambda)|,$$

up to a constant, with ℓ depending on γ. This concludes the proof of Lemma 5.3.5. □

Hence the proof of Proposition 5.3.4 is now complete.

Looking back at the proof of Proposition 5.3.4, we see that we can assume that f depends on $x \in G$ in the following way:

Corollary 5.3.7. *Let $\mathcal{R}$ be a positive Rockland operator of homogeneous degree ν. Let $m \in \mathbb{R}$ and $0 \leq \delta \leq 1$. Let*

$$f : G \times \mathbb{R}_+ \ni (x, \lambda) \mapsto f_x(\lambda) \in \mathbb{C}$$

be a smooth function. We assume that for every $\beta \in \mathbb{N}_0^n$, $X_x^\beta f_x \in \mathcal{M}_{\frac{m+\delta[\beta]}{\nu}}$. Then $\sigma(x, \pi) = f_x(\pi(\mathcal{R}))$ defines a symbol σ in $S^m_{1,\delta}$ which satisfies

$$\forall a, b, c \in \mathbb{N}_0 \qquad \exists \ell \in \mathbb{N},\ C > 0 : \qquad \|\sigma\|_{S^m_{1,\delta}, a,b,c} \leq C \sup_{[\beta]\leq b} \|X_x^\beta f_x\|_{\mathcal{M}_{\frac{m+\delta[\beta]}{\nu}}, \ell},$$

with ℓ and C independent of f.

5.3.2 Joint multipliers

To a certain extent, we can tensorise the property in Proposition 5.3.4. But we need to define the tensorisation of the space $\mathcal{M}_m$ and the multipliers of two Rockland operators.

First, we define the space $\mathcal{M}_{m_1} \otimes \mathcal{M}_{m_2}$ of functions $f \in C^\infty(\mathbb{R}_+ \times \mathbb{R}_+)$ such that

$$\|f\|_{\mathcal{M}_{m_1}\otimes\mathcal{M}_{m_2}, \ell} := \sup_{\substack{\lambda_1,\lambda_2>0\\ \ell'_1,\ell'_2=0,\ldots,\ell}} (1+\lambda_1)^{-m_1+\ell'_1}(1+\lambda_2)^{-m_2+\ell'_2} |\partial_{\lambda_1}^{\ell'_1}\partial_{\lambda_2}^{\ell'_2} f(\lambda_1, \lambda_2)|,$$

is finite for every $\ell \in \mathbb{N}_0$. It is a routine exercise to check that $\mathcal{M}_{m_1} \otimes \mathcal{M}_{m_2}$ is a Fréchet space.

Secondly, we observe that if $\mathcal{L}$ and $\mathcal{R}$ are two Rockland operators on G which commute strongly, meaning that their spectral measures $E_\mathcal{L}$ and $E_\mathcal{R}$ commute, then we can define their common spectral measure $E_{\mathcal{L},\mathcal{R}}$ via

$$E_{\mathcal{L},\mathcal{R}}(B_1 \times B_2) := E_\mathcal{L}(B_1)E_\mathcal{R}(B_2), \qquad \text{for } B_1, B_2 \text{ Borel subsets of } \mathbb{R},$$

and we can also define the multipliers in $\mathcal{L}$ and $\mathcal{R}$ by

$$f(\mathcal{L},\mathcal{R}) := \int_{\mathbb{R}_+ \times \mathbb{R}_+} f(\lambda_1, \lambda_2) dE_{\mathcal{L},\mathcal{R}}(\lambda_1, \lambda_2),$$

for any $f \in L^\infty(\mathbb{R}_+ \times \mathbb{R}_+)$.

Corollary 5.3.8. *Let $\mathcal{L}$ and $\mathcal{R}$ be two positive Rockland operators on G of respective degrees $\nu_\mathcal{L}$ and $\nu_\mathcal{R}$. We assume that $\mathcal{L}$ and $\mathcal{R}$ commute strongly, that is, their spectral measures $E_\mathcal{L}$ and $E_\mathcal{R}$ commute. If $f \in \mathcal{M}_{\frac{m_1}{\nu_\mathcal{L}}} \otimes \mathcal{M}_{\frac{m_2}{\nu_\mathcal{R}}}$ then $f(\mathcal{L},\mathcal{R})$ is in $\Psi^{m_1+m_2}$. Furthermore, we have for any $a, b, c \in \mathbb{N}_0$,*

$$\|f(\mathcal{L},\mathcal{R})\|_{\Psi^{m_1+m_2},a,b,c} \leq C\|f\|_{\mathcal{M}_{\frac{m_1}{\nu_\mathcal{L}}} \otimes \mathcal{M}_{\frac{m_2}{\nu_\mathcal{R}}},\ell},$$

where ℓ and $C > 0$ are independent of f.

Proof. By uniqueness, the spectral measure $E_{\mathcal{L},\mathcal{R}}$ is invariant under left translations. Denoting by $\pi(E_{\mathcal{L},\mathcal{R}})$ for $\pi \in \widehat{G}$ its group Fourier transform, we see that the group Fourier transform of a multiplier $f(\mathcal{L},\mathcal{R})$ for $f \in L^\infty(\mathbb{R}_+ \times \mathbb{R}_+)$ is

$$\pi(f(\mathcal{L},\mathcal{R})) = \int_{\mathbb{R}_+ \times \mathbb{R}_+} f(\lambda_1, \lambda_2) d\pi(E_{\mathcal{L},\mathcal{R}})(\lambda_1, \lambda_2),$$

since it is true for a function f of the form $f(\lambda_1, \lambda_2) = f_1(\lambda_1) f_2(\lambda_2)$ with $f_1, f_2 \in L^\infty(\mathbb{R}_+)$, by Corollary 5.3.7.

We fix $\eta \in C^\infty(\mathbb{R})$ supported in $[-\frac{1}{2}, \frac{1}{2}]$ such that

$$\forall \lambda' \in \mathbb{R} \qquad \sum_{j' \in \mathbb{Z}} \eta(\lambda' + j') = 1.$$

We also fix another function $\tilde{\eta} \in C^\infty(\mathbb{R})$ supported in $[-1, 1]$ such that $\tilde{\eta} = 1$ on $[-\frac{1}{2}, \frac{1}{2}]$. For any $j', k' \in \mathbb{Z}$, we define $\psi_{j',k'} \in C^\infty(\mathbb{R})$ by

$$\psi_{j',k'}(\lambda') := e^{-ik'(\lambda'-j')} \tilde{\eta}(\lambda' - j').$$

It is easy to show that for any $\ell' \in \mathbb{N}_0$ there exists $C = C_{\ell'} > 0$ such that

$$\forall j', k' \in \mathbb{Z} \qquad \|\psi_{j',k'}\|_{\mathcal{M}_m,\ell'} \leq C(1 + |k'|)^{\ell'} (1 + |j'|)^{-m+\ell'}.$$

Since the symbols form an algebra (see Section 5.2.5), and by Proposition 5.3.4, writing $m = m_1 + m_2$, we have for any $j_1, j_2, k_1, k_2 \in \mathbb{Z}$:

$$\begin{aligned}
&\|\psi_{j_1,k_1}(\pi(\mathcal{L}))\psi_{j_2,k_2}(\pi(\mathcal{R}))\|_{S^m,a,b,c} \\
&\quad \leq C\|\psi_{j_1,k_1}(\pi(\mathcal{L}))\|_{S^{m_1},a_1,b_1,c_1}\|\psi_{j_2,k_2}(\pi(\mathcal{R}))\|_{S^{m_2},a_2,b_2,c_2} \\
&\quad \leq C(1+|k_1|)^{\ell_1}(1+|j_1|)^{-\frac{m_1}{\nu_{\mathcal{L}}}+\ell_1}(1+|k_2|)^{\ell_2}(1+|j_2|)^{-\frac{m_2}{\nu_{\mathcal{R}}}+\ell_2} \quad (5.35)
\end{aligned}$$

for some $\ell_1, \ell_2 \in \mathbb{N}_0$.

Let f be as in the statement. We extend f to a smooth function supported in $(-1, \infty)^2$ and decompose it as a locally finite sum:

$$f = \sum_{j \in \mathbb{Z}^2} f_j \qquad \text{where} \quad f_j(\lambda) = f(\lambda)\eta(\lambda_1 - j_1')\eta(\lambda_2 - j_2'), \quad \lambda = (\lambda_1, \lambda_2).$$

For each $j \in \mathbb{Z}$, we view $f_j(\cdot + j)$ as a smooth function supported in $[-1, 1] \times [-1, 1]$ and we expand it in the Fourier series

$$f_j(\lambda + j) = \sum_{k \in \mathbb{Z}^2} c_{j,k} e^{-ik\cdot\lambda}.$$

The hypothesis on f implies that for any $\ell_1, \ell_2 \in \mathbb{N}_0$, we have

$$\begin{aligned}
|c_{j,k}| \quad \leq \quad & C_{\ell_1,\ell_2}\|f\|_{\mathcal{M}_{\frac{m_1}{\nu_{\mathcal{L}}}}\otimes\mathcal{M}_{\frac{m_2}{\nu_{\mathcal{R}}}},\ell_1+\ell_2}(1+|k_1|)^{-\ell_1}(1+|k_2|)^{-\ell_2} \times \quad (5.36) \\
& \times(1+|j_1|)^{\frac{m_1}{\nu_{\mathcal{L}}}-\ell_1}(1+|j_2|)^{\frac{m_2}{\nu_{\mathcal{R}}}-\ell_2}.
\end{aligned}$$

We have obtained that (taking different ℓ's)

$$\sum_{j,k \in \mathbb{Z}^2} |c_{j,k}|\|\psi_{j_1,k_1}\|_{\mathcal{M}_{\frac{m_1}{\nu_{\mathcal{L}}}},\ell_1}\|\psi_{j_2,k_2}\|_{\mathcal{M}_{\frac{m_2}{\nu_{\mathcal{R}}}},\ell_2} < \infty.$$

We have therefore obtained the following decomposition of f in the Fréchet space $\mathcal{M}_{\frac{m_1}{\nu_{\mathcal{L}}}} \otimes \mathcal{M}_{\frac{m_2}{\nu_{\mathcal{R}}}}$,

$$f(\lambda_1, \lambda_2) = \sum_{j,k \in \mathbb{Z}^2} c_{j,k}\psi_{j_1,k_1}(\lambda_1)\psi_{j_2,k_2}(\lambda_2).$$

And so for any a, b, c with ℓ_1, ℓ_2 as in (5.35),

$$\begin{aligned}
&\|f(\pi(\mathcal{L}), \pi(\mathcal{R}))\|_{S^m,a,b,c} \leq \sum_{j,k \in \mathbb{Z}^2} |c_{j,k}|\|\psi_{j_1,k_1}(\pi(\mathcal{L}))\psi_{j_2,k_2}(\pi(\mathcal{R}))\|_{S^m,a,b,c} \\
&\quad \leq \sum_{j,k \in \mathbb{Z}^2} |c_{j,k}|C(1+|k_1|)^{\ell_1}(1+|j_1|)^{-\frac{m_1}{\nu_{\mathcal{L}}}+\ell_1}(1+|k_2|)^{\ell_2}(1+|j_2|)^{-\frac{m_2}{\nu_{\mathcal{R}}}+\ell_2} \\
&\quad \leq C\|f\|_{\mathcal{M}_{\frac{m_1}{\nu_{\mathcal{L}}}}\otimes\mathcal{M}_{\frac{m_2}{\nu_{\mathcal{R}}}},\ell_1+\ell_2+4},
\end{aligned}$$

by (5.37) with $\ell_1 + 2$ and $\ell_2 + 2$. This shows that $f(\pi(\mathcal{L}), \pi(\mathcal{R})) \in S^m$ and the desired inequalities for the seminorms. □

Corollary 5.3.8 could be generalised by considering a finite family of positive Rockland operators which commute strongly between themselves (i.e. with commuting spectral measures), with symbols possibly depending on x in a similar way to Corollary 5.3.7.

5.4 Kernels of pseudo-differential operators

In this section we obtain estimates for the kernels of operators in the classes $\Psi^m_{\rho,\delta}$ (cf. Section 5.4.1) and some consequences for smoothing operators (cf. Section 5.4.2) and for operators of Calderón-Zygmund type in the calculus (cf. Section 5.4.4). We will also show the L^p boundedness of Ψ^0 in Section 5.4.4.

For technical reasons which will become apparent in Section 5.5.2, we will also consider the seminorms:

$$\|\sigma\|_{S^{m,R}_{\rho,\delta},a,b} := \sup_{\substack{(x,\pi)\in G\times\widehat{G}\\ [\alpha]\leq a,[\beta]\leq b}} \|\Delta^\alpha X^\beta_x \sigma(x,\pi)\pi(\mathrm{I}+\mathcal{R})^{-\frac{m-\rho[\alpha]+\delta[\beta]}{\nu}}\|_{\mathscr{L}(\mathcal{H}_\pi)}, \qquad (5.37)$$

where $\mathcal{R}$ is a positive Rockland operator of homogeneous degree ν. The superscript R indicates that the powers of $\mathrm{I}+\mathcal{R}$ are 'on the right'. As for the $S^m_{\rho,\delta}$-seminorms, this is a seminorm which is equivalent to a similar seminorm for another positive Rockland operator.

5.4.1 Estimates of the kernels

This section is devoted to describing the behaviour of the kernel of an operator with symbol in the class $S^m_{\rho,\delta}$. As usual in this chapter, G is a graded Lie group of homogeneous dimension Q. Our results in this section may be summarised in the following theorem.

Theorem 5.4.1. *Let $\sigma = \{\sigma(x,\pi)\}$ be in $S^m_{\rho,\delta}$ with $1\geq\rho\geq\delta\geq 0$, $\rho\neq 0$. Then its associated kernel $\kappa : (x,y)\mapsto\kappa_x(y)$ is smooth on $G\times(G\backslash\{0\})$. We also fix a homogeneous quasi-norm $|\cdot|$ on G.*

(i) Away from 0, κ_x has a Schwartz decay:

$$\forall M\in\mathbb{N}\quad \exists C>0,\ a,b,c\in\mathbb{N}:\quad \forall(x,y)\in G\times G$$
$$|y|>1 \Longrightarrow |\kappa_x(y)|\leq C\sup_{\pi\in\widehat{G}}\|\sigma(x,\pi)\|_{S^m_{\rho,\delta},a,b,c}|y|^{-M}.$$

(ii) Near 0, we have

– *if $Q+m>0$, κ_x behaves like $|y|^{-\frac{Q+m}{\rho}}$: there exists $C>0$ and $a,b,c\in\mathbb{N}$ such that*

$$\forall(x,y)\in G\times(G\backslash\{0\})\quad |\kappa_x(y)|\leq C\sup_{\pi\in\widehat{G}}\|\sigma(x,\pi)\|_{S^m_{\rho,\delta},a,b,c}|y|^{-\frac{Q+m}{\rho}};$$

- *if* $Q + m = 0$, κ_x *behaves like* $\ln|y|$*: there exists* $C > 0$ *and* $a, b, c \in \mathbb{N}$ *such that*

$$\forall (x, y) \in G \times (G\backslash\{0\}) \quad |\kappa_x(y)| \leq C \sup_{\pi \in \widehat{G}} \|\sigma(x, \pi)\|_{S^m_{\rho,\delta}, a,b,c} \ln|y|;$$

- *if* $Q + m < 0$, κ_x *is continuous on* G *and bounded:*

$$\sup_{z \in G} |\kappa_x(z)| \leq C \sup_{\pi \in \widehat{G}} \|\sigma(x, \pi)\|_{S^m_{\rho,\delta}, 0,0,0}.$$

Moreover, it is possible to replace the seminorm $\|\cdot\|_{S^m_{\rho,\delta},a,b,c}$ *in (i) and (ii) with a seminorm* $\|\cdot\|_{S^{m,R}_{\rho,\delta},a,b}$ *given in (5.37).*

Remark 5.4.2. Using Theorem 5.2.22 (i) Parts (3) and (2), and Corollary 5.2.25, we obtain similar properties for $X_y^{\beta_1} \tilde{X}_y^{\beta_2} (X_x^{\beta_o} \tilde{q}_\alpha(y) \kappa_x(y))$.

We start the proof of Theorem 5.4.1 with consequences of Proposition 5.2.16 as preliminary results on the right convolution kernels and then proceed to analysing the behaviour of these kernels both at zero and at infinity.

Proposition 5.2.16 has the following consequences:

Corollary 5.4.3. *Let* $\sigma = \{\sigma(x, \pi)\}$ *be in* $S^m_{\rho,\delta}$ *with* $1 \geq \rho \geq \delta \geq 0$. *Let* κ_x *denote its associated kernel.*

1. *If* $\alpha, \beta_1, \beta_2, \beta_o \in \mathbb{N}_0^n$ *are such that*

$$m - \rho[\alpha] + [\beta_1] + [\beta_2] + \delta[\beta_o] < -Q/2,$$

then the distribution $X_z^{\beta_1} \tilde{X}_z^{\beta_2} (X_x^{\beta_o} \tilde{q}_\alpha(z) \kappa_x(z))$ *is square integrable and for every* $x \in G$ *we have*

$$\int_G \left| X_z^{\beta_1} \tilde{X}_z^{\beta_2} (X_x^{\beta_o} \tilde{q}_\alpha(z) \kappa_x(z)) \right|^2 dz \leq C \sup_{\pi \in \widehat{G}} \|\sigma(x, \pi)\|^2_{S^m_{\rho,\delta}, a,b,c}$$

where $a = [\alpha]$, $b = [\beta_o]$, $c = \rho[\alpha] + [\beta_1] + [\beta_2] + \delta[\beta_o]$ *and* $C = C_{m,\alpha,\beta_1,\beta_2,\beta_o} > 0$ *is a constant independent of* σ *and* x. *If* $\beta_1 = 0$ *then we may replace the seminorm* $\|\cdot\|_{S^m_{\rho,\delta},a,b,c}$ *with a seminorm* $\|\cdot\|_{S^{m,R}_{\rho,\delta},a,b}$ *given in (5.37).*

2. *For any* $\alpha, \beta_1, \beta_2, \beta_o \in \mathbb{N}_0^n$ *satisfying*

$$m - \rho[\alpha] + [\beta_1] + [\beta_2] + \delta[\beta_o] < -Q,$$

the distribution $z \mapsto X_z^{\beta_1} \tilde{X}_z^{\beta_2} X_x^{\beta_o} \tilde{q}_\alpha(z) \kappa_x(z)$ *is continuous on* G *for every* $x \in G$ *and we have*

$$\sup_{z \in G} \left| X_z^{\beta_1} \tilde{X}_z^{\beta_2} \left\{ X_x^{\beta_o} \tilde{q}_\alpha(z) \kappa_x(z) \right\} \right| \leq C \sup_{\pi \in \widehat{G}} \|\sigma(x, \pi)\|_{S^m_{\rho,\delta}, [\alpha], [\beta_o], [\beta_2]},$$

where $C = C_{m,\alpha,\beta_1,\beta_2,\beta_o} > 0$ is a constant independent of σ and x. If $\beta_1 = 0$ then we may replace the seminorm $\|\cdot\|_{S^m_{\rho,\delta},[\alpha],[\beta_o],[\beta_2]}$ with the seminorm $\|\cdot\|_{S^{m,R}_{\rho,\delta},[\alpha],[\beta_o]}$, see (5.37).

Consequently, if $\rho > 0$ then the map $\kappa : (x,y) \mapsto \kappa_x(y)$ is smooth on $G \times (G \setminus \{0\})$.

Proof. Part (1) follows from Proposition 5.2.16 together with Theorem 5.2.22 (i) Parts (3) and (2), and Corollary 5.2.25 . Now by the Sobolev inequality in Theorem 4.4.25 (ii), if the right-hand side of the following inequality is finite:

$$\sup_{z\in G}\left|X_z^{\beta_1}\tilde{X}_z^{\beta_2}\left\{X_x^{\beta_o}\tilde{q}_\alpha(z)\kappa_x(z)\right\}\right| \leq C\left\|(\mathrm{I}+\mathcal{R}_z)^{\frac{s}{\nu}}X_z^{\beta_1}\tilde{X}_z^{\beta_2}\left\{X_x^{\beta_o}\tilde{q}_\alpha(z)\kappa_x(z)\right\}\right\|_{L^2(dz)},$$

for $s > Q/2$, then the distribution

$$z \mapsto X_z^{\beta_1}\tilde{X}_z^{\beta_2}\left\{X_x^{\beta_o}\tilde{q}_\alpha(z)\kappa_x(z)\right\}$$

is continuous and the inequality of Part (2) holds. By Theorem 4.4.16,

$$\begin{aligned}
&\left\|(\mathrm{I}+\mathcal{R}_z)^{\frac{s}{\nu}}X_z^{\beta_1}\tilde{X}_z^{\beta_2}\left\{X_x^{\beta_o}\tilde{q}_\alpha(z)\kappa_x(z)\right\}\right\|_{L^2(dz)}\\
&\quad\leq C\left\|(\mathrm{I}+\mathcal{R})^{\frac{s+[\beta_1]}{\nu}}(\mathrm{I}+\tilde{\mathcal{R}})^{\frac{[\beta_2]}{\nu}}\left\{X_x^{\beta_o}\tilde{q}_\alpha(z)\kappa_x(z)\right\}\right\|_{L^2(dz)}\\
&\quad\leq C\left\|\pi(\mathrm{I}+\mathcal{R})^{\frac{s+[\beta_1]}{\nu}}X_x^{\beta_o}\Delta^\alpha\sigma(x,\pi)\pi(\mathrm{I}+\mathcal{R})^{\frac{[\beta_2]}{\nu}}\right\|_{L^2(\widehat{G})},
\end{aligned}$$

by the Plancherel formula (1.28). By Proposition 5.2.16 (together with Theorem 5.2.22 (ii)) as long as

$$m + s + [\beta_1] - \rho[\alpha] + \delta[\beta_o] + [\beta_2] < -Q/2,$$

since

$$(\mathrm{I}+\mathcal{R})^{\frac{s+[\beta_1]}{\nu}}(\mathrm{I}+\tilde{\mathcal{R}})^{\frac{[\beta_2]}{\nu}}\left\{X_x^{\beta_o}\tilde{q}_\alpha(z)\kappa_x(z)\right\}$$

is the kernel of the symbol

$$\pi(\mathrm{I}+\mathcal{R})^{\frac{s+[\beta_1]}{\nu}}X_x^{\beta_o}\Delta^\alpha\sigma(x,\pi)\pi(\mathrm{I}+\mathcal{R})^{\frac{[\beta_2]}{\nu}},$$

we have

$$\left\|\pi(\mathrm{I}+\mathcal{R})^{\frac{s+[\beta_1]}{\nu}}X_x^{\beta_o}\Delta^\alpha\sigma(x,\pi)\pi(\mathrm{I}+\mathcal{R})^{\frac{[\beta_2]}{\nu}}\right\|_{L^2(\widehat{G})} \leq C\|\sigma(x,\pi)\|_{S^m_{\rho,\delta},[\alpha],[\beta_o],[\beta_2]},$$

if $s + [\beta_1] \leq \rho[\alpha] - m - \delta[\beta_o] - [\beta_2]$. This shows Part (2). □

Estimates at infinity

We will now prove better estimates for the kernel than the ones stated in Corollary 5.4.3. First let us show that the kernel has a Schwartz decay away from the origin.

Proposition 5.4.4. *Let $\sigma = \{\sigma(x,\pi)\}$ be in $S^m_{\rho,\delta}$ with $1 \geq \rho \geq \delta \geq 0$. Let κ_x denote its associated kernel.*

We assume that $\rho > 0$ and we fix a homogeneous quasi-norm $|\cdot|$ on G. Then for any $M \in \mathbb{R}$ and any $\alpha, \beta_1, \beta_2, \beta_o \in \mathbb{N}_0^n$ there exist $C > 0$ and $a, b, c \in \mathbb{N}$ independent of σ such that for all $x \in G$ and $z \in G$ satisfying $|z| \geq 1$, we have

$$\left| X_z^{\beta_1} \tilde{X}_z^{\beta_2} (X_x^{\beta_o} \tilde{q}_\alpha(z) \kappa_x(z)) \right| \leq C \sup_{\pi \in \widehat{G}} \|\sigma(x,\pi)\|_{S^m_{\rho,\delta},a,b,c} |z|^{-M}.$$

Furthermore, if $\beta_1 = 0$ then we may replace the seminorm $\|\cdot\|_{S^m_{\rho,\delta},a,b,c}$ with a seminorm $\|\cdot\|_{S^{m,R}_{\rho,\delta},a,b}$ given in (5.37).

Proof. We start by proving the stated result for $\alpha = \beta_1 = \beta_2 = \beta_o = 0$ and for the homogeneous quasi-norm $|\cdot|_p$ given by (3.21). Here $p > 0$ is a positive number to be chosen suitably. We also fix a number $b_o > 0$ and a function $\eta_o \in C^\infty(\mathbb{R})$ valued in $[0,1]$ with $\eta_o \equiv 0$ on $(-\infty, \frac{1}{2}]$ and $\eta_o \equiv 1$ on $[1,\infty)$. We set

$$\eta(x) := \eta_o(b_o^{-p} |x|_p^p).$$

Therefore, η is a smooth function on G such that $\eta(z) = 1$ if $|z|_p \geq b_o$. Consequently,

$$\begin{aligned} \sup_{|z|_p \geq b_o} \left| |z|_p^M \kappa_x(z) \right| &\leq \sup_{z \in G} \left| |z|_p^M \kappa_x(z) \eta(z) \right| \\ &\leq C \sum_{[\beta'] \leq \lceil Q/2 \rceil} \left\| X_z^{\beta'} \{ |z|_p^M \kappa_x(z) \eta(z) \} \right\|_{L^2(G,dz)} \end{aligned} \tag{5.38}$$

by the Sobolev inequality in Theorem 4.4.25.

We study each term separately. We assume that $p/2$ is a positive integer divisible by all the weights $\upsilon_1, \ldots, \upsilon_n$ and we introduce the polynomial

$$|z|_p^p = \sum_{j=1}^n |z_j|^{\frac{p}{\upsilon_j}}$$

and its inverse, so that

$$\begin{aligned} X_z^{\beta'} \{ |z|_p^M \kappa_x(z) \eta(z) \} &= X_z^{\beta'} \{ |z|_p^M |z|_p^{-p} |z|_p^p \kappa_x(z) \, \eta(z) \} \\ &= \overline{\sum_{[\beta_1'] + [\beta_2'] = [\beta']}} X_z^{\beta_1'} \{ |z|_p^M |z|_p^{-p} \eta(z) \} X_z^{\beta_2'} \{ |z|_p^p \kappa_x(z) \}, \end{aligned}$$

where $\overline{\sum}$ means taking a linear combination, that is, a sum involving some constants. We observe that, using a polar change of coordinates,

$$\|X_z^{\beta_1'}\left\{|z|_p^M|z|_p^{-p}\eta(z)\right\}\|_{L^2(G,dz)} < \infty$$

as long as $2(M-p-[\beta_1'])+Q-1<-1$. We assume that p has been chosen so that $2(M-p)+Q<0$. Therefore, all these L^2-norms can be viewed as constants. By the Cauchy-Schwartz inequality and the properties of Sobolev spaces, we obtain

$$\begin{aligned}\|X_z^{\beta'}\left\{|z|_p^M\kappa_x(z)\eta(z)\right\}\|_{L^2(G,dz)} &\leq C\sum_{[\beta_2']\leq[\beta']}\|X_z^{\beta_2'}\left\{|z|_p^p\kappa_x(z)\right\}\|_{L^2(G,dz)}\\ &\leq C\sum_{[\beta_2']\leq[\beta']}\sum_{[\alpha]\leq p}\|X_z^{\beta_2'}\left\{\tilde{q}_\alpha\kappa_x\right\}\|_2,\end{aligned}$$

since $|z|_p^p=\sum_{j=1}^n z_j^{\frac{p}{v_j}}$ is a polynomial of homogeneous degree p. Therefore, by Corollary 5.4.3 Part (1), we get

$$\|X_z^{\beta'}\left\{|z|_p^M\kappa_x(z)\eta(z)\right\}\|_{L^2(G,dz)} \leq C\sup_{\pi\in\widehat{G}}\|\sigma(x,\pi)\|_{S^m_{\rho,\delta},p,0,\rho p+[\beta']}$$

if $\rho p-m>Q/2+[\beta']$. We choose p accordingly. Combining this with (5.38) yields

$$\sup_{|z|_p\geq b_o}\left||z|_p^M\kappa_x(z)\right| \leq C\sup_{\pi\in\widehat{G}}\|\sigma(x,\pi)\|_{S^m_{\rho,\delta},p,0,\rho p+\lceil Q/2\rceil}.$$

Therefore, we have obtained the result for the homogeneous norm $|\cdot|_p$ and $\alpha=\beta_1=\beta_2=\beta_o=0$.

The full result follows for any homogeneous norm and indices $\alpha,\beta_1,\beta_2,\beta_o$ from the equivalence of any two homogeneous norms and by Theorem 5.2.22 (i) Parts (3) and (2), and Corollary 5.2.25. □

Remark 5.4.5.

1. During the proof of Proposition 5.4.4, we have obtained the following statement which is quantitatively more precise. We keep the setting of Proposition 5.4.4. Then for any $M\in\mathbb{R}$ and $b_o>0$, there exists $C=C_{M,b_o,m}>0$ such that

$$\sup_{|z|_p\geq b_o}\left||z|_p^M\kappa_x(z)\right| \leq C\sup_{\pi\in\widehat{G}}\|\sigma(x,\pi)\|_{S^m_{\rho,\delta},p,0,\rho p+\lceil Q/2\rceil},$$

where $p\in\mathbb{N}$ is the smallest positive integer such that $p/2$ is divisible by all the weights $v_1,\ldots,v_n$ and $p>\max(Q/2+M,\frac{1}{\rho}(m+Q+1))$.

2. Combining Part (1) above, Theorem 5.2.22 (i) Parts (3) and (2), and Corollary 5.2.25, it is possible (but not necessarily useful) to obtain a concrete expression for the numbers a,b,c appearing in Proposition 5.4.4, in terms of $m,\rho,\delta,\alpha,\beta_1,\beta_2,\beta_o$ and of Q.

Furthermore, the same statement is true for $|z|\geq b_o$ for an arbitrary lower bound $b_o>0$. However, the constant C may depend on b_o.

Estimates at the origin

We now prove a singular estimate for the kernel near the origin which is (therefore) not covered by Corollary 5.4.3 (2).

Proposition 5.4.6. *Let $\sigma = \{\sigma(x,\pi)\}$ be in $S^m_{\rho,\delta}$ with $1 \geq \rho \geq \delta \geq 0$. Let κ_x denote its associated kernel.*

We assume that $\rho > 0$ and we fix a homogeneous quasi-norm $|\cdot|$ on G. Then for any $\alpha, \beta_1, \beta_2, \beta_o \in \mathbb{N}_0^n$ with $Q + m + \delta[\beta_o] - \rho[\alpha] + [\beta_1] + [\beta_2] \geq 0$ there exist a constant $C > 0$ and computable integers $a, b, c \in \mathbb{N}_0$ independent of σ such that for all $x \in G$ and $z \in G\backslash\{0\}$, we have that if

$$Q + m + \delta[\beta_o] - \rho[\alpha] + [\beta_1] + [\beta_2] > 0,$$

then

$$\left| X_z^{\beta_1} \tilde{X}_z^{\beta_2} (X_x^{\beta_o} \tilde{q}_\alpha(z) \kappa_x(z)) \right| \leq C \sup_{\pi \in \widehat{G}} \|\sigma(x,\pi)\|_{S^m_{\rho,\delta},a,b,c} |z|^{-\frac{Q+m+\delta[\beta_o]-\rho[\alpha]+[\beta_1]+[\beta_2]}{\rho}},$$

and if

$$Q + m + \delta[\beta_o] - \rho[\alpha] + [\beta_1] + [\beta_2] = 0,$$

then

$$\left| X_z^{\beta_1} \tilde{X}_z^{\beta_2} (X_x^{\beta_o} \tilde{q}_\alpha(z) \kappa_x(z)) \right| \leq C \sup_{\pi \in \widehat{G}} \|\sigma(x,\pi)\|_{S^m_{\rho,\delta},a,b,c} \ln |z|.$$

In both estimates, if $\beta_1 = 0$ then we may replace the seminorm $\|\cdot\|_{S^m_{\rho,\delta},a,b,c}$ with a seminorm $\|\cdot\|_{S^{m,\mathcal{R}}_{\rho,\delta},a,b}$ given in (5.37).

During the proof of Proposition 5.4.6, we will need the following technical lemma which is of interest on its own.

Lemma 5.4.7. *Let $\sigma = \{\sigma(x,\pi)\}$ be in $S^m_{\rho,\delta}$ with $1 \geq \rho \geq \delta \geq 0$. Let $\eta \in \mathcal{D}(\mathbb{R})$ and $c_o > 0$. We also fix a positive Rockland operator $\mathcal{R}$ of homogeneous degree ν with corresponding seminorms for the symbol classes $S^m_{\rho,\delta}$.*

Then for any $\ell \in \mathbb{N}_0$, the symbols given by

$$\sigma_{L,\ell}(x,\pi) := \eta(2^{-\ell c_o}\pi(\mathcal{R}))\sigma(x,\pi) \quad \text{and} \quad \sigma_{R,\ell}(x,\pi) := \sigma(x,\pi)\eta(2^{-\ell c_o}\pi(\mathcal{R})),$$

are in $S^{-\infty}$. Moreover, for any $m_1 \in \mathbb{R}$ and $a, b, c \in \mathbb{N}_0$, there exists a constant $C = C_{m,m_1,\rho,\delta,a,b,c,\eta,c_o} > 0$ such that for any $\ell \in \mathbb{N}_0$ we have

$$\|\sigma_{L,\ell}(x,\pi)\|_{S^{m_1}_{\rho,\delta},a,b,c} \leq C \sup_{\pi \in \widehat{G}} \|\sigma(x,\pi)\|_{S^m_{\rho,\delta},a,b,c} 2^{\ell \frac{c_o}{\nu}(m-m_1)}.$$

The same holds for $\sigma_{R,\ell}(x,\pi)$, but with a possibly different seminorm on the right hand side.

Only for $\sigma_{R,\ell}(x,\pi)$, we also have for the seminorm $\|\cdot\|_{S^{m,\mathcal{R}}_{\rho,\delta},a,b}$ given in (5.37), the estimate

$$\|\sigma_{R,\ell}(x,\pi)\|_{S^{m_1,\mathcal{R}}_{\rho,\delta},a,b} \leq C \sup_{\pi \in \widehat{G}} \|\sigma(x,\pi)\|_{S^{m,\mathcal{R}}_{\rho,\delta},a,b} 2^{\ell \frac{c_o}{\nu}(m-m_1)}.$$

Proof of Lemma 5.4.7. For each $\ell \in \mathbb{N}_0$, the symbol $\eta(2^{-\ell c_o}\pi(\mathcal{R}))$ is in $S^{-\infty}$ by Proposition 5.3.4. Therefore, by Theorem 5.2.22 (ii) and the inclusions (5.31), $\sigma_{L,\ell}$ and $\sigma_{R,\ell}$ are in $S^{-\infty}$.

Let us fix $\alpha_o, \beta_o \in \mathbb{N}_0^n$ and $\gamma \in \mathbb{R}$. By the Leibniz formula (see (5.28)),

$$\begin{aligned}
&\pi(\mathrm{I}+\mathcal{R})^{\frac{\rho[\alpha_o]-m_1-\delta[\beta_o]+\gamma}{\nu}} X_x^{\beta_o}\Delta^{\alpha_o}\sigma_{L,\ell}\pi(\mathrm{I}+\mathcal{R})^{-\frac{\gamma}{\nu}} \\
&\quad = \pi(\mathrm{I}+\mathcal{R})^{\frac{\rho[\alpha_o]-m_1-\delta[\beta_o]+\gamma}{\nu}} X_x^{\beta_o}\Delta^{\alpha_o}\left\{\eta(2^{-\ell c_o}\pi(\mathcal{R}))\sigma(x,\pi)\right\}\pi(\mathrm{I}+\mathcal{R})^{-\frac{\gamma}{\nu}} \\
&\quad = \sum_{[\alpha_1]+[\alpha_2]=[\alpha_o]} c_{\alpha_1,\alpha_2}\pi(\mathrm{I}+\mathcal{R})^{\frac{\rho[\alpha_o]-m_1-\delta[\beta_o]+\gamma}{\nu}}\Delta^{\alpha_1}\eta(2^{-\ell c_o}\pi(\mathcal{R})) \\
&\qquad\qquad X_x^{\beta_o}\Delta^{\alpha_2}\sigma(x,\pi)\pi(\mathrm{I}+\mathcal{R})^{-\frac{\gamma}{\nu}}.
\end{aligned}$$

Therefore, taking the operator norm, we obtain

$$\begin{aligned}
&\|\pi(\mathrm{I}+\mathcal{R})^{\frac{\rho[\alpha_o]-m_1-\delta[\beta_o]+\gamma}{\nu}} X_x^{\beta_o}\Delta^{\alpha_o}\sigma_{L,\ell}\pi(\mathrm{I}+\mathcal{R})^{-\frac{\gamma}{\nu}}\| \\
&\le C\sum_{[\alpha_1]+[\alpha_2]=[\alpha_o]}\|\pi(\mathrm{I}+\mathcal{R})^{\frac{\rho[\alpha_o]-m_1-\delta[\beta_o]+\gamma}{\nu}}\Delta^{\alpha_1}\eta(2^{-\ell c_o}\pi(\mathcal{R}))\pi(\mathrm{I}+\mathcal{R})^{-\frac{\rho[\alpha_2]-m-\delta[\beta_o]+\gamma}{\nu}}\| \\
&\qquad\qquad \|\pi(\mathrm{I}+\mathcal{R})^{\frac{\rho[\alpha_2]-m-\delta[\beta_o]+\gamma}{\nu}} X_x^{\beta_o}\Delta^{\alpha_2}\sigma(x,\pi)\pi(\mathrm{I}+\mathcal{R})^{-\frac{\gamma}{\nu}}\| \\
&\le C\|\sigma(x,\pi)\|_{S^m_{\rho,\delta},[\alpha_o],[\beta_o],|\gamma|} \\
&\quad \sum_{[\alpha_1]+[\alpha_2]=[\alpha_o]}\|\pi(\mathrm{I}+\mathcal{R})^{\frac{\rho[\alpha_o]-m_1-\delta[\beta_o]+\gamma}{\nu}}\Delta^{\alpha_1}\eta(2^{-\ell c_o}\pi(\mathcal{R}))\pi(\mathrm{I}+\mathcal{R})^{-\frac{\rho[\alpha_2]-m-\delta[\beta_o]+\gamma}{\nu}}\|.
\end{aligned}$$

By Proposition 5.3.4,

$$\begin{aligned}
&\|\pi(\mathrm{I}+\mathcal{R})^{\frac{\rho[\alpha_o]-m_1-\delta[\beta_o]+\gamma}{\nu}}\Delta^{\alpha_1}\eta(2^{-\ell c_o}\pi(\mathcal{R}))\pi(\mathrm{I}+\mathcal{R})^{-\frac{\rho[\alpha_2]-m-\delta[\beta_o]+\gamma}{\nu}}\| \\
&\qquad\qquad \le C\|\eta(2^{-\ell c_o}\cdot)\|_{\mathcal{M}_{\frac{m_2}{\nu}},k},
\end{aligned}$$

for some k, where m_2 is such that

$$[\alpha_1]-m_2 = \rho[\alpha_o]-m_1-\delta[\beta_o]+\gamma-(\rho[\alpha_2]-m-\delta[\beta_o]+\gamma),$$

that is,

$$m_2 = m_1 - m + [\alpha_1](1-\rho).$$

Now, we can estimate

$$\begin{aligned}
\|\eta(2^{-\ell c_o}\cdot)\|_{\mathcal{M}_{\frac{m_2}{\nu}},k} &= \sup_{\lambda>0,\, k'=0,\dots,k}(1+\lambda)^{k'-\frac{m_2}{\nu}}\partial_\lambda^{k'}(\eta(2^{-\ell c_o}\lambda)) \\
&= \sup_{\lambda>0,\, k'=0,\dots,k}(1+\lambda)^{k'-\frac{m_2}{\nu}}2^{-\ell c_o k'}(\partial^{k'}\eta)(2^{-\ell c_o}\lambda) \\
&\le C2^{-\ell c_o\frac{m_2}{\nu}}.
\end{aligned}$$

Therefore,

$$\sum_{[\alpha_1]+[\alpha_2]=[\alpha_o]} \|\pi(\mathrm{I}+\mathcal{R})^{\frac{\rho[\alpha]-m_1+\gamma}{\nu}}\Delta^{\alpha_1}\eta(2^{-\ell c_o}\pi(\mathcal{L}))\pi(\mathrm{I}+\mathcal{R})^{-\frac{\rho[\alpha_2]-m-\delta[\beta_o]+\gamma}{\nu}}\|$$
$$\leq C\sum_{[\alpha_1]+[\alpha_2]=[\alpha_o]} 2^{-\ell c_o\frac{m_1-m+[\alpha_1](1-\rho)}{\nu}} \leq C2^{-\ell c_o\frac{m_1-m}{\nu}},$$

and we have shown that

$$\|\pi(\mathrm{I}+\mathcal{R})^{\frac{\rho[\alpha_o]-m_1-\delta[\beta_o]+\gamma}{\nu}}X_x^{\beta_o}\Delta^{\alpha_o}\sigma_{L,\ell}\pi(\mathrm{I}+\mathcal{R})^{-\frac{\gamma}{\nu}}\|$$
$$\leq C_{\alpha_o}\|\sigma(x,\pi)\|_{S^m_{\rho,\delta},[\alpha_o],[\beta_o],|\gamma|}2^{-\ell c_o\frac{m_1-m}{\nu}}.$$

The desired property for $\sigma_{L,\ell}$ follows easily. The property for $\sigma_{R,\ell}$ may be obtained by similar methods and its proof is left to the reader. □

Proof of Proposition 5.4.6. By Theorem 5.2.22 (i) Parts (3) and (2), and Corollary 5.2.25, it suffices to show the statement for $\alpha=\beta_1=\beta_2=\beta_o=0$. By equivalence of homogeneous quasi-norms (Proposition 3.1.35), we may assume that the homogeneous quasi-norm is $|\cdot|_p$ given by (3.21) where $p>0$ is such that $p/2$ is the smallest positive integer divisible by all the weights $\upsilon_1,\ldots,\upsilon_n$. Since κ_x decays faster than any polynomial away from the origin (more precisely see Proposition 5.4.4), it suffices to prove the result for $|z|_p<1$.

So let $\sigma\in S^m_{\rho,\delta}$ with $Q+m\geq 0$. By Lemma 5.4.11 (to be shown in Section 5.4.2) we may assume that the kernel $\kappa:(x,y)\mapsto\kappa_x(y)$ of σ is smooth on $G\times G$ and compactly supported in x. By Proposition 5.4.4 it is also Schwartz in y.

We fix a positive Rockland operator $\mathcal{R}$ of homogeneous degree ν and a dyadic decomposition of its spectrum: we choose two functions $\eta_0,\eta_1\in\mathcal{D}(\mathbb{R})$ supported in $[-1,1]$ and $[1/2,2]$, respectively, both valued in $[0,1]$ and satisfying

$$\forall\lambda>0\qquad \sum_{\ell=0}^{\infty}\eta_\ell(\lambda)=1,$$

where for $\ell\in\mathbb{N}$ we set

$$\eta_\ell(\lambda):=\eta_1(2^{-(\ell-1)\nu}\lambda).$$

For each $\ell\in\mathbb{N}_0$, the symbol $\eta_\ell(\pi(\mathcal{R}))$ is in $S^{-\infty}$ by Proposition 5.3.4 and its kernel $\eta_\ell(\mathcal{R})\delta_0$ is Schwartz by Corollary 4.5.2. Furthermore, by the functional calculus, $\sum_{\ell=0}^N\eta_\ell(\mathcal{R})$ converges in the strong operator topology of $\mathscr{L}(L^2(G))$ to the identity operator I as $N\to\infty$, and thus $\sum_{\ell=0}^N\eta_\ell(\mathcal{R})\delta_0$ converges in $\mathcal{K}(G)$ and in $\mathcal{S}'(G)$ to the Dirac measure δ_0 at the origin as $N\to\infty$.

By Theorem 5.2.22 (ii), the symbol σ_ℓ given by

$$\sigma_\ell(x,\pi):=\sigma(x,\pi)\eta_\ell(\pi(\mathcal{R})),\quad (x,\pi)\in G\times\widehat{G},$$

is in $S^{-\infty}$. The kernel associated with σ_ℓ is κ_ℓ given by

$$\kappa_\ell(x,y) = \kappa_{\ell,x}(y) = (\eta_\ell(\mathcal{R})\delta_0) * \kappa_x(y).$$

For each x, we have $\kappa_{\ell,x} \in \mathcal{S}(G)$. The sum $\sum_{\ell=0}^{N} \kappa_{\ell,x}$ converges in $\mathcal{S}'(G)$ to κ_x as $N \to \infty$ since

$$\sum_{\ell=0}^{N} \mathrm{Op}(\sigma_\ell(x,\cdot)) = \mathrm{Op}(\sigma(x,\cdot)) \sum_{\ell=0}^{N} \eta_\ell(\mathcal{R})$$

converges to $\mathrm{Op}(\sigma(x,\cdot))$ in the strong operator topology of $\mathscr{L}(L^2(G), L^2_{-m}(G))$. This convergence is in fact stronger. Indeed, by Lemma 5.4.7,

$$\|\sigma_\ell\|_{S^{m_1}_{\rho,\delta},a,b,c} \le C \sup_{\pi\in\widehat{G}} \|\sigma\|_{S^{m}_{\rho,\delta},a',b',c'} 2^{\ell(m-m_1)},$$

thus

$$\sum_{\ell\in\mathbb{N}} \|\sigma_\ell\|_{S^{m_1}_{\rho,\delta},a,b,c} < \infty$$

if $m_1 > m$. Consequently, the sum $\sum_\ell \sigma_\ell$ is convergent in $S^{m_1}_{\rho,\delta}$ and, fixing $x \in G$, the sum $\sum_\ell \sup_{z\in S} |\kappa_{\ell,x}(z)|$ is convergent where S is any compact subset of $G\backslash\{0\}$ by Proposition 5.4.4 or more precisely the first part in Remark 5.4.5. Necessarily, the limit of $\sum_\ell \sigma_\ell$ is σ and the limit of $\sum_\ell \kappa_{\ell,x}$ for the uniform convergence on any compact subset of $G\backslash\{0\}$ is κ_x with

$$|\kappa_x(z)| \le \sum_{\ell=0}^{\infty} |\kappa_{\ell,x}(z)|, \qquad z \in G\backslash\{0\}.$$

By Corollary 5.4.3 (2), for any $m_1 < -Q$ and $r \in \mathbb{N}_0$, we have

$$\begin{aligned} \sup_{z\in G} |z|_p^{pr} |\kappa_{\ell,x}(z)| &\le C \sum_{[\alpha]=pr} \sup_{\pi\in\widehat{G}} \|\Delta^\alpha \sigma_\ell(x,\pi)\|_{S^{m_1}_{\rho,\delta},0,0,0} \\ &\le C c_{\sigma,r} 2^{\ell(m-m_1-\rho pr)} \end{aligned} \tag{5.39}$$

by Lemma 5.4.7 and its proof, with $c_{\sigma,r} := \sup_{\pi\in\widehat{G}} \|\sigma(x,\pi)\|_{S^{m}_{\rho,\delta},pr,0,0}$.

We write $|z|_p \sim 2^{-\ell_o}$ in the sense that $\ell_o \in \mathbb{N}_0$ is the only integer satisfying $|z|_p \in (2^{-(\ell_o+1)}, 2^{-\ell_o}]$.

Let us assume that $Q+m > 0$. We use (5.39) with $r = 0$ and m_1 such that $m - m_1 = (Q+m)/\rho$. In particular,

$$m_1 = m(1-\frac{1}{\rho}) - \frac{Q}{\rho} < -Q.$$

The sum over $\ell = 0, \ldots, \ell_o - 1$, can be estimated as

$$\begin{aligned} \sum_{\ell=0}^{\ell_o-1} |\kappa_{\ell,x}(z)| &\le \sum_{\ell=0}^{\ell_o-1} C c_{\sigma,0} 2^{\ell(m-m_1)} \le c_{\sigma,0} 2^{\ell_o(m-m_1)} \\ &\le C c_{\sigma,0} |z|_p^{-\frac{Q+m}{\rho}}. \end{aligned}$$

We now choose $r \in \mathbb{N}$ and $m_1 < -Q$ such that

$$m - m_1 - \rho p r < 0 \quad \text{and} \quad pr(1-\rho) + m - m_1 = \frac{Q+m}{\rho}.$$

More precisely, we set $r := \lceil (m+Q)/(\rho p) \rceil$, that is, r is the largest integer strictly greater than $(m+Q)/(\rho p)$, while m_1 is defined by the equality just above; in particular,

$$m - m_1 > \frac{Q+m}{\rho} - (1-\rho)\frac{Q+m}{\rho} \quad \text{thus} \quad m_1 < -Q.$$

We may use (5.39) and sum over $\ell = \ell_o, \ell_o + 1 \ldots$, to get

$$\sum_{\ell=\ell_o}^{\infty} |z|_p^{pr} |\kappa_{\ell,x}(z)| \leq C c_{\sigma,r} \sum_{\ell=\ell_o}^{\infty} 2^{\ell(m-m_1-\rho p r)} \leq C c_{\sigma,r} 2^{\ell_o(m-m_1-\rho p r)}.$$

Therefore, we obtain

$$\begin{aligned}
\sum_{\ell=\ell_o}^{\infty} |\kappa_{\ell,x}(z)| &\leq C c_{\sigma,r} 2^{\ell_o(m-m_1-\rho p r)} |z|_p^{-pr} \\
&\leq C c_{\sigma,r} |z|_p^{-pr-(m-m_1-\rho p r)} = C c_{\sigma,r} |z|_p^{-\frac{Q+m}{\rho}}.
\end{aligned}$$

This yields the desired estimate for κ_x when $Q + m < 0$.

Let us assume that $Q + m = 0$. Using (5.39) with $r = 0$ and $m_1 = -m$, we obtain

$$\begin{aligned}
\sum_{\ell=0}^{\ell_o-1} |\kappa_{\ell,x}(z)| &\leq \sum_{\ell=0}^{\ell_o-1} C c_{\sigma,0} 2^{\ell(m-m_1)} \leq c_{\sigma,0} \ell_o \\
&\leq C c_{\sigma,0} \ln |z|_p.
\end{aligned}$$

Proceeding as above for the sum over $\ell \geq \ell_o$, we obtain that $\sum_{\ell=\ell_o}^{\infty} |\kappa_{\ell,x}(z)|$ is bounded. This yields the desired estimate for κ_x in the case $Q + m = 0$. □

Remark 5.4.8. It is possible to obtain a concrete expression for the numbers a, b, c appearing in Proposition 5.4.6, in terms of $m, \rho, \delta, \alpha, \beta_1, \beta_2, \beta_o$ and of Q.

5.4.2 Smoothing operators and symbols

The kernel estimates obtained in Section 5.4.1 allow us to characterise smoothing operators in terms of their kernels. Moreover they also imply that the operators in $\Psi^{-\infty}$ map the tempered distribution to smooth functions and enable the construction of sequences of smoothing operators converging in $\Psi^m_{\rho,\delta}$

Theorem 5.4.9. *1. If $T \in \Psi^{-\infty}$, then its associated kernel $\kappa : (x,y) \mapsto \kappa_x(y)$ is a smooth function on $G \times G$ such that for each $x \in G$, $y \mapsto \kappa_x(y)$ is Schwartz. Moreover, for each multi-index $\beta \in \mathbb{N}_0^n$ and each Schwartz seminorm $\|\cdot\|_{\mathcal{S}(G),N}$, there exist a constant $C > 0$ and a seminorm $\|\cdot\|_{S^m,a,b,c}$ (both independent of T) such that*

$$\sup_{x\in G} \|X_x^\beta \kappa_x\|_{\mathcal{S}(G),N} \leq C\|\sigma\|_{S^m,a,b,c}.$$

The converse is true, see Lemma 5.2.21.

2. If $T \in \Psi^{-\infty}$, then T extends to a continuous mapping from $\mathcal{S}'(G)$ to $C^\infty(G)$ via

$$Tf(x) = f * \kappa_x(x)$$

where $f \in \mathcal{S}'(G)$, $x \in G$, and κ_x is the kernel associated with T.

Furthermore, for any compact subset $K \subset G$ and any multi-index $\beta \in \mathbb{N}_0^n$, there exists a constant $C > 0$ and a seminorm $\|\cdot\|_{\mathcal{S}'(G),N}$ such that

$$\sup_{x\in K} |\partial^\beta Tf(x)| \leq C\|f\|_{\mathcal{S}'(G),N}.$$

Moreover C can be chosen as $C_1\|\sigma\|_{S^m,a,b,c}$, and $C_1 > 0$ and N can be chosen independently of f and T.

Part 1 may be rephrased as stating that the map between the smoothing operators and their associated kernels is a Fréchet isomorphism between $\Psi^{-\infty}$ and the space $C_b^\infty(G, \mathcal{S}(G))$ of functions $\kappa \in C^\infty(G \times G)$ satisfying

$$\sup_{x\in G} \|X_x^\beta \kappa_x\|_{\mathcal{S}(G),N} < \infty.$$

Here $C_b^\infty(G, \mathcal{S}(G))$ is endowed with the Fréchet structure given via the seminorms

$$\kappa \longmapsto \max_{[\beta]\leq N} \sup_{x\in G} \|X_x^\beta \kappa_x\|_{\mathcal{S}(G),N} < \infty, \qquad N \in \mathbb{N}_0.$$

Part 2 may be rephrased as stating that the mapping $T \mapsto T$ from $\Psi^{-\infty}$ to the space $\mathscr{L}(\mathcal{S}'(G), C^\infty(G))$ of linear continuous mappings from $\mathcal{S}'(G)$ to $C^\infty(G)$ is continuous (it is clearly linear).

Proof. Part 1 follows easily from Theorem 5.4.1 and Remark 5.4.2. By Lemma 3.1.55, for any tempered distribution $f \in \mathcal{S}'(G)$, the function $f * \kappa_x$ is smooth on G and the function $x \mapsto f * \kappa_x(x)$ is smooth on G. Hence T extends to $\mathcal{S}'(G)$ and $Tf \in C^\infty$ if $f \in \mathcal{S}'(G)$.

Note that Lemma 3.1.55 also implies the existence of a positive constant C and $N \in \mathbb{N}_0$ such that

$$|f * \kappa_x(z)| \leq C(1 + |z|)^N \|f\|_{\mathcal{S}'(G),N} \|\kappa_x\|_{\mathcal{S}(G),N}.$$

Using the Leibniz property for vector fields, one checks easily that for any multi-index $\beta \in \mathbb{N}_0^n$, we have

$$X^\beta(Tf)(x) = \sum_{[\beta_1]+[\beta_2]=[\beta]} c_{\beta,\beta_1,\beta_2} X^{\beta_1}_{x_1=x}(f * X^{\beta_2}_{x_2=x}\kappa_{x_2})(x_1).$$

Thus, proceeding as above, passing from left derivatives to the right, and using Lemma 3.1.55, we get

$$\begin{aligned} |X^\beta(Tf)(x)| &\leq C \sum_{[\beta_1]+[\beta_2]=[\beta]} (1+|x|)^{[\beta_1]} |(\tilde{X}^{\beta_1}_{x_1=x}(f * (X^{\beta_2}_{x_2=x}\kappa_{x_2}))(x_1)| \\ &\leq C \sum_{[\beta_1]+[\beta_2]=[\beta]} (1+|x|)^{[\beta_1]} |(\tilde{X}^{\beta_1}_{x_1=x} f) * (X^{\beta_2}_{x_2=x}\kappa_{x_2})(x_1)| \\ &\leq C \sum_{[\beta_1]+[\beta_2]=[\beta]} (1+|x|)^{[\beta_1]+N} \|\tilde{X}^{\beta_1} f\|_{\mathcal{S}'(G),N} \|X^{\beta_2}_{x_2=x}\kappa_{x_2}\|_{\mathcal{S}(G),N} \\ &\leq C(1+|x|)^{N_2} \|f\|_{\mathcal{S}'(G),N_1} \|X^{\beta_2}_{x_2=x}\kappa_{x_2}\|_{\mathcal{S}(G),N} \end{aligned}$$

with a new constant $C > 0$ and integers $N_2, N_1, N \in \mathbb{N}_0$. This shows that $f \mapsto Tf$ is continuous from $\mathcal{S}'(G)$ to $C^\infty(G)$.

Using Part 1, the inequality above also shows the continuity of $T \mapsto T$ from $\Psi^{-\infty}$ to the space of continuous mappings from $\mathcal{S}'(G)$ to $C^\infty(G)$. This concludes the proof of Theorem 5.4.9. □

Using the stability of taking the adjoint, reasoning by duality from Part 2 of Theorem 5.4.9, will yield the fact that smoothing operators map distributions with compact support to Schwartz functions, see Corollary 5.5.13.

Note that the proof of Part 2 of Theorem 5.4.9 yields the more precise result:

Corollary 5.4.10. *If $T \in \Psi^{-\infty}$ and $f \in \mathcal{S}'(G)$, then Tf is smooth and all its left-derivatives $X^\beta Tf$, $\beta \in \mathbb{N}_0^n$, have polynomial growth. More precisely, for any multi-index $\beta \in \mathbb{N}_0^n$, there exist a constant $C > 0$, and integer $M \in \mathbb{N}_0$ and a seminorm $\|\cdot\|_{\mathcal{S}'(G),N}$ such that*

$$|X^\beta Tf(x)| \leq C(1+|x|)^M \|f\|_{\mathcal{S}'(G),N}.$$

Moreover C can be chosen as $C_1\|\sigma\|_{S^m,a,b,c}$, and $C_1 > 0$ and N, M can be chosen independently of f and T.

5.4.3 Pseudo-differential operators as limits of smoothing operators

In the proof of Lemma 5.1.42, for a given symbol σ, we constructed a sequence of symbols σ_ϵ such that $\mathrm{Op}(\sigma_\epsilon)$ is a sequence of 'nice operators' converging towards $\mathrm{Op}(\sigma)$ in a certain sense. If we assume that $\sigma \in S^m_{\rho,\delta}$, then we can construct a sequence of smoothing operators with a convergence in $\Psi^m_{\rho,\delta}$ described in the next lemma and its corollary. These operators are therefore 'nice' since they have Schwartz associated kernels in the sense of Theorem 5.4.9.

Lemma 5.4.11. *Let $1 \geq \rho \geq \delta \geq 0$. If $\sigma = \{\sigma(x,\pi)\}$ is in $S^m_{\rho,\delta}$, then we can construct a family $\sigma_\epsilon = \{\sigma_\epsilon(x,\pi)\}$, $\epsilon > 0$, in $S^{-\infty}$, satisfying the following properties:*

1. *For each $\epsilon > 0$, the x-support of each σ_ϵ is compact, or in other words, the function $x \mapsto \sup_{\pi\in\widehat{G}} \|\sigma(x,\pi)\|_{\mathscr{L}(\mathcal{H}_\pi)}$ is zero outside a compact set in G. Hence the kernel $\kappa_\epsilon : (x,y) \mapsto \kappa_{\epsilon,x}(y)$ associated with each symbol σ_ϵ is Schwartz on $G \times G$ and compactly supported in x.*
2. *For any seminorm $\|\cdot\|_{S^{m_1}_{\rho,\delta},a,b,c}$, there exist a constant $C = C_{a,b,c,m,m_1\rho,\delta} > 0$ such that*
$$\forall \epsilon \in (0,1) \qquad \|\sigma_\epsilon\|_{S^{m_1}_{\rho,\delta},a,b,c} \leq C\|\sigma\|_{S^m_{\rho,\delta},a,b,c}\epsilon^{\frac{m_1-m}{\nu}},$$
and when $m \leq m_1$,
$$\forall \epsilon \in (0,1) \qquad \|\sigma_\epsilon - \sigma\|_{S^{m_1}_{\rho,\delta},a,b,c} \leq C\|\sigma\|_{S^m_{\rho,\delta},a,b,c+\rho a}\epsilon^{\frac{m_1-m}{\nu}}.$$
Here ν is the degree of homogeneity of the positive Rockland operator used to define the seminorms.

   Consequently, when $m < m_1$, the convergence $\sigma_\epsilon \to \sigma$ as $\epsilon \to 0$ holds in $S^{m_1}_{\rho,\delta}$.
3. *If $\phi \in \mathcal{S}(G)$ then $\text{Op}(\sigma_\epsilon)\phi \in \mathcal{D}(G)$ and the convergence*
$$\text{Op}(\sigma_\epsilon)\phi \underset{\epsilon\to 0}{\longrightarrow} \text{Op}(\sigma)\phi$$
holds uniformly on any compact subset of G and also in $\mathcal{S}(G)$.

Remark 5.4.12. As the construction will show, the symbols σ_ϵ are constructed independently of the order $m \in \mathbb{R}$.

Proof of Lemma 5.4.11. We consider the function χ_ϵ on G constructed in Lemma 5.1.42. Let $\eta \in \mathcal{D}(\mathbb{R})$ be such that $\eta \equiv 1$ on $[0,1]$. Let $\mathcal{R}$ be a positive Rockland operator. Let $\sigma \in S^m_{\rho,\delta}$. We set
$$\sigma_\epsilon(x,\pi) = \chi_\epsilon(x)\sigma(x,\pi)\eta(\epsilon\,\pi(\mathcal{R})).$$

Arguing as in Lemma 5.4.7 and its proof yields that
$$\{\sigma(x,\pi)\eta(\epsilon\,\pi(\mathcal{R})), (x,\pi) \in G \times \widehat{G}\}$$
is in $S^{-\infty}$. Moreover, for any $m_1 \in \mathbb{R}$ and $a,b,c \in \mathbb{N}_0$, there exists a constant $C = C_{m,m_1,\rho,\delta,a,b,c,\eta} > 0$ such that for any $\ell \in \mathbb{N}_0$ we have
$$\|\sigma(x,\pi)\eta(\epsilon\,\pi(\mathcal{R}))\|_{S^{m_1}_{\rho,\delta},a,b,c} \leq C \sup_{\pi\in\widehat{G}} \|\sigma(x,\pi)\|_{S^m_{\rho,\delta},a,b,c}\epsilon^{\frac{m_1-m}{\nu}}.$$

From this, it is clear that Property (1) and the first estimate in Property (2) hold.

Let us prove the second estimate in Property (2). We notice that

$$\begin{aligned}
&\|\pi(\mathrm{I}+\mathcal{R})^{-\frac{m_1}{\nu}}\left(\sigma(x,\pi)\eta(\epsilon\,\pi(\mathcal{R}))-\sigma(x,\pi)\right)\|_{\mathscr{L}(\mathcal{H}_\pi)}\\
&=\|\pi(\mathrm{I}+\mathcal{R})^{-\frac{m_1}{\nu}}\sigma(x,\pi)\left(\eta(\epsilon\,\pi(\mathcal{R}))-\mathrm{I}\right)\|_{\mathscr{L}(\mathcal{H}_\pi)}\\
&\leq\|\pi(\mathrm{I}+\mathcal{R})^{-\frac{m_1}{\nu}}\sigma(x,\pi)\pi(\mathrm{I}+\mathcal{R})^{\frac{m_1-m}{\nu}}\|_{\mathscr{L}(\mathcal{H}_\pi)}\\
&\qquad\|\pi(\mathrm{I}+\mathcal{R})^{\frac{m-m_1}{\nu}}\left(\eta(\epsilon\,\pi(\mathcal{R}))-\mathrm{I}\right)|_{\mathscr{L}(\mathcal{H}_\pi)},
\end{aligned}$$

and the spectral calculus properties (cf. Corollary 4.1.16) imply

$$\begin{aligned}
&\sup_{\pi\in\widehat{G}}\|\pi(\mathrm{I}+\mathcal{R})^{\frac{m-m_1}{\nu}}\left(\eta(\epsilon\,\pi(\mathcal{R}))-\mathrm{I}\right)\|_{\mathscr{L}(\mathcal{H}_\pi)}\\
&=\|(\mathrm{I}+\mathcal{R})^{\frac{m-m_1}{\nu}}\left(\eta(\epsilon\mathcal{R})-\mathrm{I}\right)\|_{\mathscr{L}(L^2(G))}\leq\sup_{\lambda>0}(1+\lambda)^{\frac{m-m_1}{\nu}}|\eta(\epsilon\lambda)-1|.
\end{aligned}$$

One checks easily that

$$\begin{aligned}
\sup_{\lambda>0}(1+\lambda)^{\frac{m-m_1}{\nu}}|\eta(\epsilon\lambda)-1| &\leq \|\eta-1\|_\infty\sup_{\lambda>\epsilon^{-1}}(1+\lambda)^{\frac{m-m_1}{\nu}}\\
&\leq t(1+\epsilon^{-1})^{\frac{m-m_1}{\nu}}\leq C\epsilon^{\frac{m_1-m}{\nu}},
\end{aligned}$$

provided that $m-m_1\leq 0$. Hence

$$\begin{aligned}
&\sup_{(x,\pi)\in G\times\widehat{G}}\|\pi(\mathrm{I}+\mathcal{R})^{-\frac{m_1}{\nu}}\left(\sigma(x,\pi)\eta(\epsilon\,\pi(\mathcal{R}))-\sigma(x,\pi)\right)\|_{\mathscr{L}(\mathcal{H}_\pi)}\\
&\qquad\qquad\leq C\|\sigma\|_{S^m_{\rho,\delta},0,0,|m_1-m|}\epsilon^{\frac{m_1-m}{\nu}}.
\end{aligned}$$

More generally, we can introduce derivatives in x and difference operators and use the Leibniz properties (cf. Proposition 5.2.10):

$$\begin{aligned}
&X_x^\beta\Delta^\alpha\left(\sigma(x,\pi)\eta(\epsilon\,\pi(\mathcal{R}))-\sigma(x,\pi)\right)\\
&=\sum_{[\alpha_1]+[\alpha_2]=[\alpha]}c_{\alpha,\alpha_1,\alpha_2}X_x^\beta\Delta^{\alpha_1}\sigma(x,\pi)\;\;\Delta^{\alpha_2}(\eta(\epsilon\,\pi(\mathcal{R}))-\mathrm{I}),
\end{aligned}$$

so that the quantity

$$\|\pi(\mathrm{I}+\mathcal{R})^{\frac{-m_1+\rho[\alpha]-\delta[\beta]-\gamma}{\nu}}X_x^\beta\Delta^\alpha\left(\sigma(x,\pi)\eta(\epsilon\,\pi(\mathcal{R}))-\sigma(x,\pi)\right)\pi(\mathrm{I}+\mathcal{R})^{\frac{\gamma}{\nu}}\|_{\mathscr{L}(\mathcal{H}_\pi)}$$

is, up to a constant, less or equal to the sum over $[\alpha_1]+[\alpha_2]=[\alpha]$ of

$$\begin{aligned}
&\|\pi(\mathrm{I}+\mathcal{R})^{\frac{-m_1+\rho[\alpha]-\delta[\beta]-\gamma}{\nu}}X_x^\beta\Delta^{\alpha_1}\sigma(x,\pi)\pi(\mathrm{I}+\mathcal{R})^{\frac{m_1-m-\rho[\alpha_2]+\gamma}{\nu}}\|_{\mathscr{L}(\mathcal{H}_\pi)}\\
&\times\|\pi(\mathrm{I}+\mathcal{R})^{-\frac{m_1-m-\rho[\alpha_2]+\gamma}{\nu}}\Delta^{\alpha_2}(\eta(\epsilon\,\pi(\mathcal{R}))-\mathrm{I})\pi(\mathrm{I}+\mathcal{R})^{\frac{\gamma}{\nu}}\|_{\mathscr{L}(\mathcal{H}_\pi)}.
\end{aligned}$$

Applying Proposition 5.3.4, we obtain

$$\|\pi(\mathrm{I}+\mathcal{R})^{-\frac{m_1-m-\rho[\alpha_2]+\gamma}{\nu}}\Delta^{\alpha_2}(\eta(\epsilon\,\pi(\mathcal{R}))-\mathrm{I})\pi(\mathrm{I}+\mathcal{R})^{\frac{\gamma}{\nu}}\|_{\mathscr{L}(\mathcal{H}_\pi)}\leq C\epsilon^{\frac{m-m_1}{\nu}}.$$

Collecting the estimates and taking the supremum over $[\alpha]\leq a$, $[\beta]\leq b$, $|\gamma|\leq c$ yield the second estimate in Property (2).

Property (3) follows from Property (2) and the continuity of $\sigma\mapsto \mathrm{Op}(\sigma)$ from $S^{m_1}_{\rho,\delta}$ to $\mathscr{L}(\mathcal{S}(G))$, see Theorem 5.2.15. □

Keeping the notation of Lemma 5.4.11, we can also show that the kernels κ_ϵ converge in some sense towards the kernel of σ. In order to make this more precise, let us define the space $C_b^\infty(G,\mathcal{S}'(G))$ as the space of functions $x\mapsto\kappa_x\in\mathcal{S}'(G)$ such that for each $x\in G$, $y\mapsto\kappa_x(y)$ is a tempered distribution and, for any $\beta\in\mathbb{N}_0^n$, the map $x\mapsto X_x^\beta\kappa_x$ is continuous and bounded on G. This definition is motivated by the following property:

Lemma 5.4.13. *If $\sigma\in S^m_{\rho,\delta}$ then its associated kernel $\kappa=\kappa^{(\sigma)}$ is in $C_b^\infty(G,\mathcal{S}'(G))$ defined above. Furthermore, the map*

$$\sigma\mapsto\kappa^{(\sigma)}$$

from $S^m_{\rho,\delta}$ to $C_b^\infty(G,\mathcal{S}'(G))$ is continuous.

Naturally, we have endowed $C_b^\infty(G,\mathcal{S}'(G))$ with the structure of Fréchet space given by the seminorms

$$\kappa\longmapsto\max_{[\beta]\leq N}\sup_{x\in G}\|X_x^\beta\kappa_x\|_{\mathcal{S}'(G),N},\qquad N\in\mathbb{N}_0.$$

Proof of Lemma 5.4.13. By Lemma 5.1.35, if σ is a symbol then its kernel is in $C^\infty(G,\mathcal{S}'(G))$. Adapting slightly its proof yields

$$\sup_{x\in G}\|X_x^\beta\kappa_x\|_{\mathcal{S}'(G)}\leq C\sup_{x\in G}\|X_x^\beta\sigma(x,\cdot)\|_{L^\infty_{0,-m-\delta[\beta]}(\widehat{G})}.$$

As the inverse Fourier transform is one-to-one and continuous from $L^\infty_{0,-m-\delta[\beta]}(\widehat{G})$ to $\mathcal{S}'(G)$, this shows the continuity of the map $\sigma\mapsto\kappa^{(\sigma)}$ from $S^m_{\rho,\delta}$ to $C_b^\infty(G,\mathcal{S}'(G))$. □

We can now express the convergence in distribution of the sequence of kernels κ_ϵ constructed in the proof of Lemma 5.4.11:

Corollary 5.4.14. *We keep the notation of Lemma 5.4.11. The sequence of kernels κ_ϵ converges towards the kernel κ associated with σ in $C_b^\infty(G,\mathcal{S}'(G))$. If $\rho>0$, the convergence is also uniform on any compact subset of $G\times(G\backslash\{0\})$.*

Proof. The statement follows from the convergence of σ_ϵ to σ in $S^{m_1}_{\rho,\delta}$ for $m_1<m$ by Part 2 of Lemma 5.4.11, together with Lemma 5.4.13 for the first part and Corollary 5.4.3 for the second part. □

5.4.4 Operators in Ψ^0 as singular integral operators

From the kernel estimates obtained in Section 5.4.1, one can show easily that the operators in Ψ^0 are Calderón-Zygmund, and generalise this to some classes $\Psi^m_{\rho,\delta}$, see Theorem 5.4.16. We are then led to study the L^2-boundedness.

First let us notice that thanks to the kernel estimates, our operators admit a representation as singular integrals in the following sense:

Lemma 5.4.15. *Let κ_x be the kernel associated with $T \in \Psi^m_{\rho,\delta}$ with $m \in \mathbb{R}$ and $1 \geq \rho \geq \delta \geq 0$ with $\rho \neq 0$. For any $f \in \mathcal{S}'(G)$ and any $x_0 \in G$ such that $f \equiv 0$ on a neighbourhood of x_0, the integral*

$$\int_G f(y)\kappa_{x_0}(y^{-1}x_0)dy$$

makes distributional sense and defines a smooth function at x_0.

This coincides with Tf if $f \in \mathcal{S}(G)$.

Proof. Let T and κ_x be as in the statement. Let $f \in \mathcal{S}'(G)$ and $x_0 \in G$. We assume that there exists a bounded open set Ω_2 containing x_0 and where $f \equiv 0$. Let $\Omega \subsetneq \Omega_1 \subsetneq \Omega_2$ be open subsets of Ω_2 such that $x_0 \in \Omega$, $\bar{\Omega} \subset \Omega_1$, and $\bar{\Omega}_1 \subset \Omega_2$. We can find $\chi_1, \chi \in \mathcal{D}(G)$ such that $\chi_1 \equiv 1$ on Ω_1 but $\chi_1 \equiv 0$ outside Ω_2, $\chi \equiv 1$ on Ω but $\chi \equiv 0$ outside Ω_1. At least formally, we have

$$\chi(x)\int_G f(y)\kappa_x(y^{-1}x)dy = \int_G f(y)\ \chi(x)(1-\chi_1)(y)\kappa_x(y^{-1}x)dy,$$

since $f \equiv 0$ on $\{\chi_1 = 1\}$. Clearly the function $(x,y) \mapsto \chi(x)(1-\chi_1)(y)$ is smooth on $G \times G$ and supported away from the diagonal $\{(x,y) \in G \times G : x = y\}$. By Theorem 5.4.1, the function

$$y \longmapsto \chi(x)(1-\chi_1)(y)\kappa_x(y^{-1}x),$$

is Schwartz and this yields a smooth mapping $G \to \mathcal{S}(G)$ (which is also compactly supported). The rest of the statement follows easily. $\square$

In Corollary 5.5.13, we will see that an operator in $\Psi^m_{\rho,\delta}$ extends naturally to $\mathcal{S}'(G)$. Lemma 5.4.15 and its proof above will then imply that the operator admits a singular representation for any tempered distribution in the sense that the following formula makes sense and holds

$$Tf(x) = \int_G f(y)\kappa_x(y^{-1}x)dy,$$

for any $f \in \mathcal{S}'(G)$ and any $x \in G$ such that $f \equiv 0$ on a neighbourhood of x. We will not use this.

We can now give sufficient condition for operator in some $\Psi^m_{\rho,\delta}$ to be Calderón-Zygmund.

Theorem 5.4.16. *1. If $T \in \Psi^0$ then the operator T is Calderón-Zygmund in the sense of Definition 3.2.15.*

2. If $T \in \Psi^m_{\rho,\delta}$ with

$$m \leq (\rho - 1)Q,$$

$1 \geq \rho \geq \delta \geq 0$ and $\rho \neq 0$, then the operator T is Calderón-Zygmund in the sense of Definition 3.2.15.

In Parts 1 and 2, the constants appearing in the Definition 3.2.15 are $\gamma = 1$ and, up to constants of the group, given by seminorms of $T \in \Psi^m_{\rho,\delta}$.

Proof. We fix a homogeneous quasi-norm $|\cdot|$ on G.

Let $T \in \Psi^0$. We denote by κ its associated kernel. Then its integral kernel κ_o is formally given via $\kappa_o(x,y) = \kappa_x(y^{-1}x)$. By Theorem 5.4.1, for any two distinct points $y, x \in G$, we have

$$|\kappa_o(x,y)| = |\kappa_x(y^{-1}x)| \leq C|y^{-1}x|^{-Q}.$$

Using Remark 5.4.2 as well and the Leibniz property for vector fields, we obtain

$$|(X_j)_x \kappa_o(x,y)| \leq |(X_j)_{x_1=x}\kappa_{x_1}(y^{-1}x)| + |(X_j)_{x_2=x}\kappa_x(y^{-1}x_2)| \leq C|y^{-1}x|^{-(Q+v_j)},$$

and

$$|(X_j)_y \kappa_o(x,y)| \leq |(\tilde{X}_j)_{z=y^{-1}x}\kappa_x(z)| \leq C|y^{-1}x|^{-(Q+v_j)}.$$

Hence κ_o satisfies the hypotheses of Lemma 3.2.19. This shows Part 1.

Let us now assume that $T \in \Psi^m_{\rho,\delta}$. Again, let κ be its associated kernel. Let $\chi \in C^\infty(G)$ be supported in the unit ball $\{x \in G : |x| \leq 1\}$ and such that $\chi \equiv 1$ on $\{x \in G : |x| \leq 1/2\}$. By Theorem 5.4.1 and Remark 5.4.2 together with Lemma 5.2.21, the operator given by $\phi \mapsto \phi * \{(1-\chi)\kappa\}$ is smoothing (as $\rho \neq 0$) hence it is a Calderón-Zygmund operator by Part 1. Thus we just have to study the operator $\phi \mapsto \phi * \{\chi\kappa\}$. Its integral kernel is κ_o given via

$$\kappa_o(x,y) = \chi(y^{-1}x)\kappa_x(y^{-1}x).$$

Proceeding as above, in particular by Theorem 5.4.1, we have

$$\begin{aligned} |\kappa_o(x,y)| &= |(\chi\kappa_x)(y^{-1}x)| \lesssim |y^{-1}x|^{-\frac{Q+m}{\rho}}, \\ |(X_j)_y \kappa_o(x,y)| &= |(\tilde{X}_j)_{z=y^{-1}x}\kappa_x(z)| \lesssim |y^{-1}x|^{-\frac{Q+m+v_j}{\rho}}, \end{aligned}$$

and κ_o is supported on $\{(x,y) \in G : |y^{-1}x| \leq 1\}$ where we have

$$\begin{aligned} |(X_j)_x \kappa_o(x,y)| &\leq |(X_j)_{x_1=x}\kappa_{x_1}(y^{-1}x)| + |(X_j)_{x_2=x}\kappa_x(y^{-1}x_2)| \\ &\lesssim |y^{-1}x|^{-\frac{Q+m+\delta v_j}{\rho}} + |y^{-1}x|^{-\frac{Q+m+v_j}{\rho}} \lesssim |y^{-1}x|^{-\frac{Q+m}{\rho} - \frac{\delta}{\rho}v_j} \\ &\lesssim |y^{-1}x|^{-\frac{Q+m+v_j}{\rho}}, \end{aligned}$$

since $|y^{-1}x| \leq 1$. Hence if $(Q+m)/\rho \leq Q$, we can apply Lemma 3.2.19. □

In order to apply the singular integrals theorem (Theorem A.4.4), we still need to show that the operators are L^2-bounded. In the case $(\rho,\delta)=(1,0)$, it is not very difficult to adapt the Euclidean case to show that the operators in Ψ^0 are L^2-bounded.

Theorem 5.4.17. *If $T\in\Psi^0$ then T extends to a bounded operator on $L^2(G)$. Furthermore, there exist constants $C>0$ and $a,b,c\in\mathbb{N}_0$ of the group such that*

$$\forall f\in\mathcal{S}(G)\qquad \|Tf\|_{L^2(G)}\leq C\|T\|_{\Psi^m,a,b,c}\|f\|_{L^2(G)}.$$

During the proof of Theorem 5.4.17, we will need the following observation:

Lemma 5.4.18. *The collection of operators Ψ^0 is invariant under left translations in the sense that*

$$T\in\Psi^0\Longrightarrow\forall x_o\in G\quad \tau_{x_o}T\tau_{x_o}^{-1}\in\Psi^0,\qquad \textit{where}\qquad \tau_{x_o}:f\mapsto f(x_o\,\cdot).$$

Furthermore, if κ_x is the kernel of T and $\sigma=\mathrm{Op}^{-1}(T)$ is its symbol, then the operator $\tau_{x_o}T\tau_{x_o}^{-1}$ has κ_{x_ox} as kernel and $\sigma(x_ox,\pi)$ as symbol, and

$$\|T\|_{\Psi^0,a,b,c}=\|\tau_{x_o}T\tau_{x_o}^{-1}\|_{\Psi^0,a,b,c}.$$

Proof of Lemma 5.4.18. Let $T\in\Psi^0$ and let κ_x be its kernel. Then

$$\begin{aligned}\tau_{x_o}T\tau_{x_o}^{-1}f(x) &= T(\tau_{x_o}^{-1}f)(x_ox)=(\tau_{x_o}^{-1}f)*\kappa_{x_ox}(x_ox)\\ &= \int_G f(x_o^{-1}y)\kappa_{x_ox}(y^{-1}x_ox)dy\\ &= \int_G f(z)\kappa_{x_ox}(z^{-1}x)dz\end{aligned}$$

after the change of variable $z=x_o^{-1}y$. Therefore

$$\tau_{x_o}T\tau_{x_o}^{-1}f(x)=f*\kappa_{x_ox}(x).$$

Since $\mathcal{F}_G(\kappa_{x_ox})(\pi)=\sigma(x_ox,\pi)$ if σ denotes the symbol of T, we see that κ_{x_ox} is the kernel associated to the symbol $\{\sigma(x_ox,\pi),(x,\pi)\in G\times\widehat{G}\}$ and the corresponding operator is $\tau_{x_o}T\tau_{x_o}^{-1}$. The rest of the statement follows easily. □

Proof of Theorem 5.4.17. The proof follows the Euclidean case as given in [Ste93, ch. VI §2]. Let $T\in\Psi^0$ and let $\sigma=\mathrm{Op}^{-1}(T)$ be its symbol. We claim that it suffices to show Theorem 5.4.17 under the additional assumption that the kernel κ associated with σ is smooth in x and Schwartz in y, and such that $G\ni x\mapsto\kappa_x\in\mathcal{S}(G)$ is smooth. Indeed, this would imply that Theorem 5.4.17 is proved for each operator $T_\epsilon=\mathrm{Op}(\sigma_\epsilon)$ where σ_ϵ is as in Lemma 5.4.11. The properties (2) and (3) in Lemma 5.4.11 allow to pass through the limit as $\epsilon\to 0$ and imply then the theorem. This shows our earlier claim and hence we may assume that $G\ni x\mapsto\kappa_x\in\mathcal{S}(G)$ is smooth.

We fix $|\cdot|$ to be the homogeneous quasi-norm $|\cdot|_p$ given by (3.21), where $p>0$ is such that $p/2$ is the smallest positive integer divisible by all the weights $\upsilon_1,\ldots,\upsilon_n$. The balls are defined by $B(x_o,r):=\{x\in G:|x^{-1}x_o|<r\}$. We denote by $C_o\geq 1$ a constant such that for all $x,y\in G$, we have

$$|xy|\leq C_o(|x|+|y|) \quad \text{and} \quad |y|\leq\frac{|x|}{2}\Longrightarrow ||xy|-|x||\leq C_o|y|,$$

see the triangle inequality in Proposition 3.1.38 and its converse (3.26).

Let $f\in\mathcal{S}(G)$ and let us write it as

$$f=f_1+f_2,$$

where f_1 and f_2 are two smooth functions supported in $B(0,4C_o)$ and outside of $B(0,2C_o)$, respectively, and satisfying $|f_1|,|f_2|\leq|f|$.

First, we claim that there exists a constant $C>0$ of the group such that

$$\int_{B(0,1)}|Tf_1(x)|^2dx\leq C\|\sigma\|^2_{S^0,0,\lceil Q/2\rceil,0}\ \|f_1\|^2_{L^2(G)}. \tag{5.40}$$

Let us prove this. We fix a function $\chi\in\mathcal{D}(G)$ which is identically 1 on $B(0,1)$. Then

$$\begin{aligned}\int_{B(0,1)}|Tf_1(x)|^2dx &\leq \int_{B(0,1)}|\chi(x)\ f_1*\kappa_x(x)|^2dx\\ &\leq \int_{B(0,1)}\sup_{z\in G}|\chi(z)\ f_1*\kappa_z(x)|^2dx.\end{aligned}$$

We now use the Sobolev inequality in Theorem 4.4.25 to get

$$\sup_{z\in G}|\chi(z)\ f_1*\kappa_z(x)|^2\leq C\sum_{[\alpha]\leq\lceil Q/2\rceil}\int_G\left|X_z^\alpha\{\chi(z)\ f_1*\kappa_z(x)\}\right|^2dz.$$

Since

$$X_z^\alpha\{\chi(z)\ f_1*\kappa_z(x)\}=f_1*X_z^\alpha\{\chi(z)\kappa_z\}(x),$$

we have obtained

$$\begin{aligned}\int_{B(0,1)}|Tf_1(x)|^2dx &\leq \int_{B(0,1)}C\sum_{[\alpha]\leq\lceil Q/2\rceil}\int_G\left|f_1*X_z^\alpha\{\chi(z)\kappa_z\}(x)\right|^2dzdx\\ &=C\sum_{[\alpha]\leq\lceil Q/2\rceil}\int_G\int_{B(0,1)}\left|f_1*X_z^\alpha\{\chi(z)\kappa_z\}(x)\right|^2dxdz,\end{aligned}$$

by Fubini's property. But the integral over $B(0,1)$ can be estimated using Plancherel's Theorem (see Theorem 1.8.11) by

$$\begin{aligned}\int_{B(0,1)}\left|f_1*X_z^\alpha\{\chi(z)\kappa_z\}(x)\right|^2dx &\leq \|f_1*X_z^\alpha\{\chi(z)\kappa_z\}\|_2^2\\ &\leq \|\pi(X_z^\alpha\{\chi(z)\kappa_z\})\|^2_{L^\infty(\widehat{G})}\|f_1\|_2^2.\end{aligned}$$

Now the Leibniz formula for X_z^α gives

$$\begin{aligned}\|\pi(X_z^\alpha\{\chi(z)\kappa_z\})\|_{\mathscr{L}(L^2(G))} &\leq \sum_{[\alpha_1]+[\alpha_2]=[\alpha]} c_{\alpha_1,\alpha_2}\|\pi\left(X^{\alpha_1}\chi(z)X_z^{\alpha_2}\kappa_z\}\right)\|_{\mathscr{L}(L^2(G))}\\ &\leq C_\alpha \max_{[\beta]\leq[\alpha]}\|\pi\left(X_z^\beta\kappa_z\}\right)\|_{\mathscr{L}(L^2(G))}\sum_{[\alpha_1]\leq[\alpha]}|X^{\alpha_1}\chi(z)|.\end{aligned}$$

Since $\pi\left(X_z^\beta\kappa_z\right) = X_z^\beta\sigma(z,\pi)$, we have obtained

$$\begin{aligned}&\int_{B(0,1)} |f_1 * X_z^\alpha\{\chi(z)\kappa_z\}(x)|^2\,dx\\ &\qquad\leq C\max_{[\beta]\leq[\alpha]}\|X_z^\beta\sigma(z,\pi)\|^2_{L^\infty(\widehat{G})}\|f_1\|_2^2\sum_{[\alpha_1]\leq[\alpha]}|X^{\alpha_1}\chi(z)|^2.\end{aligned}$$

Therefore,

$$\begin{aligned}\int_{B(0,1)}|Tf_1(x)|^2dx &\leq C\sum_{[\alpha]\leq\lceil Q/2\rceil}\int_G\int_{B(0,1)}|f_1 * X_z^\alpha\{\chi(z)\kappa_z\}(x)|^2\,dxdz\\ &\leq C\max_{[\beta]\leq\lceil Q/2\rceil}\sup_{z\in G}\|X_z^\beta\sigma(z,\pi)\|^2_{L^\infty(\widehat{G})}\ \|f_1\|_2^2.\end{aligned}$$

This concludes the proof of Claim (5.40).

Secondly, we claim that for any $r\in\mathbb{N}$, there exists a constant $C = C_r > 0$ such that

$$\int_{B(0,1)}|Tf_2(x)|^2dx \leq C\|\sigma\|^2_{S^0,pr,0,pr}\ \|(1+|\cdot|)^{-pr}f_2\|^2_{L^2(G)}. \tag{5.41}$$

Let us prove this. We write

$$Tf_2(x) = \int_{y\notin B(0,2C_o)} f_2(y)|y^{-1}x|^{-pr}(|\cdot|^{pr}\kappa_x)(y^{-1}x)dy.$$

If $x\in B(0,1)$ and $y\notin B(0,2C_o)$, then

$$|y^{-1}| - |y^{-1}x| \leq C_o|x| \leq C_o \quad \text{thus} \quad |y^{-1}x| \geq |y| - C_o \geq \frac{1}{2}|y| \geq \frac{1}{4}(1+|y|),$$

and

$$\begin{aligned}|Tf_2(x)| &\leq \int_{y\notin B(0,2C_o)}|f_2(y)|\left(\frac{1}{4}(1+|y|)\right)^{-pr}\left|(|\cdot|^{pr}\kappa_x)(y^{-1}x)\right|dy\\ &\leq 4^{pr}\|(1+|\cdot|)^{-pr}f_2\|_{L^2(G)}\ \|(|\cdot|^{pr}\kappa_x)\|_{L^2(G)},\end{aligned}$$

after having used the Cauchy-Schwartz inequality. Integrating the square of the left-hand side over $x \in B(0,1)$, and taking the supremum over $x \in B(0,1)$ of the right-hand side, we obtain

$$\int_{B(0,1)} |Tf_2(x)|^2 dx \leq 4^{2pr} \sup_{x\in B(0,1)} \| |\cdot|^{pr} \kappa_x \|^2_{L^2(G)} \, \|(1+|\cdot|)^{-pr} f_2\|^2_{L^2(G)}. \quad (5.42)$$

Now writing $|z|_p^{pr} = \sum_{[\alpha]=pr} c_\alpha \tilde{q}_\alpha(z)$, we have

$$\| |\cdot|^{pr} \kappa_x \|^2_{L^2(G)} \leq C_r \sum_{[\alpha]=pr} \|\tilde{q}_\alpha \kappa_x\|^2_{L^2(G)}$$

and since by Corollary 5.4.3 (1), if $[\alpha] > Q/2$,

$$\|\tilde{q}_\alpha \kappa_x\|^2_{L^2(G)} \leq C_\alpha \sup_{\pi\in\widehat{G}} \|\sigma(x,\pi)\|^2_{S^m_{\rho,\delta},[\alpha],0,[\alpha]},$$

we have obtained that if $pr > Q/2$, then

$$\sup_{x\in B(0,1)} \| |\cdot|^{pr} \kappa_x \|^2_{L^2(G)} \leq C_r \|\sigma\|^2_{S^0,pr,0,pr}.$$

This and (5.42) show Claim (5.41).

Now, combining together Claims (5.40) and (5.41), we obtain

$$\int_{B(0,1)} |Tf(x)|^2 dx \leq C_r \|T\|^2_{\Psi^0,pr,\lceil Q/2\rceil,pr} \, \|(1+|\cdot|)^{-pr} f\|^2_{L^2(G)},$$

and this is so for any $f \in \mathcal{S}(G)$. Therefore, by Lemma 5.4.18 (and its notation), we have for any $x_o \in G$, that

$$\begin{aligned}
\int_{B(x_o,1)} |Tf(x)|^2 dx &= \int_{|x_o^{-1}x|<1} |Tf(x)|^2 dx = \int_{B(0,1)} |Tf(x_o x')|^2 dx' \\
&= \int_{B(0,1)} |\tau_{x_o}(Tf)(x')|^2 dx' = \int_{B(0,1)} |(\tau_{x_o} T \tau_{x_o}^{-1})(\tau_{x_o} f)(x')|^2 dx' \\
&\leq C_r \|\tau_{x_o} T \tau_{x_o}^{-1}\|^2_{\Psi^0,pr,\lceil Q/2\rceil,pr} \, \|(1+|\cdot|)^{-pr} \tau_{x_o} f\|^2_{L^2(G)} \\
&= C_r \|T\|^2_{\Psi^0,pr,\lceil Q/2\rceil,pr} \, \|(1+|\cdot|)^{-pr} \tau_{x_o} f\|^2_{L^2(G)}.
\end{aligned}$$

Integrating over $x_o \in G$, we obtain for the left hand side,

$$\begin{aligned}
\int_G \int_{B(x_o,1)} |Tf(x)|^2 dx dx_o &= \int_G \int_G 1_{|x_o^{-1}x|<1} |Tf(x)|^2 dx dx_o \\
&= \int_G \int_G 1_{|y|<1} |Tf(x)|^2 dx dy = |B(0,1)| \|Tf\|_2^2,
\end{aligned}$$

and for the last term in the right hand side,

$$\begin{aligned}\int_G \|(1+|\cdot|)^{-pr}\tau_{x_o} f\|^2_{L^2(G)} dx_o &= \int_G\int_G \left|(1+|x|)^{-pr} f(x_o x)\right|^2 dx dx_o \\ &= \|f\|_2^2 \int_G (1+|x|)^{-2pr} dx.\end{aligned}$$

Assuming $-2pr+Q<0$, this last integral is finite.

We have obtained that if $r>Q/2p$ (for instance $r=\lceil Q/2p\rceil$) then $pr>Q/2$ and

$$|B(0,1)|\|Tf\|_2^2 \le C\|T\|^2_{\Psi^0,pr,\lceil Q/2\rceil,pr}\|f\|_2^2.$$

This concludes the proof of Theorem 5.4.17. □

Remark 5.4.19. More precisely we have obtained that if $T\in\Psi^0$, then

$$\|Tf\|_2 \le C\|T\|_{\Psi^0,pr,\lceil Q/2\rceil,pr}\|f\|_2,$$

where $r:=\lceil \frac{Q}{2p}\rceil$, and $p\in\mathbb{R}$ is such that $p/2$ is the smallest positive integer divisible by all the weights $\upsilon_1,\ldots,\upsilon_n$.

Theorem 5.4.16 and Theorem 5.4.17 show that any operator of order 0 and of type (1,0) satisfies the hypotheses of the singular integrals theorem, see Sections 3.2.3 and A.4. Therefore, we have the following corollary:

Corollary 5.4.20. *If $T\in\Psi^0$ then T extends to a bounded operator on $L^p(G)$ for any $p\in(1,\infty)$. Furthermore, there exist constants $a,b,c\in\mathbb{N}_0$ such that*

$$\forall p\in(1,\infty)\quad \exists C>0\quad \forall f\in\mathcal{S}(G)\qquad \|Tf\|_{L^p(G)}\le C\|T\|_{\Psi^0,a,b,c}\|f\|_{L^p(G)}.$$

5.5 Symbolic calculus

In this section we present elements of the symbolic calculus of operators with symbols in the classes $S^m_{\rho,\delta}$. In particular, we will discuss asymptotic sums of symbols, adjoints, and compositions.

5.5.1 Asymptotic sums of symbols

We now establish a nilpotent analogue of the asymptotic sum of symbols of decreasing orders going to $-\infty$.

Theorem 5.5.1. *We assume $1\ge\rho\ge\delta\ge0$. Let $\{\sigma_j\}_{j\in\mathbb{N}_0}$ be a sequence of symbols such that $\sigma_j\in S^{m_j}_{\rho,\delta}$ with m_j strictly decreasing to $-\infty$. Then there exists $\sigma\in S^{m_0}_{\rho,\delta}$, unique modulo $S^{-\infty}$, such that*

$$\forall M\in\mathbb{N}\qquad \sigma-\sum_{j=0}^{M}\sigma_j\in S^{m_{M+1}}_{\rho,\delta}. \tag{5.43}$$

Definition 5.5.2. Under the hypotheses and conclusions of Theorem 5.5.1, we write

$$\sigma \sim \sum_j \sigma_j.$$

Proof. We keep the notation of the statement. We also fix a positive Rockland operator $\mathcal{R}$ of homogeneous degree ν on G. Let $\chi \in C^\infty(\mathbb{R})$ with $\chi_{|(-\infty,1/2)} = 0$ and $\chi_{|[1,\infty)} = 1$. We fix $t \in (0,1)$.

Let us check that for any seminorm $\|\cdot\|_{S^{m_0}_{\rho,\delta},a,b,c}$, there exists a constant $C = C_{a,b,c} > 0$ such that for any $t \in (0,1)$ and any $j \in \mathbb{N}$, we have

$$\|\sigma_j(x,\pi)\chi(t\pi(\mathcal{R}))\|_{S^{m_0}_{\rho,\delta},a,b,c} \le C\|\sigma_j(x,\pi)\|_{S^{m_0}_{\rho,\delta},a,b,c+\rho a+m_0-m_j} t^{\frac{m_0-m_j}{\nu}}. \quad (5.44)$$

Indeed, from the Leibniz formula (see Formula (5.28)), we obtain easily

$$\begin{aligned}
&\|\pi(\mathrm{I}+\mathcal{R})^{\frac{\rho[\alpha_o]-m_0-\delta[\beta_o]+\gamma}{\nu}} X_x^{\beta_o}\Delta^{\alpha_o}\left(\sigma_j(x,\pi)\chi(t\pi(\mathcal{R}))\right)\pi(\mathrm{I}+\mathcal{R})^{-\frac{\gamma}{\nu}}\|_{\mathscr{L}(\mathcal{H}_\pi)} \\
&\lesssim \sum_{[\alpha_1]+[\alpha_2]=[\alpha_o]} \|\pi(\mathrm{I}+\mathcal{R})^{\frac{\rho[\alpha_o]-m_0-\delta[\beta_o]+\gamma}{\nu}} X_x^{\beta_o}\Delta^{\alpha_1}\sigma_j(x,\pi) \\
&\qquad\qquad \Delta^{\alpha_2}\chi(t\pi(\mathcal{R}))\ \pi(\mathrm{I}+\mathcal{R})^{-\frac{\gamma}{\nu}}\|_{\mathscr{L}(\mathcal{H}_\pi)} \\
&\lesssim \sum_{[\alpha_1]+[\alpha_2]=[\alpha_o]} \|\sigma_j(x,\pi)\|_{S^{m_0}_{\rho,\delta},[\alpha_1],[\beta_o],\rho([\alpha_o]-[\alpha_1])+m_0-m_j+|\gamma|} \\
&\qquad\qquad \|\pi(\mathrm{I}+\mathcal{R})^{\frac{\rho[\alpha_2]-m_0+m_j+\gamma}{\nu}}\Delta^{\alpha_2}\chi(t\pi(\mathcal{R}))\pi(\mathrm{I}+\mathcal{R})^{-\frac{\gamma}{\nu}}\|_{\mathscr{L}(\mathcal{H}_\pi)}.
\end{aligned}$$

By the functional calculus, we have

$$\begin{aligned}
&\|\pi(\mathrm{I}+\mathcal{R})^{\frac{\rho[\alpha_2]-m_0+m_j+\gamma}{\nu}}\Delta^{\alpha_2}\chi(t\pi(\mathcal{R}))\pi(\mathrm{I}+\mathcal{R})^{-\frac{\gamma}{\nu}}\|_{\mathscr{L}(\mathcal{H}_\pi)} \\
&\le \|\pi(\mathrm{I}+\mathcal{R})^{\frac{[\alpha_2]-m_0+m_j+\gamma}{\nu}}\Delta^{\alpha_2}\chi(t\pi(\mathcal{R}))\pi(\mathrm{I}+\mathcal{R})^{-\frac{\gamma}{\nu}}\|_{\mathscr{L}(\mathcal{H}_\pi)} \\
&\lesssim \sup_{\substack{k'\le k\\ \lambda>0}} (1+\lambda)^{\frac{-m_0+m_j}{\nu}+k'}|\partial_\lambda^{k'}\{\chi(t\lambda)\}| \lesssim t^{\frac{m_0-m_j}{\nu}},
\end{aligned}$$

by Proposition 5.3.4 for some $k \in \mathbb{N}_0$. This shows (5.44).

Let us choose strictly increasing sequences $\{a_\ell\}$, $\{b_\ell\}$ and $\{c_\ell\}$ of positive integers. For each ℓ there exists $C_\ell > 0$ such that for any $j \in \mathbb{N}$ and $t \in (0,1)$, we have

$$\|\sigma_j(x,\pi)\chi(t\pi(\mathcal{R}))\|_{S^{m_0}_{\rho,\delta},a_\ell,b_\ell,c_\ell} \le C_\ell\|\sigma_j(x,\pi)\|_{S^{m_0}_{\rho,\delta},a_\ell,b_\ell,c_\ell+\rho a_\ell+m_0-m_j} t^{\frac{m_0-m_j}{\nu}}.$$

We may assume that the constants C_ℓ are increasing with ℓ.

We now choose a decreasing sequence of numbers $\{t_j\}$ such that for any $j \in \mathbb{N}$,

$$t_j \in (0,2^{-j}) \quad \text{and} \quad C_j \sup_{\substack{x\in G\\ \pi\in\widehat{G}}} \|\sigma_j(x,\pi)\|_{S^{m_0}_{\rho,\delta},a_j,b_j,c_j+\rho a_j+m_0-m_j} t_j^{\frac{m_0-m_j}{\nu}} \le 2^{-j}.$$

For any $j \in \mathbb{N}$, we define the symbols

$$\tilde{\sigma}_j(x,\pi) := \sigma_j(x,\pi)\chi(t_j\pi(\mathcal{R})).$$

For any $\ell \in \mathbb{N}$, the sum

$$\sum_{j=0}^{\infty} \|\tilde{\sigma}_j\|_{S^{m_0}_{\rho,\delta},a_\ell,b_\ell,c_\ell} \leq \sum_{j=0}^{\ell} \|\tilde{\sigma}_j\|_{S^{m_0}_{\rho,\delta},a_\ell,b_\ell,c_\ell} + \sum_{j=\ell+1}^{\infty} 2^{-j},$$

is finite. Since $S^{m_0}_{\rho,\delta}$ is a Fréchet space, we obtain that

$$\sigma := \sum_{j=0}^{\infty} \tilde{\sigma}_j,$$

is a symbol in $S^{m_0}_{\rho,\delta}$.

Starting the sequence at m_{M+1}, the same proof gives

$$\sum_{j=M+1}^{\infty} \tilde{\sigma}_j \in S^{m_{M+1}}_{\rho,\delta}.$$

By Proposition 5.3.4, each symbol given by $(1-\chi)(t_j\pi(\mathcal{R}))$ is in $S^{-\infty}$. Thus by Theorem 5.2.22 (ii) and the inclusions (5.31), each symbol given by $\sigma_j(x,\pi)(1-\chi)(t_j\pi(\mathcal{R}))$ is in $S^{-\infty}$. Therefore, the symbol given by

$$\sigma(x,\pi) - \sum_{j=0}^{M} \sigma_j(x,\pi) = \sum_{j=0}^{M} \sigma_j(x,\pi)(1-\chi)(t_j\pi(\mathcal{R})) + \sum_{j=M+1}^{\infty} \tilde{\sigma}_j(x,\pi),$$

is in $S^{m_{M+1}}_{\rho,\delta}$. This shows (5.43) for σ.

If τ is another symbol as in the statement of the theorem, then for any $M \in \mathbb{N}$,

$$\sigma - \tau = \left(\sigma - \sum_{j=0}^{M} \sigma_j\right) - \left(\tau - \sum_{j=0}^{M} \sigma_j\right)$$

is in $S^{m_{M+1}}$. Thus $\sigma - \tau \in S^{-\infty}$. □

We note that the proof above does not produce a symbol σ depending continuously on $\{\sigma_j\}$, the same as in the abelian case.

5.5.2 Composition of pseudo-differential operators

In this section, we show that the class of operators $\cup_{m\in\mathbb{R}} \Psi^m_{\rho,\delta}$ is an algebra:

Theorem 5.5.3. *Let $1 \geq \rho \geq \delta \geq 0$ with $\delta \neq 1$ and $m_1, m_2 \in \mathbb{R}$. If $T_1 \in \Psi^{m_1}_{\rho,\delta}$ and $T_2 \in \Psi^{m_2}_{\rho,\delta}$ are two pseudo-differential operators of type (ρ,δ), then their composition T_1T_2 is in $\Psi^{m_1+m_2}_{\rho,\delta}$. Moreover, the mapping*

$$(T_1, T_2) \mapsto T_1T_2$$

is continuous from $\Psi^{m_1}_{\rho,\delta} \times \Psi^{m_2}_{\rho,\delta}$ to $\Psi^{m_1+m_2}_{\rho,\delta}$.

Since any operator in $\Psi^{m}_{\rho,\delta}$ maps $\mathcal{S}(G)$ to itself continuously (see Theorem 5.2.15), the composition of any two operators in $\Psi^{m_1}_{\rho,\delta}$ and $\Psi^{m_2}_{\rho,\delta}$ defines an operator in $\mathscr{L}(\mathcal{S}(G))$.

Let us start the proof of Theorem 5.5.3 with observing that the symbol of T_1T_2 is necessarily known and unique at least formally or under favourable conditions such as between smoothing operators:

Lemma 5.5.4. *Let σ_1 and σ_2 be two symbols in $S^{-\infty}$ and let κ_1 and κ_2 be their associated kernels. We set*

$$\kappa_x(y) := \int_G \kappa_{2,xz^{-1}}(yz^{-1})\kappa_{1,x}(z)dz, \qquad x, y \in G.$$

Then $\sigma(x,\pi) = \pi(\kappa_x)$ defines a smooth symbol σ in the sense of Definition 5.1.34. Furthermore, it satisfies

$$\mathrm{Op}(\sigma_1)\mathrm{Op}(\sigma_2) = \mathrm{Op}(\sigma).$$

and

$$\sigma(x,\pi) = \int_G \kappa_{1,x}(z)\pi(z)^*\sigma_2(xz^{-1},\pi)\, dz, \tag{5.45}$$

In particular, if $\sigma_2(x,\pi)$ is independent of x then $\sigma_1 \circ \sigma_2 = \sigma_1\sigma_2$.

We will often write

$$\sigma := \sigma_1 \circ \sigma_2.$$

Proof of Lemma 5.5.4. We keep the notation of the statement. Clearly $\kappa : (x,y) \mapsto \kappa_x(y)$ is smooth on $G \times G$, compactly supported in x. Furthermore, κ_x is integrable in y since

$$\begin{aligned}
\int_G |\kappa_x(y)|dy &\leq \int_G\int_G |\kappa_2(xz^{-1}, yz^{-1})\kappa_1(x,z)|dzdy \\
&\leq \int_G\int_G |\kappa_{2,xz^{-1}}(w)|dw\ |\kappa_1(x,z)|dz \\
&\leq \max_{x' \in G} \int_G |\kappa_{2,x'}(w)|dw \int_G |\kappa_{1,x}(z)|dz.
\end{aligned}$$

Therefore, $\sigma(x,\pi) = \pi(\kappa_x)$ defines a symbol σ in the sense of Definition 5.1.33.

Using the Leibniz formula iteratively, one obtains easily that for any $\beta_o \in \mathbb{N}_0^n$, $\tilde{X}_x^{\beta_o}\kappa_x(y)$ is a linear combination of

$$\int_G \tilde{X}_{x_2=xz^{-1}}^{\beta_2}\kappa_{2,x_2}(yz^{-1})\tilde{X}_{x_1=x}^{\beta_1}\kappa_{1,x_1}(z)dz, \qquad [\beta_1]+[\beta_2]=[\beta_o].$$

Hence proceeding as above

$$\int_G |\tilde{X}_x^{\beta_o}\kappa_x(y)|dy \lesssim \sum_{[\beta_1]+[\beta_2]=[\beta_o]} \max_{x_2\in G}\int_G |\tilde{X}_{x_2}^{\beta_2}\kappa_{2,x_2}(w)|dw \int_G |\tilde{X}_x^{\beta_1}\kappa_{1,x}(z)|dz.$$

This together with the link between abelian and right-invariant derivatives (see Section 3.1.5, especially 3.17) implies easily that σ is a smooth symbol in the sense of Definition 5.1.34.

The properties of κ_1 and κ_2 (see Theorem 5.4.9) justify the equalities

$$\begin{aligned}
\mathrm{Op}(\sigma_1)\mathrm{Op}(\sigma_2)\phi(x) &= \int_G T_2\phi(y)\kappa_{1,x}(y^{-1}x)dy \\
&= \int_G\int_G \phi(z)\kappa_{2,y}(z^{-1}y)\kappa_{1,x}(y^{-1}x)dzdy \\
&= \int_G\int_G \phi(z)\kappa_{2,xw^{-1}}(z^{-1}xw^{-1})\kappa_{1,x}(w)dzdw \\
&= \int_G \phi(z)\kappa_x(z^{-1}x)dz = \phi * \kappa_x(x),
\end{aligned}$$

with the change of variables $y^{-1}x = w$. This yields $T_1T_2 = \mathrm{Op}(\sigma)$. We have then finally

$$\begin{aligned}
\sigma(x,\pi) &= \widehat{\kappa}_x(\pi) = \int_G \kappa_x(y)\pi(y)^*dy \\
&= \int_G\int_G \kappa_{2,xz^{-1}}(yz^{-1})\kappa_{1,x}(z)\pi(z)^*\pi(yz^{-1})^*dydz \\
&= \int_G \kappa_{1,x}(z)\pi(z)^*\sigma_2(xz^{-1},\pi)\ dz,
\end{aligned}$$

after an easy change of variable. □

From Lemma 5.5.4 and its proof, we see that if $T = \mathrm{Op}(\sigma_1)\mathrm{Op}(\sigma_2)$ then the symbol σ of T is not $\sigma_1\sigma_2$ in general, unless the symbol $\{\sigma_2(x,\pi)\}$ does not depend on $x \in G$ for instance. However, we can link formally σ with σ_1 and σ_2 in the following way: using the vector-valued Taylor expansion (see (5.27)) for $\sigma_2(x,\pi)$ in the variable x, we have

$$\sigma_2(xz^{-1},\pi) \approx \sum_\alpha q_\alpha(z^{-1})X_x^\alpha\sigma_2(x,\pi),$$

Thus, implementing this in the expression (5.45), we obtain informally

$$\begin{aligned}\sigma(x,\pi) &\approx \int_G \kappa_{1,x}(z)\pi(z)^* \sum_\alpha q_\alpha(z^{-1}) X_x^\alpha \sigma_2(x,\pi)\, dz \\ &= \sum_\alpha \int_G q_\alpha(z^{-1})\kappa_{1,x}(z)\pi(z)^* dz\; X_x^\alpha \sigma_2(x,\pi) \\ &= \sum_\alpha \Delta^\alpha \sigma_1(x,\pi)\; X_x^\alpha \sigma_2(x,\pi).\end{aligned}$$

We will show that in fact these formal manipulations effectively give the asympotitcs, see Corollary 5.5.8. From Theorem 5.2.22, we know that if $\sigma_1 \in S^{m_1}_{\rho,\delta}$, $\sigma_2 \in S^{m_2}_{\rho,\delta}$ then

$$\Delta^\alpha \sigma_1\; X_x^\alpha \sigma_2 \in S^{m_1+m_2-(\rho-\delta)[\alpha]}_{\rho,\delta}. \tag{5.46}$$

The main problem with the informal approach above is that one needs to estimate the remainder

$$\sigma_1 \circ \sigma_2 - \sum_{[\alpha]\le M} \Delta^\alpha \sigma_1\; X_x^\alpha \sigma_2.$$

We will first show how to estimate this remainder in the case of $\rho > \delta$ using the following property.

Lemma 5.5.5. *We fix a positive Rockland operator of homogeneous degree ν. Let $m_1, m_2 \in \mathbb{R}$, $1 \ge \rho \ge \delta \le 0$ with $\rho \ne 0$ and $\delta \ne 1$, $\beta_0 \in \mathbb{N}_0^n$, and $M, M_1 \in \mathbb{N}_0$. We assume that*

$$\begin{cases} \frac{m_2+\delta(c_{\beta_0}+\upsilon_n)}{1-\delta} \le \nu M_1 < M - Q - m_1 - \delta[\beta_0] + \rho(Q+\upsilon_1), \\ m_2 + \delta(c_{\beta_0} + \upsilon_n + M) \le \nu M_1 < -Q - m_1 - \delta[\beta_0] + \rho(Q+M), \end{cases} \tag{5.47}$$

where

$$c_{\beta_0} := \max_{\substack{[\beta_{02}]\le[\beta_0] \\ [\beta']\ge[\beta_{02}],\ |\beta'|\ge|\beta_{02}|}} [\beta'].$$

If $M \ge \nu M_1$, only the second condition may be assumed.

Then there exist a constant $C > 0$, and two pseudo-norms $\|\cdot\|_{S^{m_1,R}_{\rho,\delta},a_1,b_1}$, $\|\cdot\|_{S^{m_2}_{\rho,\delta},0,b_2,0}$, such that for any $\sigma_1, \sigma_2 \in S^{-\infty}$ and any $(x,\pi) \in G \times \widehat{G}$ we have

$$\begin{aligned}&\|X_x^{\beta_0}(\sigma_1 \circ \sigma_2(x,\pi) - \sum_{[\alpha]\le M} \Delta^\alpha \sigma_1(x,\pi)\; X_x^\alpha \sigma_2(x,\pi))\|_{\mathscr{L}(\mathcal{H}_\pi)} \\ &\le C\|\sigma_1\|_{S^{m_1,R}_{\rho,\delta},a_1,b_1} \|\sigma_2\|_{S^{m_2}_{\rho,\delta},0,b_2,0}.\end{aligned}$$

In the proof of Lemma 5.5.5, we will use the following easy consequence of the estimates of the kernels given in Theorem 5.2.22.

Lemma 5.5.6. *Let $\sigma \in S^m_{\rho,\delta}$ with $1 \geq \rho \geq \delta \geq 0$ with $\rho \neq 0$. We denote by κ_x its associated kernel. For any $\gamma \in \mathbb{R}$, if $\gamma + Q > \max(\frac{m+Q}{\rho}, 0)$ then there exist a constant $C > 0$ and a seminorm $\|\cdot\|_{S^m_{\rho,\delta},a,b,c}$ such that*

$$\int_G |z|^\gamma |\kappa_x(z)| dz \leq C \|\sigma\|_{S^m_{\rho,\delta},a,b,c}.$$

We may replace $\|\cdot\|_{S^m_{\rho,\delta},a,b,c}$ with $\|\cdot\|_{S^{m,R}_{\rho,\delta},a,b}$.

Proof of Lemma 5.5.6. We keep the notation and the statement and write

$$\int_G |z|^\gamma |\kappa_x(z)| dz = \int_{|z|\geq 1} + \int_{|z|<1}.$$

The estimate for large $|z|$ given in Theorem 5.4.1 easily implies that the integral $\int_{|z|\geq 1}$ is bounded up to a constant of γ, m, ρ, δ, by a seminorm of σ. The estimate for small $|z|$ yield

$$\int_{|z|\leq 1} |z|^\gamma |\kappa_x(z)| dz \lesssim \begin{cases} \int_{|z|\leq 1} |z|^{\gamma - \frac{m+Q}{\rho}} dz & \text{if } m+Q > 0, \\ \int_{|z|\leq 1} |z|^\gamma |\ln |z|| dz & \text{if } m+Q = 0, \\ \int_{|z|\leq 1} |z|^\gamma dz & \text{if } m+Q < 0. \end{cases}$$

Using the polar change of coordinates yields the result. □

Proof of Lemma 5.5.5, case $\beta_0 = 0$. By Lemma 5.5.4 and the observations that follow, we have

$$\begin{aligned} &\sigma(x,\pi) - \sum_{[\alpha]\leq M} \Delta^\alpha \sigma_1(x,\pi)\, X_x^\alpha \sigma_2(x,\pi) \\ &= \int_G \kappa_{1,x}(z)\pi(z)^* \left(\sigma_2(xz^{-1},\pi) - \sum_{[\alpha]\leq M} q_\alpha(z^{-1}) X_x^\alpha \sigma_2(x,\pi) \right) dz \\ &= \int_G \kappa_{1,x}(z)\pi(z)^* R_{x,M}^{\sigma_2(\cdot,\pi)}(z^{-1}) dz, \end{aligned}$$

where $R_{x,M}^{\sigma_2(\cdot,\pi)}$ denotes the remainder of the (vector-valued) Taylor expansion of $v \mapsto \sigma_2(xv,\pi)$ of order M at 0. We now introduce powers of $\pi(\mathrm{I}+\mathcal{R})$ near $\pi(z)^*$

$$\pi(z)^* = \pi(z)^* \pi(\mathrm{I}+\mathcal{R})^{M_1} \pi(\mathrm{I}+\mathcal{R})^{-M_1} = \overline{\sum_{[\beta]\leq \nu M_1}} \pi(z)^* \pi(X)^\beta \pi(\mathrm{I}+\mathcal{R})^{-M_1}$$

and we notice that

$$\pi(z)^* \pi(X)^\beta = (-1)^{|\beta|} \left(\pi(X)^\beta \pi(z)\right)^* = (-1)^{|\beta|} \left(\tilde{X}_z^\beta \pi(z)\right)^*. \tag{5.48}$$

We integrate by parts and obtain

$$\begin{aligned}\sigma(x,\pi) &- \sum_{[\alpha]\leq M} \Delta^\alpha\sigma_1(x,\pi)\ X_x^\alpha\sigma_2(x,\pi)\\ &= \overline{\sum_{[\beta_1]+[\beta_2]\leq\nu M_1}} \int_G \tilde{X}_{z_1=z}^{\beta_1}\kappa_{1,x}(z_1)\pi(z)^*\tilde{X}_{z_2=z}^{\beta_2} R_{x,M}^{\pi(\mathrm{I}+\mathcal{R})^{-M_1}\sigma_2(\cdot,\pi)}(z_2^{-1})dz\\ &= \overline{\sum_{[\beta_1]+[\beta_2]\leq\nu M_1}} \int_G \tilde{X}_{z_1=z}^{\beta_1}\kappa_{1,x}(z_1)\pi(z)^* R_{x,M-[\beta_2]}^{\pi(\mathrm{I}+\mathcal{R})^{-M_1}X^{\beta_2}\sigma_2(\cdot,\pi)}(z^{-1})dz\end{aligned}$$

by Lemma 3.1.50. Taking the operator norm, we have

$$\begin{aligned}\|\sigma(x,\pi) &- \sum_{[\alpha]\leq M} \Delta^\alpha\sigma_1(x,\pi)\ X_x^\alpha\sigma_2(x,\pi)\|_{\mathscr{L}(\mathcal{H}_\pi)}\\ &\lesssim \sum_{[\beta_1]+[\beta_2]\leq\nu M_1} \int_G |\tilde{X}_{z_1=z}^{\beta_1}\kappa_{1,x}(z_1)|\ \|R_{x,M-[\beta_2]}^{\pi(\mathrm{I}+\mathcal{R})^{-M_1}X^{\beta_2}\sigma_2(\cdot,\pi)}(z^{-1})\|_{\mathscr{L}(\mathcal{H}_\pi)}dz.\end{aligned}$$

The adapted statement of Taylor's estimates remains valid for vector-valued function, see Theorem 3.1.51 and Remark 3.1.52 (3), so we have

$$\begin{aligned}&\|R_{x,M-[\beta_2]}^{\pi(\mathrm{I}+\mathcal{R})^{-M_1}X^{\beta_2}\sigma_2(\cdot,\pi)}(z^{-1})\|_{\mathscr{L}(\mathcal{H}_\pi)}\\ &\lesssim \sum_{\substack{|\gamma|\leq\lceil(M-[\beta_2])_+\rfloor+1\\ [\gamma]>(M-[\beta_2])_+}} |z|^{[\gamma]} \sup_{x_1\in G} \|\pi(\mathrm{I}+\mathcal{R})^{-M_1}X_{x_1}^\gamma X_{x_1}^{\beta_2}\sigma_2(x_1,\pi)\|_{\mathscr{L}(\mathcal{H}_\pi)}.\end{aligned}$$

We have obtained that

$$\begin{aligned}\|\sigma(x,\pi) &- \sum_{[\alpha]\leq M} \Delta^\alpha\sigma_1(x,\pi)\ X_x^\alpha\sigma_2(x,\pi)\|_{\mathscr{L}(\mathcal{H}_\pi)}\\ &\lesssim \sum_{\substack{[\gamma]>(M-[\beta_2])_+\\ |\gamma|\leq\lceil(M-[\beta_2])_+\rfloor+1}} \int_G |z|^{[\gamma]}|\tilde{X}_{z_1=z}^{\beta_1}\kappa_{1,x}(z_1)|dz\\ &\qquad \sup_{x_1\in G} \|\pi(\mathrm{I}+\mathcal{R})^{-M_1}X_{x_1}^\gamma X_{x_1}^{\beta_2}\sigma_2(x_1,\pi)\|_{\mathscr{L}(\mathcal{H}_\pi)}.\end{aligned}$$

If $M-[\beta_2]\leq 0$, the integrals above are finite by Lemma 5.5.6 and the suprema are bounded by a $S_{\rho,\delta}^{m_2}$-seminorm in σ_2 when

$$\begin{cases} m_1+[\beta_1]+Q<\rho(Q+v_1)\\ -\nu M_1+m_2+\delta(v_n+[\beta_2])\leq 0\end{cases},$$

and it suffices

$$\begin{cases} m_1+\nu M_1-M+Q<\rho(Q+v_1)\\ -\nu M_1+m_2+\delta(v_n+\nu M_1)\leq 0\end{cases}.$$

If $M-[\beta_2]>0$, the integrals above are finite by Lemma 5.5.6 and the suprema are bounded by a $S^{m_2}_{\rho,\delta}$-seminorm in σ_2 when

$$\begin{cases} m_1+[\beta_1]+Q<\rho(Q+[\gamma]) \\ -\nu M_1+m_2+\delta([\gamma]+[\beta_2])\leq 0 \end{cases},$$

and it suffices

$$\begin{cases} m_1+\nu M_1+Q<\rho(Q+M) \\ -\nu M_1+m_2+\delta(\upsilon_n+M)\leq 0 \end{cases}.$$

Our conditions on M and M_1 ensure that the sufficient conditions above are satisfied. Collecting the various estimates yields the statement in the case $\rho\neq 0$ and $\beta_0=0$. □

Proof of Lemma 5.5.5, general case. Using Formula (5.45), the Leibniz property for left invariant vector fields easily implies that

$$X_x^{\beta_0}\sigma_1\circ\sigma_2(x,\pi)=\overline{\overline{\sum_{[\beta_{01}]+[\beta_{02}]=[\beta_0]}}}\int_G X_x^{\beta_{01}}\kappa_{1,x}(z)\pi(z)^* X_{x_2=x}^{\beta_{02}}\sigma_2(x_2z^{-1},\pi)\,dz.$$

Proceeding as in the case $\beta_0=0$, we have

$$\begin{aligned} X_x^{\beta_0}&\left(\sigma_1\circ\sigma_2(x,\pi)-\sum_{[\alpha]\leq M}\Delta^\alpha\sigma_1(x,\pi)\ X_x^\alpha\sigma_2(x,\pi)\right) \\ &=\overline{\overline{\sum_{[\beta_{01}]+[\beta_{02}]=[\beta_0]}}}\int_G X_x^{\beta_{01}}\kappa_{1,x}(z)\pi(z)^* R_{0,M}^{X_{x_2=x}^{\beta_{02}}\sigma_2(x_2\,\cdot,\pi)}(z^{-1})\,dz. \end{aligned}$$

Introducing the powers of $\pi(\mathrm{I}+\mathcal{R})$, each integral on the right-hand side above is equal to

$$\begin{aligned} \overline{\overline{\sum_{[\beta_1]+[\beta_2]\leq\nu M_1}}}\int_G &\tilde{X}_{z_1=z}^{\beta_1}X_x^{\beta_{01}}\kappa_{1,x}(z_1)\pi(z)^* \\ &R_{0,M-[\beta_2]}^{\pi(\mathrm{I}+\mathcal{R})^{-M_1}X_{x_2=x}^{\beta_{02}}X^{\beta_2}\sigma_2(x_2\,\cdot,\pi)}(z^{-1})\,dz, \end{aligned} \tag{5.49}$$

by Corollary 3.1.53. We use a more precise version for the Taylor remainder than in the proof of the case $\beta_0=0$:

$$\begin{aligned} &\|R_{0,M-[\beta_2]}^{\pi(\mathrm{I}+\mathcal{R})^{-M_1}X_{x_2=x}^{\beta_{02}}X^{\beta_2}\sigma_2(x_2\,\cdot,\pi)}(z^{-1})\|_{\mathscr{L}(\mathcal{H}_\pi)} \\ &\qquad\leq C_M\sum_{\substack{[\gamma]>(M-[\beta_2])_+\\|\gamma|\leq\lceil(M-[\beta_2])_+\rfloor+1}}|z|^{[\gamma]}S(z,M_1,\gamma,\beta_{02},\beta_2), \end{aligned}$$

where $S(z,M_1,\gamma,\beta_{02},\beta_2)$ denotes the supremum

$$S(z,M_1,\gamma,\beta_{02},\beta_2):=\sup_{|y|\leq\eta^{\lceil M\rfloor+1}|z|}\|\pi(\mathrm{I}+\mathcal{R})^{-M_1}X_y^\beta X_{x_2=x}^{\beta_{02}}X_y^{\beta_2}\sigma_2(x_2y,\pi)\|_{\mathscr{L}(\mathcal{H}_\pi)}.$$

For any reasonable function $f : G \to \mathbb{C}$, the definitions of left and right-invariant vector fields imply

$$X_x^\beta f(xy) = \tilde{X}_y^\beta f(xy) \tag{5.50}$$

and the properties of left or right-invariant vector fields (see Section 3.1.5) then yield

$$X_x^\beta f(xy) = \tilde{X}_y^\beta f(xy) = \sum_{\substack{|\beta'|\leq|\beta| \\ [\beta']\geq[\beta]}} Q_{\beta,\beta'}(y) X_y^{\beta'} f(xy), \tag{5.51}$$

where $Q_{\beta,\beta'}$ are $([\beta'] - [\beta])$-homogeneous polynomials. Therefore

$$S(z, M_1, \gamma, \beta_{02}, \beta_2) \lesssim \sum_{\substack{[\beta'_{02}]\geq[\beta_{02}] \\ |\beta'_{02}|\leq|\beta_{02}|}} |z|^{[\beta'_{02}]-[\beta_{02}]} \tilde{S}(M_1, [\gamma] + [\beta'_{02}] + [\beta_2]),$$

where $\tilde{S}(M_1, [\beta_0])$ denotes the supremum

$$\tilde{S}(M_1, [\beta_0]) := \sup_{[\gamma']=[\beta_0]} \sup_{x_1\in G} \|\pi(\mathrm{I} + \mathcal{R})^{-M_1} X_{x_1}^{\gamma'} \sigma_2(x_1, \pi)\|_{\mathscr{L}(\mathcal{H}_\pi)}.$$

We then obtain that (5.49) is bounded up to a constant by

$$\sum_{[\beta_1]+[\beta_2]\leq\nu M_1} \int_G |\tilde{X}_{z_1=z}^{\beta_1} X_x^{\beta_{01}} \kappa_{1,x}(z_1)| \sum_{\substack{[\gamma]>(M-[\beta_2])_+ \\ |\gamma|\leq\lceil(M-[\beta_2])_+\rfloor+1}} |z|^{[\gamma]}$$
$$\sum_{\substack{[\beta'_{02}]\geq[\beta_{02}] \\ |\beta'_{02}|\leq|\beta_{02}|}} |z|^{[\beta'_{02}]-[\beta_{02}]} \tilde{S}(M_1, [\gamma] + [\beta'_{02}] + [\beta_2])\, dz.$$

We conclude in the same way as in the case $\beta_0 = 0$. □

To take into account the difference operator, we will use the following observation.

Lemma 5.5.7. *Let $\sigma_1, \sigma_2 \in S^{-\infty}$. For any $\alpha \in \mathbb{N}_0^n$, $\Delta^\alpha(\sigma_1 \circ \sigma_2)$ is a linear combination independent of σ_1, σ_2 of $(\Delta^{\alpha_1}\sigma_1) \circ (\Delta^{\alpha_2}\sigma_2)$, over $\alpha_1, \alpha_2 \in \mathbb{N}_0^n$ satisfying $[\alpha_1] + [\alpha_2] = [\alpha]$. It is the same linear combination as in the Leibniz rule (5.28).*

Proof of Lemma 5.5.7. We keep the notation of Lemma 5.5.4 and adapt the proof of the Leibniz rule for Δ^α given in Proposition 5.2.10. By Proposition 5.2.3 (4), we have

$$\begin{aligned} \tilde{q}_\alpha(y)\kappa_x(y) &= \int_G \tilde{q}_\alpha(yz^{-1}z)\kappa_{2,xz^{-1}}(yz^{-1})\kappa_{1,x}(z)dz \\ &= \overline{\sum_{[\alpha_1]+[\alpha_2]=[\alpha]}} \int_G \tilde{q}_{\alpha_2}(yz^{-1})\kappa_{2,xz^{-1}}(yz^{-1})\, \tilde{q}_{\alpha_1}(z)\kappa_{1,x}(z)dz, \end{aligned}$$

where $\overline{\sum}$ denotes a linear combination. Lemma 5.5.4 implies easily the statement. □

Proof of Theorem 5.5.3 with $\rho > \delta$. We assume $\rho > \delta$. We fix a positive Rockland operator $\mathcal{R}$ of homogeneous degree ν. Let us show that for any $\alpha_0, \beta_0 \in \mathbb{N}_0^n$, and $M_0 \in \mathbb{N}$, there exists $M \geq M_0$, a constant $C > 0$ and seminorms $\|\cdot\|_{S^{m_1,R}_{\rho,\delta},a_1,b_1}$, $\|\cdot\|_{S^{m_2}_{\rho,\delta},a_2,b_2,c_2}$ such that for any $\sigma_1, \sigma_2 \in S^{-\infty}$ we have

$$\begin{aligned}\|X_x^{\beta_0}\Delta^{\alpha_0}\tau_M(x,\pi)\ \pi(\mathrm{I}+\mathcal{R})^{-\frac{m-(\rho-\delta)M_0-\rho[\alpha_0]+\delta[\beta_0]}{\nu}}\|_{\mathscr{L}(\mathcal{H}_\pi)}\\ \leq C\|\sigma_1\|_{S^{m_1,R}_{\rho,\delta},a_1,b_1}\|\sigma_2\|_{S^{m_2}_{\rho,\delta},a_2,b_2,c_2},\end{aligned} \tag{5.52}$$

where we have denoted $m = m_1 + m_2$ and

$$\tau_M := \sigma_1 \circ \sigma_2 - \sum_{[\alpha]\leq M} \Delta^\alpha \sigma_1 X_x^\alpha \sigma_2.$$

By Lemma 5.5.7, it suffices to show (5.52) only for $\alpha_0 = 0$.

Let $\beta_0 \in \mathbb{N}_0$ and $M_0 \in \mathbb{N}$. We fix $m_2' := -m_1 + (\rho-\delta)M_0 - \delta[\beta_0]$. As $\rho > \delta$, we can find $M \geq \max(M_0, \upsilon_1)$ such that

$$(-Q - m_1 - \delta[\beta_0] + \rho(Q+M)) - (m_2' + \delta(c_{\beta_0} + \upsilon_n + M)) \geq \nu.$$

This shows that we can find M_1 satisfying the second condition in (5.47) for m_1, m_2' and therefore also the first. Hence we can apply Lemma 5.5.5 to M, M_1 and the symbols σ_1 and $\sigma_2\pi(\mathrm{I}+\mathcal{R})^{-\frac{m-(\rho-\delta)M_0+\delta[\beta_0]}{\nu}}$, with orders m_1 and m_2'. The left-hand side of (5.52) is then bounded up to a constant by

$$\begin{aligned}\|\sigma_1\|_{S^{m_1,R}_{\rho,\delta},a_1,b_1}\|\sigma_2\pi(\mathrm{I}+\mathcal{R})^{-\frac{m-(\rho-\delta)M_0+\delta[\beta_0]}{\nu}}\|_{S^{m_2'}_{\rho,\delta},0,b_2,0}\\ \lesssim \|\sigma_1\|_{S^{m_1,R}_{\rho,\delta},a_1,b_1}\|\sigma_2\|_{S^{m_2}_{\rho,\delta},0,b_2,c_2}.\end{aligned}$$

Hence (5.52) is proved.

Using (5.46), classical considerations imply that (5.52) yield that for any $M_0 \in \mathbb{N}_0$, and any seminorm $\|\cdot\|_{S^{m-M_0(\rho-\delta),R}_{\rho,\delta},a,b}$, there exist a constant $C > 0$ and two seminorms $\|\cdot\|_{S^{m_1,R}_{\rho,\delta},a_1,b_1}$, $\|\cdot\|_{S^{m_2}_{\rho,\delta},a_2,b_2,c_2}$ such that for any $\sigma_1, \sigma_2 \in S^{-\infty}$ we have

$$\|\tau_{M_0}\|_{S^{m-M_0(\rho-\delta),R}_{\rho,\delta},a,b} \leq C\|\sigma_1\|_{S^{m_1,R}_{\rho,\delta},a_1,b_1}\|\sigma_2\|_{S^{m_2}_{\rho,\delta},a_2,b_2,c_2}. \tag{5.53}$$

In Section 5.5.4, we will see that for any seminorm $\|\cdot\|_{S^{\tilde{m}}_{\rho,\delta},\tilde{a},\tilde{b},\tilde{c}}$ there exist a constant $C > 0$ and a seminorm $\|\cdot\|_{S^{m,R}_{\rho,\delta},a,b}$ such that

$$\forall \sigma \in S^{-\infty} \qquad \|\sigma\|_{S^{\tilde{m}}_{\rho,\delta},\tilde{a},\tilde{b},\tilde{c}} \leq C\|\sigma\|_{S^{\tilde{m},R}_{\rho,\delta},a,b}. \tag{5.54}$$

Inequalities (5.54) together with (5.53) and Lemma 5.4.11 (to pass from $S^{-\infty}$ to $S^{m_1}_{\rho,\delta}, S^{m_2}_{\rho,\delta}$) conclude the proof of Theorem 5.5.3 in the case $\rho > \delta$. □

Note that the proof of the case $\rho > \delta$ above also shows:

Corollary 5.5.8. *We assume* $1 \geq \rho > \delta \geq 0$. *If* $\sigma_1 \in S^{m_1}_{\rho,\delta}$ *and* $\sigma_2 \in S^{m_2}_{\rho,\delta}$, *then there exists a unique symbol* σ *in* $S^m_{\rho,\delta}$, $m = m_1 + m_2$, *such that*

$$\mathrm{Op}(\sigma) = \mathrm{Op}(\sigma_1)\mathrm{Op}(\sigma_2). \tag{5.55}$$

Moreover, for any $M \in \mathbb{N}_0$, *we have*

$$\{\sigma - \sum_{[\alpha]\leq M} \Delta^\alpha \sigma_1 \; X_x^\alpha \sigma_2\} \in S^{m-(\rho-\delta)M}_{\rho,\delta}. \tag{5.56}$$

Furthermore, the mapping

$$\left\{ \begin{array}{ccc} S^m_{\rho,\delta} & \longrightarrow & S^{m-(\rho-\delta)M}_{\rho,\delta} \\ \sigma & \longmapsto & \{\sigma - \sum_{[\alpha]\leq M} \Delta^\alpha \sigma_1 \; X_x^\alpha \sigma_2\} \end{array} \right. ,$$

is continuous.

Consequently, we can also write

$$\sigma \sim \sum_{j=0}^{\infty} \left(\sum_{[\alpha]=j} \Delta^\alpha \sigma_1 \; X_x^\alpha \sigma_2 \right), \tag{5.57}$$

in the sense of an asymptotic expansion as in Definition 5.5.2.

The case $\rho = \delta$ is more delicate to prove but relies on the same kind of arguments as above. If $\rho = \delta$, the asymptotic formula (5.56) does not bring any improvement and, in this sense, is not interesting.

We will need the following variation of the properties given in Lemma 5.5.6 obtained using Corollary 5.4.3 instead of Theorem 5.4.1.

Lemma 5.5.9. *Let* $\sigma \in S^m_{\rho,\delta}$ *with* $1 \geq \rho \geq \delta \geq 0$. *We denote by* κ_x *its associated kernel. Let* $\gamma \geq 0$ *and* $m < -Q$. *Then there exist a constant* $C > 0$ *and a seminorm* $\|\cdot\|_{S^m_{\rho,\delta},a,b,c}$ *such that*

$$\int_G |z|^\gamma |\kappa_x(z)| dz \leq C \|\sigma\|_{S^m_{\rho,\delta},a,b,c}.$$

We may replace $\|\cdot\|_{S^m_{\rho,\delta},a,b,c}$ *with* $\|\cdot\|_{S^{m,R}_{\rho,\delta},a,b}$

Proof of Lemma 5.5.9. By Part 2 of Corollary 5.4.3, $z \mapsto |\kappa_x(z)|$ is a continuous bounded function if $m - \rho\gamma < -Q$ hence the integral $\int_{|z|<1} |z|^\gamma |\kappa_x(z)| dz$ is finite. By the Cauchy-Schwartz inequality, we have

$$\begin{aligned} \int_{|z|>1} |z|^\gamma |\kappa_x(z)| dz &\leq \sqrt{\int_{|z|>1} |z|^{-Q-\frac{1}{2}}} \sqrt{\int_{|z|>1} |z|^{2\gamma+Q+\frac{1}{2}} |\kappa_x(z)|^2 dz} \\ &\lesssim \sum_{[\alpha]=M} \|\tilde{q}_\alpha \kappa_x\|_{L^2(G)}, \end{aligned}$$

where $M/2 \in \mathbb{N}$ is the smallest integer divisible by $\upsilon_1, \ldots, \upsilon_n$ satisfying $M \geq 2\gamma + Q + \frac{1}{2}$, having chosen (3.21) with $p = M$ for quasi-norm. By Part 1 of Corollary 5.4.3, the sum above is finite when $m - \rho M < -Q/2$, which holds true. $\square$

Using Lemma 5.5.9 instead of Lemma 5.5.6 in the proof of Lemma 5.5.10 produces the following result.

Lemma 5.5.10. *We fix a positive Rockland operator of homogeneous degree ν. Let $m_1 \in \mathbb{R}$, $1 \geq \rho \geq \delta \leq 0$ with $\delta \neq 1$, $\beta_0 \in \mathbb{N}_0^n$, and $M, M_1 \in \mathbb{N}_0$. We assume that*

$$\begin{cases} m_1 + \nu M_1 < -Q \\ -\nu M_1 + m_2 + \delta(c_{\beta_0} + \upsilon_n + \max(\nu M_1, M)) \leq 0 \end{cases},$$

where

$$c_{\beta_0} := \max_{\substack{[\beta_{02}] \leq [\beta_0] \\ [\beta'] \geq [\beta_{02}],\ |\beta'| \geq |\beta_{02}|}} [\beta'].$$

Then there exist a constant $C > 0$, and two seminorms $\|\cdot\|_{S^{m_1,R}_{\rho,\delta},a_1,b_1}$, $\|\cdot\|_{S^{m_2}_{\rho,\delta},0,b_2,0}$, such that for any $\sigma_1, \sigma_2 \in S^{-\infty}$ and any $(x,\pi) \in G \times \widehat{G}$ we have

$$\begin{aligned} &\|X_x^{\beta_0}\big(\sigma_1 \circ \sigma_2(x,\pi) - \sum_{[\alpha] \leq M} \Delta^\alpha \sigma_1(x,\pi)\ X_x^\alpha \sigma_2(x,\pi)\big)\|_{\mathscr{L}(\mathcal{H}_\pi)} \\ &\leq C \|\sigma_1\|_{S^{m_1,R}_{\rho,\delta},a_1,b_1} \|\sigma_2\|_{S^{m_2}_{\rho,\delta},0,b_2,0}. \end{aligned}$$

The details of the proof of Lemma 5.5.10 are left to the reader. The first inequality in the statement just above shows that we will require the ability to choose m_1 as negative as one wants. We can do this thanks to the following remark:

Lemma 5.5.11. *Let $\sigma_1, \sigma_2 \in S^{-\infty}$. For any $X \in \mathfrak{g}$ and any $\sigma_1, \sigma_2 \in S^{-\infty}$, we have*

$$(\sigma_1 \pi(X)) \circ \sigma_2 = \sigma_1 \circ (X_x \sigma_2) + \sigma_1 \circ (\pi(X)\sigma_2).$$

More generally, for any $\beta \in \mathbb{N}_0^n$, we have

$$\{\sigma_1 \pi(X)^\beta\} \circ \sigma_2 = \overline{\sum_{[\beta_1]+[\beta_2]=[\beta]}} \sigma_1 \circ \{\pi(X)^{\beta_1} X_x^{\beta_2} \sigma_2\},$$

where $\overline{\sum}$ denotes a linear combination independent of σ_1, σ_2.

Note that in the expression above, $\pi(X)^{\beta_1}$ and $X_x^{\beta_2}$ commute.

Proof of Lemma 5.5.7. We keep the notation of Lemma 5.5.4. Using integration by parts and the Leibniz formula, we obtain

$$\begin{aligned} (\sigma_1 \pi(X)) \circ \sigma_2\ (x,\pi) &= \int_G \tilde{X}_{z_1=z} \kappa_{1,x}(z_1) \pi(z)^* \sigma_2(xz^{-1},\pi)\, dz \\ &= -\int_G \kappa_{1,x}(z) \left(\tilde{X}_{z_1=z} \pi(z_1)^* \sigma_2(xz^{-1},\pi) + \pi(z)^* \tilde{X}_{z_2=z} \sigma_2(xz_2^{-1},\pi)\right) dz \\ &= \int_G \kappa_{1,x}(z) \left(\pi(z)^* \pi(X) \sigma_2(xz^{-1},\pi) + \pi(z)^* X_{x_2=xz^{-1}} \sigma_2(x_2,\pi)\right) dz. \end{aligned}$$

This shows the first formula. The next formula is obtained recursively. □

We can now sketch the proof of Theorem 5.5.3 in the case $\rho = \delta$.

Sketch of the proof of Theorem 5.5.3 with $\rho = \delta$. We assume $\rho = \delta \in [0,1)$. Writing $\sigma_1 = \sigma_1 \pi(\mathrm{I}+\mathcal{R})^{-N}\pi(\mathrm{I}+\mathcal{R})^N$ and using Lemma 5.5.11, it suffices to prove (5.52) for m_1 as negative as one wants. We proceed as in the proof of the case $\rho > \delta$ replacing Lemma 5.5.5 with Lemma 5.5.10. The details are left to the reader. □

5.5.3 Adjoint of a pseudo-differential operator

Here we prove that the classes $\Psi^m_{\rho,\delta}$ are stable under taking the formal adjoints of operators.

Theorem 5.5.12. *We assume $1 \geq \rho \geq \delta \geq 0$ with $\delta \neq 1$ and $m \in \mathbb{R}$. If $T \in \Psi^m_{\rho,\delta}$ then its formal adjoint T^* is also in $\Psi^m_{\rho,\delta}$. Moreover, the mapping $T \mapsto T^*$ is continuous on $\Psi^m_{\rho,\delta}$.*

Recall that the formal adjoint of an operator $T : \mathcal{S}(G) \to \mathcal{S}'(G)$ is the operator $T^* : \mathcal{S}(G) \to \mathcal{S}'(G)$ defined by

$$\forall \phi, \psi \in \mathcal{S}(G) \qquad \int_G T\phi(x)\,\overline{\psi(x)}\,dx = \int_G \phi(x)\,\overline{T^*\psi(x)}\,dx.$$

We observe that the operator $T = \mathrm{Op}(\sigma) \in \Psi^m_{\rho,\delta}$ maps $\mathcal{S}(G)$ to itself continuously (see Theorem 5.2.15) and therefore has a formal adjoint T^*.

Before beginning the proof of Theorem 5.5.12, let us point out some of its consequences.

Corollary 5.5.13. *1. We assume $1 \geq \rho \geq \delta \geq 0$ with $\delta \neq 1$, and $m \in \mathbb{R}$.*

Any $T \in \Psi^m_{\rho,\delta}$ extends uniquely to a continuous operator on $\mathcal{S}'(G)$. Furthermore the mapping $T \mapsto T$ from $\Psi^m_{\rho,\delta}$ to the space $\mathscr{L}(\mathcal{S}'(G))$ of continuous operators on $\mathcal{S}'(G)$ is linear and continuous.

2. Any smoothing operator $T \in \Psi^{-\infty}$ maps continuously the space $\mathcal{E}'(G)$ of compactly supported distributions to the Schwartz space $\mathcal{S}(G)$. Furthermore the mapping $T \mapsto T$ from $\Psi^{-\infty}$ to the space $\mathscr{L}(\mathcal{E}'(G), \mathcal{S}(G))$ of continuous mappings from $\mathcal{E}'(G)$ to $\mathcal{S}(G)$ is linear and continuous.

Proof of Corollary 5.5.13. We admit Theorem 5.5.12 (whose proof is given below). The statement then follows by classical arguments of duality and Theorem 5.2.15 for Part 1, and Part 2 of Theorem 5.4.9 for Part 2. □

Let us start the proof of Theorem 5.5.12 by observing that the symbol $\sigma^{(*)}$ of the adjoint T^* of $T = \mathrm{Op}(\sigma)$ is necessarily known and unique at least formally or under favourable conditions such as in the case of a smoothing operator:

Lemma 5.5.14. *Let $\sigma \in S^{-\infty}$ and let $\kappa : (x,y) \mapsto \kappa_x(y)$ be its associated kernel. We set*

$$\kappa_x^{(*)}(y) := \bar{\kappa}_{xy^{-1}}(y^{-1}), \quad x, y \in G.$$

Then $\kappa^{(*)} : (x,y) \mapsto \kappa_x^{(*)}(y)$ is smooth on $G \times G$ and for every $\alpha \in \mathbb{N}_0^n$, $x \mapsto X^\alpha \kappa_x^{(*)}$ is continuous from G to $\mathcal{S}(G)$.

The symbol $\sigma^{(*)}$ defined via

$$\sigma^{(*)}(x,\pi) := \mathcal{F}_G(\kappa_x^{(*)})(\pi), \quad (x,\pi) \in G \times \widehat{G},$$

is a smooth symbol in the sense of Definition 5.1.34 and satisfies

$$(\mathrm{Op}(\sigma))^* = \mathrm{Op}(\sigma^{(*)}).$$

In particular, if σ does not depend on x, then $\sigma^{(*)} = \sigma^*$.

Note that this operation is an involution since

$$\kappa_x(y) = \bar{\kappa}^{(*)}_{xy^{-1}}(y^{-1}).$$

Recall that if $\sigma = \{\sigma(x,\pi), (x,\pi) \in G \times \widehat{G}\}$ then we have defined the adjoint symbol

$$\sigma^* = \{\sigma(x,\pi)^*, (x,\pi) \in G \times \widehat{G}\},$$

(see Theorem 5.2.22). Hence we may write

$$\sigma^*(x,\pi) := \sigma(x,\pi)^*.$$

Proof of Lemma 5.5.14. By Corollary 3.1.30, we have

$$\begin{aligned}
X_x^{\beta_o}\{\kappa_x^{(*)}(y)\} &= X_x^{\beta_o}\{\bar{\kappa}_{xy^{-1}}(y^{-1})\} = (-1)^{|\beta_o|}\tilde{X}^{\beta_o}_{y_1=y^{-1}}\{\bar{\kappa}_{xy_1}(y^{-1})\} \\
&= (-1)^{|\beta_o|} \sum_{|\beta|\leq|\beta_o|,\, [\beta]\geq[\beta_o]} Q_{\beta_o,\beta}(y^{-1}) X^{\beta}_{y_1=y^{-1}}\{\bar{\kappa}_{xy_1}(y^{-1})\} \\
&= (-1)^{|\beta_o|} \sum_{|\beta|\leq|\beta_o|,\, [\beta]\geq[\beta_o]} Q_{\beta_o,\beta}(y^{-1}) X^{\beta}_{x_1=xy^{-1}}\{\bar{\kappa}_{x_1}(y^{-1})\},
\end{aligned}$$

where the $Q_{\beta_o,\beta}$'s are $([\beta_o]-[\beta])$-homogeneous polynomials. The regularity of κ described in Theorem 5.4.9 implies that $\kappa^{(*)} : (x,y) \mapsto \kappa_x^{(*)}(y)$ is smooth in x and y (but maybe not compactly supported in x), and it is also Schwartz in y in such a way that all the mappings $G \ni x \mapsto X_x^\alpha \kappa_x^{(*)} \in \mathcal{S}(G)$ are continuous. Clearly $\sigma^{(*)}(x,\pi) = \pi(\kappa_x^{(*)})$ defines a smooth symbol $\sigma^{(*)}$.

Let $\phi, \psi \in \mathcal{S}(G)$ and let $x \in G$. The regularity of κ described in Theorem 5.4.9 justifies easily the following computations:

$$\begin{aligned}
\int_G (\mathrm{Op}(\sigma)\phi)(x)\overline{\psi(x)}dx &= \int_G \phi * \kappa_x(x)\bar{\psi}(x)dx = \int_G\int_G \phi(z)\kappa_x(z^{-1}x)\bar{\psi}(x)dzdx \\
&= \int_G\int_G \phi(z)\bar{\kappa}^{(*)}_{x(z^{-1}x)^{-1}}((z^{-1}x)^{-1})\bar{\psi}(x)dzdx \\
&= \int_G\int_G \phi(z)\overline{\kappa^{(*)}_z(x^{-1}z)\psi(x)}dzdx \\
&= \int_G \phi(z)\overline{\psi * \kappa^{(*)}_z(z)}dz.
\end{aligned}$$

This shows that $\mathrm{Op}(\sigma)^*\psi(z) = \psi * \kappa^{(*)}_z(z)$. □

In general, $\sigma^{(*)}$ is not the adjoint σ^* of the symbol σ, unless for instance it does not depend on $x \in G$. However, we can perform formal considerations to link $\sigma^{(*)}$ with σ^* in the following way: using the Taylor expansion for κ^*_x in x (see equality (5.27)), we obtain

$$\kappa^{(*)}_x(y) = \kappa^*_{xy^{-1}}(y) \approx \sum_\alpha q_\alpha(y^{-1})X^\alpha_x\kappa^*_x(y) = \sum_\alpha \tilde{q}_\alpha(y)X^\alpha_x\kappa^*_x(y).$$

Thus, taking the group Fourier transform at $\pi \in \widehat{G}$, we get

$$\sigma^{(*)}(x,\pi) = \pi(\kappa^{(*)}_x) \approx \sum_\alpha \pi(\tilde{q}_\alpha(y)X^\alpha_x\kappa^*_x(y)) = \sum_\alpha \Delta^\alpha X^\alpha_x\sigma(x,\pi)^*.$$

From Theorem 5.2.22 we know that if $\sigma \in S^m_{\rho,\delta}$ then

$$\Delta^\alpha X^\alpha_x\sigma(x,\pi)^* \in S^{m-(\rho-\delta)[\alpha]}_{\rho,\delta}. \tag{5.58}$$

From these formal computations we see that the main problem is to estimate the remainder coming from the use of the Taylor expansion. This is the purpose of the following technical lemma.

Lemma 5.5.15. *We fix a positive Rockland operator of homogeneous degree ν. Let $m \in \mathbb{R}$, $1 \geq \rho \geq \delta \geq 0$ with $\rho \neq 0$ and $\delta \neq 1$, $\beta_0 \in \mathbb{N}_0^n$, and $M, M_1 \in \mathbb{N}_0$. We assume that $M \geq \nu M_1$ and $(\rho-\delta)M + \rho Q > m + \delta[\beta_0] + \nu M_1 + Q$. Then there exist a constant $C > 0$, and a seminorm $\|\cdot\|_{S^m_{\rho,\delta},a,b,0}$, such that for any $\sigma \in S^{-\infty}$ and any $(x,\pi) \in G \times \widehat{G}$ we have*

$$\|X^{\beta_0}_x(\sigma^{(*)}(x,\pi) - \sum_{[\alpha]\leq M}\Delta^\alpha X^\alpha_x\sigma^*(x,\pi))\pi(\mathrm{I}+\mathcal{R})^{M_1}\|_{\mathscr{L}(\mathcal{H}_\pi)} \leq C\|\sigma\|_{S^m_{\rho,\delta},a,b,0}.$$

Proof of Lemma 5.5.15, case $\beta_0 = 0$. By Lemma 5.5.14 and the observations that follow, we have

$$\begin{aligned}
\sigma^{(*)}(x,\pi) &- \sum_{[\alpha]\leq M} \Delta^\alpha X_x^\alpha \sigma^*(x,\pi) \\
&= \int_G \left(\kappa^*_{xz^{-1}}(z) - \sum_{[\alpha]\leq M} q_\alpha(z^{-1}) X_x^\alpha \kappa_x^*(z) \right) \pi(z)^* dz \\
&= \int_G R_{x,M}^{\kappa_x^*(z)}(z^{-1})\pi(z)^* dz,
\end{aligned}$$

where $R_{x,M}^{\kappa_x^*(z)}$ denotes the remainder of the (vector-valued) Taylor expansion of $v \mapsto \kappa^*_{xv}(z)$ of order M at 0. Using (5.48), we can integrate by parts to obtain

$$\begin{aligned}
\big(\sigma^{(*)}(x,\pi) &- \sum_{[\alpha]\leq M} \Delta^\alpha X_x^\alpha \sigma^*(x,\pi)\big)\pi(\mathrm{I}+\mathcal{R})^{M_1} \\
&= \overline{\sum_{[\beta_1]+[\beta_2]\leq \nu M_1}} \int_G \tilde{X}^{\beta_1}_{z_1=z} R_{x,M}^{\tilde{X}^{\beta_2}_{z_2=z}\kappa_x^*(z_2)}(z_1^{-1})\pi(z)^* dz \\
&= \overline{\sum_{[\beta_1]+[\beta_2]\leq \nu M_1}} \int_G R_{x_1=x,M-[\beta_1]}^{\tilde{X}^{\beta_2}_{z_2=z} X^{\beta_1}_{x_1}\kappa^*_{x_1}(z_2)}(z^{-1})\pi(z)^* dz.
\end{aligned}$$

Taking the operator norm, we have

$$\begin{aligned}
\|\big(\sigma^{(*)}(x,\pi) &- \sum_{[\alpha]\leq M} \Delta^\alpha X_x^\alpha \sigma^*(x,\pi)\big)\pi(\mathrm{I}+\mathcal{R})^{M_1}\|_{\mathscr{L}(\mathcal{H}_\pi)} \\
&\lesssim \sum_{[\beta_1]+[\beta_2]\leq \nu M_1} \int_G |R_{x_1=x,M-[\beta_1]}^{\tilde{X}^{\beta_2}_{z_2=z} X^{\beta_1}_{x_1}\kappa^*_{x_1}(z_2)}(z^{-1})| dz.
\end{aligned}$$

For $|z| < 1$, we will use Taylor's theorem, see Theorem 3.1.51:

$$|R_{x_1=x,M-[\beta_1]}^{\tilde{X}^{\beta_2}_{z_2=z} X^{\beta_1}_{x_1}\kappa^*_{x_1}(z_2)}(z^{-1})| \lesssim \sum_{\substack{|\gamma|\leq \lceil (M-[\beta_1])_+ \rfloor +1 \\ [\gamma]>(M-[\beta_1])_+}} |z|^{[\gamma]} \sup_{x_1\in G} |X_z^\gamma \tilde{X}^{\beta_2}_{z_2=z} X^{\beta_1}_{x_1}\kappa^*_{x_1}(z_2)|,$$

together with the estimate for z near the origin given in Theorem 5.4.1. The link between left and right derivatives, see (1.11), implies

$$\sup_{x_1\in G} |X_z^\gamma \tilde{X}^{\beta_2}_{z_2=z} X^{\beta_1}_{x_1}\kappa^*_{x_1}(z_2)| = \sup_{x_1\in G} |X_z^\gamma X^{\beta_2}_{z_2=z} X^{\beta_1}_{x_1}\kappa_{x_1}(z_2)|.$$

Proceeding as in the proof of Lemma 5.5.6, we obtain that the integral

$$\begin{aligned}
\int_{|z|<1} &|R_{x,M-[\beta_1]}^{\tilde{X}^{\beta_2}_{z_2=z} X^{\beta_1}_{x}\kappa^*_{x}(z_2)}(z^{-1})| dz \\
&\lesssim \sum_{\substack{|\gamma|\leq \lceil (M-[\beta_1])_+ \rfloor +1 \\ [\gamma]>(M-[\beta_1])_+}} \int_{|z|<1} |z|^{[\gamma]} \sup_{x_1\in G} |X_{x_1}^\gamma X^{\beta_2}_{z_2=z} X^{\beta_1}_{x_1}\kappa_{x_1}(z_2)| dz
\end{aligned}$$

is finite whenever $[\gamma] + Q > (m + [\beta_2] + \delta([\gamma] + [\beta_1]) + Q)/\rho$ with the indices as above. These conditions are implied by the hypotheses of the statement. The estimates for z large given in Theorem 5.4.1 show directly that the integral

$$\int_{|z|>1} |R_{x_1=x,M-[\beta_1]}^{\tilde{X}_{z_2=z}^{\beta_2} X_{x_1}^{\beta_1}\kappa_{x_1}^*(z_2)}(z^{-1})|dz,$$

is finite. Collecting the various estimates yields the statement in the case $\rho \neq 0$ and $\beta_0 = 0$. □

Proof of Lemma 5.5.15, general case. We proceed as above and introduce the derivatives with respect to x. We obtain

$$X_x^{\beta_0}\big(\sigma^{(*)}(x,\pi) - \sum_{[\alpha]\leq M} \Delta^\alpha X_x^\alpha \sigma^*(x,\pi)\big) = \int_G R_{0,M}^{X_x^{\beta_0}\kappa_x^*.(z)}(z^{-1})\pi(z)^* dz.$$

And adding $(\mathrm{I}+\mathcal{R})^{M_1}$, we have

$$\begin{aligned}
X_x^{\beta_0}\big(\sigma^{(*)}(x,\pi) - &\sum_{[\alpha]\leq M} \Delta^\alpha X_x^\alpha \sigma^*(x,\pi)\big)(\mathrm{I}+\mathcal{R})^{M_1} \\
&= \overline{\sum_{[\beta_1]+[\beta_2]\leq \nu M_1}} \int_G R_{x_1=0,M-[\beta_1]}^{\tilde{X}_{z_2=z}^{\beta_2} X_{x_1}^{\beta_1} X_x^{\beta_0}\kappa_{xx_1}^*(z_2)}(z^{-1})\pi(z)^* dz.
\end{aligned}$$

Taking the operator norm, we have

$$\begin{aligned}
\|X_x^{\beta_0}\big(\sigma^{(*)}(x,\pi) - &\sum_{[\alpha]\leq M} \Delta^\alpha X_x^\alpha \sigma^*(x,\pi)\big)\pi(\mathrm{I}+\mathcal{R})^{M_1}\|_{\mathscr{L}(\mathcal{H}_\pi)} \\
&\lesssim \sum_{[\beta_1]+[\beta_2]\leq \nu M_1} \int_G |R_{x_1=0,M-[\beta_1]}^{\tilde{X}_{z_2=z}^{\beta_2} X_{x_1}^{\beta_1} X_x^{\beta_0}\kappa_{xx_1}^*(z_2)}(z^{-1})|dz.
\end{aligned}$$

For $|z| < 1$, we use the more precise version of Taylor's theorem than in the case $\beta_0 = 0$:

$$\begin{aligned}
&|R_{x,M-[\beta_1]}^{\tilde{X}_{z_2=z}^{\beta_2} X_{x_1}^{\beta_1} X_x^{\beta_0}\kappa_{xx_1}^*(z_2)}(z^{-1})| \\
&\quad\lesssim \sum_{\substack{|\gamma|\leq \lceil (M-[\beta_1])_+ \rfloor +1 \\ [\gamma]>(M-[\beta_1])_+}} |z|^{[\gamma]} \sup_{|y|\leq \eta^{\lceil (M-[\beta_1])_+ \rfloor +1}|z|} |X_y^\gamma \tilde{X}_{z_2=z}^{\beta_2} X_y^{\beta_1} X_x^{\beta_0}\kappa_{xy}^*(z_2)|.
\end{aligned}$$

We proceed as in the proof of Lemma 5.5.5, that is, we use (5.51) to obtain

$$\begin{aligned}
&\sup_{|y|\leq \eta^{\lceil (M-[\beta_1])_+ \rfloor +1}|z|} |X_y^\gamma \tilde{X}_{z_2=z}^{\beta_2} X_y^{\beta_1} X_x^{\beta_0}\kappa_{xy}^*(z_2)| \\
&\qquad\lesssim \sum_{\substack{[\beta_0']\geq[\beta_0] \\ |\beta_0'|\leq|\beta_0|}} |z|^{[\beta_0']-[\beta_0]} \sup_{\substack{x_1\in G \\ [\gamma_0]=[\gamma]+[\beta_0']}} |X_{x_1}^{\gamma_0}\tilde{X}_{z_2=z}^{\beta_2}\kappa_{x_1}^*(z_2)|.
\end{aligned}$$

We conclude by adapting the case $\beta_0 = 0$. □

To take into account the difference operator, we will use the following observation.

Lemma 5.5.16. *For any* $\alpha \in \mathbb{N}_0^n$ *and* $\sigma \in S^{-\infty}$, $\Delta^\alpha \sigma^{(*)}$ *can be written as a linear combination (independent of* σ*) of* $\{\Delta^{\alpha'}\sigma\}^{(*)}$ *over* $\alpha' \in \mathbb{N}_0^n$, $[\alpha'] = [\alpha]$. *This is the same linear combination as when writing* $\Delta^\alpha\sigma^*$ *as a linear combination of* $\{\Delta^{\alpha'}\sigma\}^*$.

Proof of Lemma 5.5.16. For $\sigma \in S^{-\infty}$, let κ_σ be the kernel associated with the symbol σ and similarly for any other symbol.

Let us prove Part 1. We have

$$\{\tilde{q}_\alpha \kappa_{\sigma^{(*)},x}\}(y) = \tilde{q}_\alpha(y)\bar{\kappa}_{\sigma,xy^{-1}}(y^{-1}).$$

As $\tilde{q}_\alpha$ is a $[\alpha]$-homogeneous polynomial, by Proposition 5.2.3, $\overline{\tilde{q}_\alpha}$ is a linear combination of $\tilde{q}_{\alpha'}$ over multi-indices $\alpha' \in \mathbb{N}_0^n$ satisfying $[\alpha'] = [\alpha]$. Hence

$$\{\tilde{q}_\alpha \kappa_{\sigma^{(*)},x}\}(y) = \overline{\sum_{[\alpha']=[\alpha]}} \overline{\tilde{q}_{\alpha'}\kappa_{\sigma,xy^{-1}}}(y^{-1}) = \overline{\sum_{[\alpha']=[\alpha]}} \{\tilde{q}_{\alpha'}\kappa_\sigma\}^{(*)}(y),$$

where $\overline{\sum}$ means taking a linear combination. Taking the Fourier transform, we obtain

$$\mathcal{F}_G\{\tilde{q}_\alpha \kappa_{\sigma^{(*)},x}\}(\pi) = \Delta^\alpha \sigma^{(*)}(x,\pi) = \overline{\sum_{[\alpha']=[\alpha]}} \{\Delta^{\alpha'}\sigma\}^{(*)}.$$

□

We can now prove Theorem 5.5.12 in the case $\rho > \delta$.

Proof of Theorem 5.5.12 with $\rho > \delta$. We assume $\rho > \delta$. We fix a positive Rockland operator of homogeneous degree ν. Let us show that for any $\alpha_0, \beta_0 \in \mathbb{N}_0^n$, and $M_0 \in \mathbb{N}$, there exists $M \geq M_0$, a constant $C > 0$ and a seminorm $\|\cdot\|_{S^m_{\rho,\delta},a_1,b_1,0}$, such that for any $\sigma \in S^{-\infty}$ we have

$$\begin{aligned}\|X_x^{\beta_0}\Delta^{\alpha_0}\tau_M(x,\pi)\ \pi(\mathrm{I}+\mathcal{R})^{-\frac{m-(\rho-\delta)M_0-\rho[\alpha_0]+\delta[\beta_0]}{\nu}}\|_{\mathscr{L}(\mathcal{H}_\pi)}& \\ \leq C\|\sigma\|_{S^m_{\rho,\delta},a_1,b_1,0},&\end{aligned} \tag{5.59}$$

where we have denoted $\tau_M := \sigma^{(*)} - \sum_{[\alpha]\leq M}\Delta^\alpha X_x^\alpha \sigma^*$. By Lemma 5.5.16, it suffices to show (5.59) only for $\alpha_0 = 0$.

Let $\beta_0 \in \mathbb{N}_0$ and $M_0 \in \mathbb{N}$. Let $M_1 \in \mathbb{N}_0$ be the smallest non-negative integer such that

$$-\frac{m-(\rho-\delta)M_0+\delta[\beta_0]}{\nu} \leq M_1.$$

We choose $M \geq \max(M_0, \nu M_1)$ such that $(\rho-\delta)M + \rho Q > m + \delta[\beta_0] + \nu M_1 + Q$. This is possible as $\rho > \delta$. Then (5.59) follows from the application of Lemma 5.5.15 to M, M_1 and the symbol σ.

Using (5.58), classical considerations imply that (5.59) yields that for any $M_0 \in \mathbb{N}_0$, and any seminorm $\|\cdot\|_{S^{m-M_0(\rho-\delta),R}_{\rho,\delta},a,b}$, there exist a constant $C > 0$ and a seminorm $\|\cdot\|_{S^m_{\rho,\delta},a_1,b_1,0}$, such that for any $\sigma_1, \sigma_2 \in S^{-\infty}$ we have

$$\|\tau_{M_0}\|_{S^{m-M_0(\rho-\delta),R}_{\rho,\delta},a,b} \leq C\|\sigma\|_{S^m_{\rho,\delta},a_1,b_1,0}.$$

We can then conclude as in the proof of Theorem 5.5.3 in the case $\rho > \delta$. □

In fact, we have obtained a much more precise result:

Corollary 5.5.17. *We assume $1 \geq \rho > \delta \geq 0$. If $\sigma \in S^m_{\rho,\delta}$, then there exists a unique symbol $\sigma^{(*)}$ in $S^m_{\rho,\delta}$ such that*

$$(\mathrm{Op}(\sigma))^* = \mathrm{Op}(\sigma^{(*)}).$$

Furthermore, for any $M \in \mathbb{N}_0$,

$$\{\sigma^{(*)}(x,\pi) - \sum_{[\alpha]\leq M} X_x^\alpha \Delta^\alpha \sigma^*(x,\pi)\} \in S^{m-(\rho-\delta)M}_{\rho,\delta}.$$

Moreover, the mapping

$$\left\{ \begin{array}{ccc} S^m_{\rho,\delta} & \longrightarrow & S^{m-(\rho-\delta)M}_{\rho,\delta} \\ \sigma & \longmapsto & \{\sigma^{(*)}(x,\pi) - \sum_{[\alpha]\leq M} X_x^\alpha \Delta^\alpha \sigma^*(x,\pi)\} \end{array} \right. ,$$

is continuous.

Consequently, we can also write

$$\sigma^{(*)} \sim \sum_{j=0}^{\infty} \left(\sum_{[\alpha]=j} X_x^\alpha \Delta^\alpha \sigma^* \right), \tag{5.60}$$

where the asymptotic was defined in Definition 5.5.2.

As for composition, in the case $\rho = \delta$, the asymptotic formula does not bring any improvement and, in this sense, is not interesting. The proof of this case is more delicate to prove but relies on the same kind of arguments as above. Using Lemma 5.5.9 instead of Lemma 5.5.6 in the proof of Lemma 5.5.15 produces the following result:

Lemma 5.5.18. *We fix a positive Rockland operator of homogeneous degree ν. Let $m \in \mathbb{R}$, $1 \leq \rho \leq \delta \leq 0$ with $\delta \neq 1$, $\beta_0 \in \mathbb{N}_0^n$, and $M, M_1 \in \mathbb{N}_0$. We assume that*

$$M \geq \nu M_1 \quad \text{and} \quad m + \delta(M + c_{\beta_0}) + \nu M_1 < -Q,$$

where

$$c_{\beta_0} := \max_{\substack{[\beta_0'] \leq [\beta_0] \\ [\beta'] \geq [\beta_0'],\ |\beta'| \geq |\beta_0'|}} [\beta'].$$

Then there exist a constant $C > 0$, and a seminorm $\|\cdot\|_{S^{m,R}_{\rho,\delta},a,b}$, such that for any $\sigma \in S^{-\infty}$ and any $(x,\pi) \in G \times \widehat{G}$ we have

$$\|X_x^{\beta_0}(\sigma^{(*)}(x,\pi) - \sum_{[\alpha]\leq M} \Delta^\alpha X_x^\alpha \sigma^*(x,\pi))\pi(\mathrm{I}+\mathcal{R})^{M_1}\|_{\mathscr{L}(\mathcal{H}_\pi)} \leq C\|\sigma\|_{S^{m,R}_{\rho,\delta},a,b}.$$

The details of the proof of Lemma 5.5.18 are left to the reader. The conditions in the statement just above show that we will require the ability to choose m as negative as one wants. We can do this thanks to the following remark.

Lemma 5.5.19. *For any $\sigma \in S^{-\infty}$ and any $X \in \mathfrak{g}$, we have*

$$\{\pi(X)\sigma\}^{(*)} = -\sigma^{(*)}(x,\pi)\ \pi(X) - \{X_x\sigma\}^{(*)}(x,\pi).$$

More generally, for any $\beta \in \mathbb{N}_0^n$, we have

$$\{\pi(X)^\beta\sigma\}^{(*)} = \overline{\sum_{[\beta_1]+[\beta_2]=[\beta]}} \{X_x^{\beta_1}\sigma\}^{(*)}\pi(X)^{\beta_2},$$

where $\overline{\sum}$ denotes a linear combination independent of σ_1, σ_2.

Proof of Lemma 5.5.19. We keep the notation of Lemma 5.5.14. The kernel of $\sigma^{(*)}\pi(X)$ is given via

$$\tilde{X}_y\kappa_x^{(*)}(y) = \tilde{X}_y\{\bar{\kappa}_{xy^{-1}}(y^{-1})\} = -X_{x_1=xy^{-1}}\bar{\kappa}_{x_1}(y^{-1}) - X_{y_2=y^{-1}}\bar{\kappa}_{xy^{-1}}(y_2),$$

having used (5.50) and the Leibniz property for vector fields. Hence we recognise:

$$\tilde{X}_y\kappa_x^{(*)}(y) = -(X_x\kappa_x)^{(*)}(y) - (X\kappa_x)^{(*)}(y),$$

and

$$\sigma^{(*)}\pi(X) = -(X_x\sigma)^{(*)} - (\pi(X)\sigma)^{(*)}.$$

This shows the first formula. The second formula is obtained recursively. □

We can now show sketch the proof of Theorem 5.5.3 in the case $\rho = \delta$.

Sketch of the proof of Theorem 5.5.3 with $\rho = \delta$. We assume $\rho = \delta \in [0,1)$. Writing $\sigma = \pi(\mathrm{I}+\mathcal{R})^N\pi(\mathrm{I}+\mathcal{R})^{-N}\sigma$ and using Lemma 5.5.19, it suffices to prove (5.59) for m as negative as one wants. We proceed as in the proof of the case $\rho > \delta$ replacing Lemma 5.5.15 with Lemma 5.5.18. The details are left to the reader. □

5.5.4 Simplification of the definition of $S^m_{\rho,\delta}$

In this section, we show that it is possible to choose $\gamma = 0$ in the definition of symbols as it was pointed out in Remark 5.2.13 Part (3). This simplifies the definition of the symbol classes $S^m_{\rho,\delta}$ given in Definition 5.2.11. We will also show a pivotal argument in the proof of Theorems 5.5.3 and 5.5.12, namely Inequalities (5.54).

Theorem 5.5.20. *Let $m, \rho, \delta \in \mathbb{R}$ with $1 \geq \rho \geq \delta \geq 0$ and $\delta \neq 1$.*

(L) *A symbol $\sigma = \{\sigma(x,\pi), (x,\pi) \in G \times \widehat{G}\}$ is in $S^m_{\rho,\delta}$ if and only if for each $\alpha, \beta \in \mathbb{N}_0^n$, the field of operators*

$$X_x^\beta \Delta^\alpha \sigma = \{X_x^\beta \Delta^\alpha \sigma(x,\pi) : \mathcal{H}_\pi^\infty \to \mathcal{H}_\pi, (x,\pi) \in G \times \widehat{G}\}$$

is in $L^\infty_{0,\rho[\alpha]-m-\delta[\beta]}(\widehat{G})$ uniformly in $x \in G$, that is,

$$\sup_{x\in G} \|X_x^\beta \Delta^\alpha \sigma(x,\cdot)\|_{L^\infty_{0,\rho[\alpha]-m-\delta[\beta]}(\widehat{G})} < \infty. \tag{5.61}$$

Furthermore, the family of seminorms

$$\sigma \longmapsto \|\sigma\|_{S^m_{\rho,\delta},a,b,0} = \sup_{\substack{[\alpha]\leq a\\ [\beta]\leq b}} \sup_{x\in G} \|X_x^\beta \Delta^\alpha \sigma(x,\cdot)\|_{L^\infty_{0,\rho[\alpha]-m-\delta[\beta]}(\widehat{G})}, \quad a,b \in \mathbb{N}_0,$$

yields the topology of $S^m_{\rho,\delta}$.

(R) *A symbol $\sigma = \{\sigma(x,\pi), (x,\pi) \in G \times \widehat{G}\}$ is in $S^m_{\rho,\delta}$ if and only if for each $\alpha, \beta \in \mathbb{N}_0^n$, the field of operators*

$$X_x^\beta \Delta^\alpha \sigma = \{X_x^\beta \Delta^\alpha \sigma(x,\pi) : \mathcal{H}_\pi^\infty \to \mathcal{H}_\pi, (x,\pi) \in G \times \widehat{G}\}$$

is in $L^\infty_{m+\delta[\beta]-\rho[\alpha],0}(\widehat{G})$ uniformly in $x \in G$, that is,

$$\sup_{x\in G} \|X_x^\beta \Delta^\alpha \sigma(x,\cdot)\|_{L^\infty_{m+\delta[\beta]-\rho[\alpha],0}(\widehat{G})} < \infty. \tag{5.62}$$

Furthermore, the family of seminorms

$$\sigma \longmapsto \|\sigma\|_{S^{m,R}_{\rho,\delta},a,b} = \sup_{\substack{[\alpha]\leq a\\ [\beta]\leq b}} \sup_{x\in G} \|X_x^\beta \Delta^\alpha \sigma(x,\cdot)\|_{L^\infty_{m+\delta[\beta]-\rho[\alpha],0}(\widehat{G})}, \quad a,b \in \mathbb{N}_0,$$

yields the topology of $S^m_{\rho,\delta}$.

In other words,

(R) a symbol $\sigma = \{\sigma(x,\pi), (x,\pi) \in G \times \widehat{G}\}$ is in $S^m_{\rho,\delta}$ if and only if for each $\alpha, \beta \in \mathbb{N}_0^n$, the field of operators

$$X_x^\beta \Delta^\alpha \sigma = \{X_x^\beta \Delta^\alpha \sigma(x,\pi) : \mathcal{H}_\pi^\infty \to \mathcal{H}_\pi, (x,\pi) \in G \times \widehat{G}\}$$

is defined on smooth vectors and satisfy

$$\sup_{x\in G, \pi\in\widehat{G}} \|X_x^\beta \Delta^\alpha \sigma(x,\cdot)\pi(\mathrm{I}+\mathcal{R})^{\frac{\rho[\alpha]-m-\delta[\beta]}{\nu}}\|_{\mathscr{L}(\mathcal{H}_\pi)} < \infty$$

for one (and then any) positive Rockland operator $\mathcal{R}$ of homogeneous degree ν (as the symbol is given by a field of operators defined on smooth vectors, and since $\pi(\mathrm{I}+\mathcal{R})^{\frac{s}{\nu}}$ acts on smooth vectors, this condition makes sense);

(L) a symbol $\sigma = \{\sigma(x,\pi), (x,\pi) \in G \times \widehat{G}\}$ is in $S^m_{\rho,\delta}$ if and only if for each $\alpha, \beta \in \mathbb{N}_0^n$, the field of operators

$$X_x^\beta \Delta^\alpha \sigma = \{X_x^\beta \Delta^\alpha \sigma(x,\pi) : \mathcal{H}_\pi^\infty \to \mathcal{H}_\pi^{\rho[\alpha]-m-\delta[\beta]}, (x,\pi) \in G \times \widehat{G}\}$$

is defined on smooth vectors and has range in $\mathcal{H}_\pi^{\rho[\alpha]-m-\delta[\beta]}$, and satisfies

$$\sup_{x\in G, \pi\in\widehat{G}} \|\pi(\mathrm{I}+\mathcal{R})^{\frac{\rho[\alpha]-m-\delta[\beta]}{\nu}} X_x^\beta \Delta^\alpha \sigma(x,\cdot)\|_{\mathscr{L}(\mathcal{H}_\pi)} < \infty$$

for one (and then any) positive Rockland operator $\mathcal{R}$ of homogeneous degree ν. The notion of a field having range in a Sobolev space $\mathcal{H}_\pi^s$ is described in Definition 5.1.10 and allows us to compose on the left with $\pi(\mathrm{I}+\mathcal{R})^{\frac{s}{\nu}}$ with $s = \rho[\alpha] - m - \delta[\beta]$ here, see (5.4).

Naturally, the condition does not depend on the choice of the positive Rockland operator $\mathcal{R}$.

Theorem 5.5.20 makes it considerably easier to check whether a symbol is in one of our symbol classes. However using the definition 'with any γ' has the advantages

1. that we see easily that the symbols are fields of operators acting on smooth vectors,
2. that we see easily that the symbols in $S^m_{\rho,\delta}$, $m \in \mathbb{R}$, form an algebra (cf. Theorem 5.2.22),
3. and that the properties for the multipliers in $\mathcal{R}$ in Proposition 5.3.4 are for the definition 'with any γ'.

While showing Theorem 5.5.20, we will also finish the proofs of Theorems 5.5.3 and 5.5.12. Indeed, an important argument used in the proof of Theorems 5.5.3 and 5.5.12 (i.e. the properties of stability under composition and taking the adjoint) is Inequality (5.54) which can easily be seen as equivalent to Part 2 of Theorem 5.5.20.

Before showing Theorem 5.5.20, let us summarise what has been shown in the proofs of Theorems 5.5.3 and 5.5.12 up to before the use of Inequality (5.54):

$$\|\sigma_1 \circ \sigma_2\|_{S^{m_1+m_2,R}_{\rho,\delta},a,b} \lesssim \|\sigma_1\|_{S^{m_1,R}_{\rho,\delta},a_1,b_1} \|\sigma_2\|_{S^{m_2}_{\rho,\delta},a_2,b_2,c_2}, \tag{5.63}$$

$$\|\sigma^{(*)}\|_{S^{m,R}_{\rho,\delta},a,b} \lesssim \|\sigma\|_{S^m_{\rho,\delta},a',b',0}; \tag{5.64}$$

these estimates are valid for any $\sigma, \sigma_1, \sigma_2 \in S^{-\infty}$ in the sense that for any seminorm on the left hand side, one can find seminorms on the right.

Proof of Theorem 5.5.20. Using Estimate (5.64) together with the properties of taking the adjoint and of the difference operators together, one checks easily that the two families of seminorms $\{\|\cdot\|_{S^{m,R}_{\rho,\delta},a,b}, a,b\in\mathbb{N}\}$ and $\{\|\cdot\|_{S^m_{\rho,\delta},a,b,0}, a,b\in\mathbb{N}\}$ yield the same topology on $S^{-\infty}$ and that taking the adjoint of a symbol is continuous for this topology. Consequently, for any $\gamma\in\mathbb{R}$, any symbol $\sigma\in S^{-\infty}$ and any seminorm $\|\cdot\|_{S^{m,R}_{\rho,\delta},a,b}$, we have

$$\|\pi(\mathrm{I}+\mathcal{R})^{\frac{\gamma}{\nu}}\sigma\|_{S^{m+\gamma,R}_{\rho,\delta},a,b} \lesssim \|\sigma^*\pi(\mathrm{I}+\mathcal{R})^{\frac{\gamma}{\nu}}\|_{S^{m+\gamma,R}_{\rho,\delta},a_1,b_1} \lesssim \|\sigma^*\|_{S^{m,R}_{\rho,\delta},a_2,b_2},$$

having used (5.63) and the fact that $\pi(\mathrm{I}+\mathcal{R})^{\frac{\gamma}{\nu}}\in S^\gamma$. As taking the adjoint is a continuous operator for the $S^{m,R}$-topology, we have obtained

$$\|\pi(\mathrm{I}+\mathcal{R})^{\frac{\gamma}{\nu}}\sigma\|_{S^{m+\gamma,R}_{\rho,\delta},a,b} \lesssim \|\sigma\|_{S^{m,R}_{\rho,\delta},a_3,b_3}.$$

One checks easily that

$$\forall a,b,c\in\mathbb{N}_0 \qquad \|\sigma\|_{S^m_{\rho,\delta},a,b,c} \leq \max_{|\gamma|\leq c}\|\pi(\mathrm{I}+\mathcal{R})^{\frac{\gamma}{\nu}}\sigma\|_{S^{m+\gamma,R}_{\rho,\delta},a,b},$$

whereas

$$\forall a,b\in\mathbb{N}_0 \qquad \|\sigma\|_{S^{m,R}_{\rho,\delta},a,b} \leq \|\sigma\|_{S^m_{\rho,\delta},a,b,|m|+\rho a+\delta b}.$$

This easily implies that the topologies on $S^{-\infty}$ coming from the two families of seminorms $\{\|\cdot\|_{S^m_{\rho,\delta},a,b,c}, a,b,c\in\mathbb{N}_0\}$ and $\{\|\cdot\|_{S^{m,R}_{\rho,\delta},a,b}, a,b\in\mathbb{N}_0\}$ coincide. This together with Lemma 5.4.11 (to pass from $S^{-\infty}$ to $S^m_{\rho,\delta}$) concludes the proof of Theorem 5.5.20. □

5.6 Amplitudes and amplitude operators

In this section, we discuss the notion of an amplitude extending that of the symbol, to functions/operators depending on both space variables x and y. This allows for another way of writing pseudo-differential operators as amplitude operators, analogous to Formula (2.27) in the case of compact groups. However, as in the classical theory, or as in Theorem 2.2.15 in the case of compact groups, we can show that amplitude operators with symbols in suitable amplitude classes reduce to pseudo-differential operator with symbols in corresponding symbol classes, with asymptotic formulae relating amplitudes to symbols.

5.6.1 Definition and quantization

Following the Euclidean and compact cases, it is natural to define amplitudes in the following way, extending the notion of symbols from Definitions 5.1.33 and 5.1.34:

Definition 5.6.1. An *amplitude* is a field of operators

$$\{\mathcal{A}(x,y,\pi):\mathcal{H}^\infty_\pi\to\mathcal{H}_\pi, \pi\in\widehat{G}\}$$

depending on $x, y \in G$, satisfying for each $x, y \in G$

$$\exists a, b \in \mathbb{R} \qquad \mathcal{A}(x, y, \cdot) := \{\mathcal{A}(x, y, \pi) : \mathcal{H}_\pi^\infty \to \mathcal{H}_\pi, \pi \in \widehat{G}\} \in L_{a,b}^\infty(\widehat{G}).$$

- An amplitude $\{\mathcal{A}(x, y, \pi) : \mathcal{H}_\pi^\infty \to \mathcal{H}_\pi, \pi \in \widehat{G}\}$ is said to be *continuous* in $x, y \in G$ whenever there exists $a, b \in \mathbb{R}$ such that

$$\forall x, y \in G \quad \mathcal{A}(x, y, \cdot) := \{\mathcal{A}(x, y, \pi) : \mathcal{H}_\pi^\infty \to \mathcal{H}_\pi, \pi \in \widehat{G}\} \in L_{a,b}^\infty(\widehat{G}),$$

  and the map $(x, y) \mapsto \mathcal{A}(x, y, \cdot)$ is continuous from $G \times G \sim \mathbb{R}^n \times \mathbb{R}^n$ to the Banach space $L_{a,b}^\infty(\widehat{G})$.

- An amplitude $\mathcal{A} = \{\mathcal{A}(x, y, \pi) : \mathcal{H}_\pi^\infty \to \mathcal{H}_\pi, \pi \in \widehat{G}\}$ is said to be *smooth* in $x, y \in G$ whenever it is a field of operators depending smoothly on $(x, y) \in G \times G$ (see Remark 1.8.16) and, for every $\beta_1, \beta_2 \in \mathbb{N}_0^n$, the field $\{\partial_x^{\beta_1} \partial_y^{\beta_2} \mathcal{A}(x, y, \pi) : \mathcal{H}_\pi^\infty \to \mathcal{H}_\pi, \pi \in \widehat{G}\}$ is continuous.

Clearly if an amplitude $\mathcal{A} = \{\mathcal{A}(x, y, \pi)\}$ does not depend on y, that is, $\mathcal{A}(x, y, \pi) = \sigma(x, \pi)$, then it defines a symbol $\sigma = \{\sigma(x, \pi)\}$. More generally any amplitude $\mathcal{A} = \{\mathcal{A}(x, y, \pi)\}$ defines a symbol σ given by $\sigma(x, \pi) = \mathcal{A}(x, x, \pi)$. In Section 5.6.2, we will define amplitude classes and give other examples of amplitudes.

Similarly to the symbol case, one can associate a kernel with an amplitude:

Definition 5.6.2. Let $\mathcal{A}$ be an amplitude. For each $(x, y) \in G \times G$, let $\kappa_{x,y} \in \mathcal{S}'(G)$ be the unique distribution such that

$$\mathcal{F}_G(\kappa_{x,y})(\pi) = \mathcal{A}(x, y, \pi).$$

The map $G \times G \ni (x, y) \mapsto \kappa_{x,y} \in \mathcal{S}'(G)$ is called its *kernel.*

As in the symbol case, the map $G \times G \ni (x, y) \mapsto \kappa_{x,y} \in \mathcal{S}'(G)$ is smooth, see Lemma 5.1.35 for the proof of this as well as for the existence and uniqueness of $\kappa_{x,y}$ in the case of symbols.

Before defining the amplitude quantization, we need to open a (quick) parenthesis to describe the following property from distribution theory:

Lemma 5.6.3. *Let $G \times G \ni (x, y) \mapsto \kappa_{x,y} \in \mathcal{S}'(G)$ be a continuous mapping. For each x, we consider the distribution $\tilde{\kappa}_x$ defined by*

$$\int_G \tilde{\kappa}_x(y) \phi(y) dy = \lim_{\epsilon \to 0} \int_{G \times G} \kappa_{x,w}(y^{-1}x) \phi(y) \psi_\epsilon(wy^{-1}) dy dw,$$

where $\phi \in \mathcal{D}(G)$, $\psi_1 \in \mathcal{D}(G)$, $\int_G \psi_1 = 1$ and $\psi_\epsilon(z) = \epsilon^{-Q} \psi(\epsilon^{-1} z)$, $\epsilon > 0$.

Indeed this limit exists and is independent of the choice of ψ_1.

This defines a continuous map $G \ni x \longmapsto \tilde{\kappa}_x \in \mathcal{D}'(G)$.

Proof of Lemma 5.6.3. Since $\kappa_{x,y} \in \mathcal{S}'(G)$, there exists a seminorm $\|\cdot\|_{\mathcal{S}(G),N}$ such that

$$\forall \phi \in \mathcal{S}(G) \qquad |\langle \kappa_{x,y}, \phi\rangle| \leq C_{x,y,N} \|\phi\|_{\mathcal{S}(G),N}.$$

Furthermore, since the map $G \times G \ni (x,y) \mapsto \kappa_{x,y} \in \mathcal{S}'(G)$ is smooth, we obtain that the constant $C_{x,y,N} = \|\kappa_{x,y}\|_{\mathcal{S}'(G),N}$ can be chosen locally uniform with respect to x and y. Furthermore, fixing two compacts K_1 and K_2 of G, there exists a seminorm $\|\cdot\|_{\mathcal{S}(G),N}$ (depending on K_1 and K_2) such that the map

$$((x,y),(x',y')) \in (K_1 \times K_2) \times (K_1 \times K_2) \mapsto \|\kappa_{x,y} - \kappa_{x',y'}\|_{\mathcal{S}'(G),N},$$

is uniformly continuous. This is easily proved using a cover of the compacts $K_1 \times K_2$ by balls of sufficiently small radius, and the continuity at each centre of these balls.

For any $\psi_1 \in \mathcal{D}(G)$, $\epsilon > 0$ and $x \in G$, we define the distribution $T_{\psi_1,\epsilon,x}$ by

$$T_{\psi_1,\epsilon,x}(\phi) := \int_{G\times G} \kappa_{x,w}(y^{-1}x)\phi(y)\psi_\epsilon(wy^{-1})dydw,$$

where $\phi \in \mathcal{D}(G)$ is supported in a fixed compact $K \subset G$. Using the change of variable from w to z with $z = \epsilon^{-1}(wy^{-1})$, so that $w = (\epsilon z)y$, we obtain

$$T_{\psi_1,\epsilon,x}(\phi) = \int_{G\times G} \kappa_{x,(\epsilon z)y}(y^{-1}x)\phi(y)\psi_1(z)dydz.$$

Therefore, for any $\epsilon_1, \epsilon_2 \in (0,1)$, we get

$$\begin{aligned}
&|(T_{\psi_1,\epsilon_1,x} - T_{\psi_1,\epsilon_2,x})(\phi)| \\
&= \left|\int_{G\times G} \left(\kappa_{x,(\epsilon_1 z)y}(y^{-1}x) - \kappa_{x,(\epsilon_2 z)y}(y^{-1}x)\right)\phi(y)\psi_1(z)dydz\right| \\
&\leq \sup_{\substack{z\in \mathrm{supp}\psi_1 \\ y \in \mathrm{supp}\phi}} \|\kappa_{x,(\epsilon_1 z)y} - \kappa_{x,(\epsilon_2 z)y}\|_{\mathcal{S}'(G),N} \|\phi\|_{\mathcal{S}(G),N} \|\psi_1\|_{L^1(G)},
\end{aligned}$$

where $\|\cdot\|_{\mathcal{S}(G),N}$ is chosen with respect to the compact sets

$$\{x\} \quad \text{and} \quad \{(\epsilon z)y, \epsilon \in [0,1],\ z \in \mathrm{supp}\psi_1,\ y \in K_2\}.$$

This shows that the scalar sequence $(T_{\psi_1,\epsilon,x}(\phi))$ converges as $\epsilon \to 0$ and that the linear map

$$\psi_1 \in \mathcal{D}(G) \longmapsto \lim_{\epsilon\to 0} T_{\psi_1,\epsilon,x}(\phi), \tag{5.65}$$

extends continuously to $L^1(K_o) \to \mathbb{C}$ for any compact $K_o \subset G$. Thus the map given in (5.65) is given by integration against a locally bounded function on G.

Let us show that the map given in (5.65) is invariant under left or right translation. Indeed, modifying the argument above we obtain

$$\begin{aligned}
&\left|T_{\psi_1,\epsilon,x}(\phi) - T_{\psi_1(\cdot y_o^{-1}),\epsilon,x}(\phi)\right| \\
&= \left|\int_{G\times G} \left(\kappa_{x,(\epsilon z)y} - \kappa_{x,(\epsilon(zy_o))y}\right)(y^{-1}x)\phi(y)\psi_1(z)dydz\right| \\
&\leq \sup_{\substack{z\in\text{supp}\psi_1\\ y\in\text{supp}\phi}} \|\kappa_{x,(\epsilon z)y} - \kappa_{x,(\epsilon(zy_o))y}\|_{\mathcal{S}'(G),N}\|\phi\|_{\mathcal{S}(G),N}\|\psi_1\|_{L^1(G)}
\end{aligned}$$

for a suitable seminorm $\|\cdot\|_{\mathcal{S}(G),N}$, (depending locally on y_o). Since the two sequences $((\epsilon z)y)_{\epsilon>0}$ and $((\epsilon(zy_o))y)_{\epsilon>0}$ converge to y in G, we see that

$$\lim_{\epsilon\to 0} T_{\psi_1,\epsilon,x}(\phi) = \lim_{\epsilon\to 0} T_{\psi_1(\cdot y_o^{-1}),\epsilon,x}(\phi),$$

and the same is true for right translation. Therefore, the locally bounded function given by the mapping (5.65) is a constant which we denote by $T_{0,x}(\phi)$:

$$\lim_{\epsilon\to 0} T_{\psi_1,\epsilon,x}(\phi) = T_{0,x}(\phi)\int_G \psi_1.$$

One checks easily that $T_{0,x}(\phi)$, $\phi \in \mathcal{D}(G)$, $\text{supp}\,\phi \subset K$, defines a distribution $\tilde{\kappa}_x \in \mathcal{D}'(G)$ which is therefore independent of ψ_1. Refining the argument given above shows that $\tilde{\kappa}_x \in \mathcal{D}'(G)$ depends continuously on $x \in G$. □

If $G \times G \ni (x,y) \mapsto \kappa_{x,y} \in \mathcal{S}'(G)$ is a continuous mapping, we will allow ourselves to denote the distribution defined in Lemma 5.6.3 by

$$\tilde{\kappa}_x(y) := \kappa_{x,y}(y^{-1}x).$$

This closes our parenthesis about distribution theory.

We can now define the operator

$$T = \text{AOp}(\mathcal{A})$$

associated with an amplitude $\mathcal{A} = \{\mathcal{A}(x,y,\pi)\}$ with amplitude kernel $\kappa_{x,y}$, by

$$T\phi(x) := \int_G \phi(y)\kappa_{x,y}(y^{-1}x)dy, \quad \phi \in \mathcal{D}(G),\ x \in G. \tag{5.66}$$

The quantization defined by formula (5.66) makes sense for any amplitude $\mathcal{A} = \{\mathcal{A}(x,y,\pi)\}$. Clearly the quantization mapping $\mathcal{A} \mapsto \text{AOp}(\mathcal{A})$ is linear. However, as in the Euclidean or compact cases, it is injective but not necessarily 1-1 since different amplitudes may lead to the same operator, in contrast to the situation for symbols, cf. Theorem 5.1.39.

Remark 5.6.4. If an amplitude $\mathcal{A} = \{\mathcal{A}(x, y, \pi)\}$ does not depend on y, that is, $\mathcal{A}(x, y, \pi) = \sigma(x, \pi)$, then the corresponding symbol $\sigma = \{\sigma(x, \pi)\}$ yield the same operator:

$$\mathrm{AOp}(\mathcal{A}) = \mathrm{Op}(\sigma)$$

since in this case the amplitude $\kappa_{x,y}$ is a function/distribution κ_x independent of y which coincides with the kernel of the symbol σ.

As in the symbol case in Lemma 5.1.42, we may see $\mathrm{AOp}(\mathcal{A})$ as a limit of nice operators in the following sense:

Lemma 5.6.5. *If $\mathcal{A} = \{\mathcal{A}(x, y, \pi)\}$ is an amplitude, we can construct explicitly a family of amplitudes $\mathcal{A}_\epsilon = \{\mathcal{A}_\epsilon(x, y, \pi)\}$, $\epsilon > 0$, in such a way that*

1. *the kernel $\kappa_{\epsilon,x,y}(z)$ of $\mathcal{A}_\epsilon$ is smooth in both x, y and z, and compactly supported in x and y,*
2. *the associated kernel $\tilde{\kappa}_{\epsilon,x}(y) = \kappa_{\epsilon,x,y}(y^{-1}x)$ is smooth and compactly supported in both x, y,*
3. *if $\phi \in \mathcal{S}(G)$ then $\mathrm{AOp}(\mathcal{A}_\epsilon)\phi \in \mathcal{D}(G)$, and*
4. *$\mathrm{AOp}(\mathcal{A}_\epsilon)\phi \underset{\epsilon \to 0}{\longrightarrow} \mathrm{AOp}(\mathcal{A})\phi$ uniformly on any compact subset of G.*

Proof of Lemma 5.6.5. We use the same notation $\chi_\epsilon \in \mathcal{D}(G)$, $|\pi|$ and $\mathrm{proj}_{\epsilon,\pi}$ as in the proof of Lemma 5.1.42. We consider for any $\epsilon \in (0, 1)$ the amplitude given by

$$\mathcal{A}_\epsilon(x, y, \pi) := \chi_\epsilon(x)\chi_\epsilon(y)1_{|\pi| \leq \epsilon^{-1}}\mathcal{A}(x, y, \pi) \circ \mathrm{proj}_{\epsilon,\pi}.$$

By Definition 5.6.2 and the Fourier inversion formula (1.26), the corresponding kernel is

$$\kappa_{\epsilon,x,y}(z) = \chi_\epsilon(x)\chi_\epsilon(y) \int_{|\pi| \leq \epsilon^{-1}} \mathrm{Tr}\left(\mathcal{A}(x, y, \pi)\ \mathrm{proj}_{\epsilon,\pi}\pi(z)\right) d\mu(\pi),$$

which is smooth in x, y and z and compactly supported in x and y. The rest follows easily. □

There is a simple relation between the amplitudes of an operator and its adjoint, much simpler than in the symbol case:

Proposition 5.6.6. *Let $\mathcal{A}$ be an amplitude. Then $\mathcal{B}$ given by*

$$\mathcal{B}(x, y, \pi) := \mathcal{A}(y, x, \pi)^*$$

is also an amplitude. Furthermore, the formal adjoint of the operator $T = \mathrm{AOp}(\mathcal{A})$ is $T^* = \mathrm{AOp}(\mathcal{B})$. If $\{\kappa_{x,y}(z)\}$ is the kernel of $\mathcal{A}$, then the kernel of $\mathcal{B}$ is given via $(x, y, z) \mapsto \bar{\kappa}_{y,x}(z^{-1})$.

Proof. On one hand, from the amplitude quantization in (5.66), we compute for $\phi, \psi \in \mathcal{D}(G)$, that

$$(T\phi, \psi) = \int_G \int_G \phi(y)\kappa_{x,y}(y^{-1}x)\bar{\psi}(x)dy\, dx = (\phi, T^*\psi),$$

therefore

$$T^*\psi(y) = \int_G \bar{\kappa}_{x,y}(y^{-1}x)\psi(x)dx$$

or, equivalently,

$$T^*\psi(x) = \int_G \bar{\kappa}_{y,x}(x^{-1}y)\psi(y)dy.$$

One the other hand, the amplitude kernel for $\mathcal{B}$ is $\kappa'_{x,y}$ satisfying

$$\pi(\kappa'_{x,y}) = \mathcal{B}(x,y,\pi) = \mathcal{A}(y,x,\pi)^* = \pi(\kappa_{y,x})^* = \pi(\kappa^*_{y,x}),$$

with $\kappa^*_{y,x}(z) = \bar{\kappa}_{y,x}(z^{-1})$, and therefore,

$$\kappa'_{x,y}(z) = \kappa^*_{y,x}(z) = \bar{\kappa}_{y,x}(z^{-1}).$$

By (5.66), this implies that $T^* = \mathrm{AOp}(\mathcal{B})$. □

5.6.2 Amplitude classes

Again similarly to the symbol case, we may define the amplitude classes $AS^m_{\rho,\delta}$. This is done in analogy to Definition 5.2.11 for symbols and its equivalent reformulation in (5.29).

Definition 5.6.7. Let $m, \rho, \delta \in \mathbb{R}$ with $1 \geq \rho \geq \delta \geq 1$. An amplitude $\mathcal{A}$ is called an *amplitude of order* m *and of type* (ρ, δ) whenever, for each $\alpha, \beta \in \mathbb{N}_0^n$ and $\gamma \in \mathbb{R}$, the field $\{X_x^{\beta_1} X_y^{\beta_2} \Delta^\alpha \mathcal{A}(x,y,\pi)\}$ is in $L^\infty_{\gamma,\rho[\alpha]-m-\delta([\beta_1]+[\beta_2])+\gamma}(\widehat{G})$ uniformly in $(x,y) \in G$, i.e. if

$$\sup_{x,y\in G} \|X_x^{\beta_1} X_y^{\beta_2} \Delta^\alpha \mathcal{A}(x,y,\cdot)\|_{L^\infty_{\gamma,\rho[\alpha]-m-\delta([\beta_1]+[\beta_2])+\gamma}(\widehat{G})} < \infty. \tag{5.67}$$

In this case, proceeding in a similar way to $S^m_{\rho,\delta}$ in Section 5.2.2, we see that the fields of operators $X_x^{\beta_1} X_y^{\beta_2} \Delta^\alpha \mathcal{A}(x,y,\cdot)$ act on smooth vectors and (5.67) implies

$$\sup_{\substack{x,y\in G\\ \pi\in\widehat{G}}} \|\pi(\mathrm{I}+\mathcal{R})^{\frac{\rho[\alpha]-m-\delta([\beta_1]+[\beta_2])+\gamma}{\nu}} X_x^{\beta_1} X_y^{\beta_2} \Delta^\alpha \mathcal{A}(x,y,\cdot)\pi(\mathrm{I}+\mathcal{R})^{-\frac{\gamma}{\nu}}\|_{\mathscr{L}(\mathcal{H}_\pi)} < \infty. \tag{5.68}$$

The converse also holds.

The *amplitude class* $AS^m_{\rho,\delta} = AS^m_{\rho,\delta}(G)$ is the set of amplitudes of order m and of type (ρ, δ). We also define

$$AS^{-\infty} := \bigcap_{m\in\mathbb{R}} AS^m_{\rho,\delta},$$

the class of smoothing amplitudes. As in the case of symbols, the class $AS^{-\infty}$ is independent of ρ and δ and can be denoted just by $AS^{-\infty}$.

It is a routine exercise to check that each amplitude class $AS^m_{\rho,\delta}$ is a vector space and that we have the inclusions

$$m_1 \le m_2, \quad \delta_1 \le \delta_2, \quad \rho_1 \ge \rho_2 \quad \Longrightarrow \quad AS^{m_1}_{\rho_1,\delta_1} \subset AS^{m_2}_{\rho_2,\delta_2}. \tag{5.69}$$

We assume that a positive Rockland operator $\mathcal{R}$ of degree ν is fixed. If $\mathcal{A}$ is an amplitude and $a, b, c \in [0, \infty)$, we set

$$\|\mathcal{A}(x, y, \pi)\|_{AS^m_{\rho,\delta},a,b,c} := \sup_{\substack{|\gamma|\le c \\ [\alpha]\le a,\, [\beta_1],[\beta_2]\le b}} \|\pi(\mathrm{I}+\mathcal{R})^{\frac{\rho[\alpha]-m-\delta([\beta_1]+[\beta_2])+\gamma}{\nu}} X_x^{\beta_1} X_y^{\beta_2} \Delta^\alpha \mathcal{A}(x, y, \pi)\pi(\mathrm{I}+\mathcal{R})^{-\frac{\gamma}{\nu}}\|_{\mathscr{L}(\mathcal{H}_\pi)},$$

and

$$\|\mathcal{A}\|_{AS^m_{\rho,\delta},a,b,c} := \sup_{(x,y)\in G\times G,\, \pi\in\widehat{G}} \|\mathcal{A}(x, y, \pi)\|_{AS^m_{\rho,\delta},a,b,c}.$$

Again, one checks easily that the resulting maps $\|\cdot\|_{S^m_{\rho,\delta},a,b,c}$, $a, b, c \in [0, \infty)$, are seminorms over the vector space $AS^m_{\rho,\delta}$. Furthermore, taking a, b, c as non-negative integers, they endow $AS^m_{\rho,\delta}$ with the structure of a Fréchet space. The class of smoothing amplitudes $AS^{-\infty}$ is then equipped with the topology of projective limit. Similarly to the case of symbols in Proposition 5.2.12, two different positive Rockland operators give equivalent families of seminorms.

The inclusions given in (5.69) are continuous for these topologies.

Symbols in $S^m_{\rho,\delta}$ are examples of amplitudes in $AS^m_{\rho,\delta}$ which do not depend on y. Conversely, if an amplitude $\mathcal{A} = \{\mathcal{A}(x, y, \pi)\}$ in $AS^m_{\rho,\delta}$ does not depend on y, that is, $\mathcal{A}(x, y, \pi) = \sigma(x, \pi)$, then it defines a symbol $\sigma = \{\sigma(x, \pi)\}$ in $S^m_{\rho,\delta}$. More generally we check easily:

Lemma 5.6.8. *If $\mathcal{A} = \{\mathcal{A}(x, y, \pi)\}$ is in $AS^m_{\rho,\delta}$, then the symbol σ given by*

$$\sigma(x, \pi) := \mathcal{A}(x, x, \pi)$$

is in $S^m_{\rho,\delta}$.

A wider class of examples is given by the following property which can be shown by an easy adaption of Proposition 5.3.4 and Corollary 5.3.7:

Corollary 5.6.9. *Let $\mathcal{R}$ be a positive Rockland operator of degree ν. Let $m \in \mathbb{R}$ and $0 \leq \delta < 1$. Let $f : G \times G \times \mathbb{R}^+ \ni (x, y, \lambda) \mapsto f_{x,y}(\lambda) \in \mathbb{C}$ be a smooth function. We assume that for every $\beta_1, \beta_2 \in \mathbb{N}_0^n$, we have*

$$X_x^{\beta_1} X_y^{\beta_2} f_{x,y} \in \mathcal{M}_{\frac{m+\delta([\beta_1]+[\beta_2])}{\nu}},$$

where $\mathcal{M}$ is as in Definition 5.3.1. Then

$$\mathcal{A}(x, y, \pi) = f_{x,y}(\pi(\mathcal{R}))$$

defines an amplitude $\mathcal{A}$ in $AS^m_{1,\delta}$ which satisfies

$$\forall a, b, c \in \mathbb{N}_0 \qquad \exists \ell \in \mathbb{N},\ C > 0$$
$$\|\mathcal{A}\|_{AS^m_{1,\delta},a,b,c} \leq C \sup_{[\beta_1],[\beta_2]\leq b} \|X_x^{\beta_1} X_y^{\beta_2} f_{x,y}\|_{\mathcal{M}_{\frac{m+\delta[\beta_1+\beta_2]}{\nu}},\ell},$$

with ℓ and C independent of f.

This can also be generalised easily to multipliers in a finite family of strongly commuting positive Rockland operators.

5.6.3 Properties of amplitude classes and kernels

One can readily prove properties for the amplitudes similar to the ones already established for symbols. Here we note that although the subsequent properties would follow also from Theorem 5.6.14 in the sequel and from the corresponding properties of symbols in Section 5.2.5, we now indicate what can be shown concerning amplitudes and their classes by a simple adaptation of proofs of the corresponding properties for symbols.

Proceeding as in Section 5.2.5, we also have the following properties for the amplitude classes:

Proposition 5.6.10. *Let $1 \geq \rho \geq \delta \geq 0$ and $\delta \neq 1$.*

(i) Let $\mathcal{A} \in AS^m_{\rho,\delta}$ have kernel $\kappa_{x,y}$. Then we have the following properties.

1. *For every $x, y \in G$ and $\gamma \in \mathbb{R}$, $\tilde{q}_\alpha X_x^{\beta_1} X_y^{\beta_2} \kappa_{x,y} \in \mathcal{K}_{\gamma,\rho[\alpha]-m-\delta[\beta_1+\beta_2]+\gamma}$, where we recall the notation $\tilde{q}_\alpha(x) = q_\alpha(x^{-1})$.*

2. *If $\beta_1, \beta_2 \in \mathbb{N}_0^n$ then the amplitude $\{X_x^{\beta_1} X_y^{\beta_2} \mathcal{A}(x, y, \pi), (x, y, \pi) \in G \times G \times \widehat{G}\}$ is in $AS^{m+\delta[\beta_1+\beta_2]}_{\rho,\delta}$ with kernel $X_x^{\beta_1} X_y^{\beta_2} \kappa_{x,y}$, and*

$$\|X_x^{\beta_1} X_y^{\beta_2} \mathcal{A}(x, y, \pi)\|_{AS^{m+\delta[\beta_1+\beta_2]}_{\rho,\delta},a,b,c} \leq C \|\mathcal{A}(x, y, \pi)\|_{AS^m_{\rho,\delta},a,b+[\beta_1+\beta_2],c},$$

with $C = C_{b,\beta_1,\beta_2}$.

3. *If* $\alpha_o \in \mathbb{N}_0^n$ *then the amplitude* $\{\Delta^{\alpha_o}\mathcal{A}(x,y,\pi),(x,y,\pi)\in G\times G\times\widehat{G}\}$ *is in* $AS^{m-\rho[\alpha_o]}_{\rho,\delta}$ *with kernel* $\tilde{q}_{\alpha_o}\kappa_{x,y}$*, and*

$$\|\Delta^{\alpha_o}\mathcal{A}(x,\pi)\|_{S^{m-\rho[\alpha_o]}_{\rho,\delta},a,b,c} \leq C_{a,\alpha_o}\|\mathcal{A}(x,\pi)\|_{S^m_{\rho,\delta},a+[\alpha_o],b,c}.$$

4. *The symbol* $\{\mathcal{A}(x,y,\pi)^*,(x,\pi)\in G\times G\times\widehat{G}\}$ *is in* $AS^m_{\rho,\delta}$ *with kernel* $\kappa^*_{x,y}$ *given by* $\kappa^*_{x,y}(z)=\bar{\kappa}_{y,x}(z^{-1})$*, and*

$$\|\mathcal{A}(x,y,\pi)^*\|_{AS^m_{\rho,\delta},a,b,c} = \sup_{\substack{|\gamma|\leq c\\ [\alpha]\leq a,\,[\beta_1],[\beta_2]\leq b}} \|\pi(\mathrm{I}+\mathcal{R})^{-\frac{\gamma}{\nu}}X_x^{\beta_1}X_y^{\beta_2}\Delta^\alpha\mathcal{A}(x,y,\pi)\pi(\mathrm{I}+\mathcal{R})^{\frac{\rho[\alpha]-m-\delta([\beta_1]+[\beta_2])+\gamma}{\nu}}\|_{\mathscr{L}(\mathcal{H}_\pi)}.$$

(ii) *Let* $\mathcal{A}_1\in AS^{m_1}_{\rho,\delta}$ *and* $\mathcal{A}_2\in AS^{m_2}_{\rho,\delta}$ *have kernels* $\kappa_{1,x,y}$ *and* $\kappa_{2,x,y}$*, respectively. Then*

$$\mathcal{A}(x,y,\pi):=\mathcal{A}_1(x,y,\pi)\mathcal{A}_2(x,y,\pi)$$

defines the amplitude $\mathcal{A}$ *in* $S^m_{\rho,\delta}$*,* $m=m_1+m_2$*, with kernel* $\kappa_{2,x,y}*\kappa_{1,x,y}$ *with the convolution in the sense of Definition 5.1.19. Furthermore,*

$$\|\mathcal{A}(x,y,\pi)\|_{S^m_{\rho,\delta},a,b,c}\leq C\|\mathcal{A}_1(x,y,\pi)\|_{S^{m_1}_{\rho,\delta},a,b,c+\rho a+|m_2|+\delta b}\|\mathcal{A}_2(x,y,\pi)\|_{S^{m_2}_{\rho,\delta},a,b,c},$$

where the constant $C=C_{a,b,c}>0$ *does not depend on* $\mathcal{A}_1,\mathcal{A}_2$*.*

A direct consequence of Part (ii) of Proposition 5.6.10 is that the amplitudes in the introduced amplitude classes form an algebra:

Corollary 5.6.11. *Let* $1\geq\rho\geq\delta\geq 0$ *and* $\delta\neq 1$*. The collection of symbols* $\bigcup_{m\in\mathbb{R}}AS^m_{\rho,\delta}$ *forms an algebra.*

Furthermore, if $\mathcal{A}_0\in AS^{-\infty}$ *is smoothing and* $\mathcal{A}\in AS^m_{\rho,\delta}$ *is of order* $m\in\mathbb{R}$*, then* $\mathcal{A}_0\mathcal{A}$ *and* $\mathcal{A}\mathcal{A}_0$ *are also in* $AS^{-\infty}$*.*

Another consequence of Part (ii) together with Lemma 5.2.17 gives the following property:

Corollary 5.6.12. *Let* $1\geq\rho\geq\delta\geq 0$ *and* $\delta\neq 1$*. Let* $\mathcal{A}\in AS^m_{\rho,\delta}$ *have kernel* $\kappa_{x,y}$*. If* β *and* $\tilde{\beta}$ *are in* $\mathbb{N}_0^n$*, then*

$$\{\pi(X)^\beta\mathcal{A}\,\pi(X)^{\tilde{\beta}},(x,\pi)\in G\times\widehat{G}\}\in AS^{m+[\beta]+[\tilde{\beta}]}_{\rho,\delta}$$

with kernel $X_z^\beta\tilde{X}_z^{\tilde{\beta}}\kappa_{x,y}(z)$*. Furthermore, for any* a,b,c *there exists* $C=C_{a,b,c}$ *independent of* $\mathcal{A}$ *such that*

$$\|\pi(X)^\beta\mathcal{A}\pi(X)^{\tilde{\beta}}\|_{AS^m_{\rho,\delta},a,b,c}\leq C\|\mathcal{A}\|_{AS^m_{\rho,\delta},a,b,c+\rho a+[\beta]+[\tilde{\beta}]+\delta b}.$$

Proceeding as in Section 5.4.1, taking into account the dependence in x and y, we obtain

Proposition 5.6.13. *Let $\mathcal{A} = \{\mathcal{A}(x, y, \pi)\}$ be in $AS^m_{\rho,\delta}$ with $1 \geq \rho \geq \delta \geq 0$. Let $\kappa_{x,y}$ denote its associated kernel.*

1. *If $\alpha, \beta_1, \beta_2, \beta_o, \beta'_o \in \mathbb{N}_0^n$ are such that*

$$m - \rho[\alpha] + [\beta_1] + [\beta_2] + \delta([\beta_o] + [\beta'_o]) < -Q/2,$$

then the distribution $X_z^{\beta_1} \tilde{X}_z^{\beta_2} (X_x^{\beta_o} X_y^{\beta'_o} \tilde{q}_\alpha(z)\kappa_{x,y}(z))$ is square integrable and for every $x \in G$ we have

$$\int_G \left| X_z^{\beta_1} \tilde{X}_z^{\beta_2} (X_x^{\beta_o} X_y^{\beta'_o} \tilde{q}_\alpha(z)\kappa_{x,y}(z)) \right|^2 dz \leq C \sup_{\pi \in \widehat{G}} \|\mathcal{A}(x, \pi)\|^2_{AS^m_{\rho,\delta}, a, b, c}$$

where $a = [\alpha]$, $b = [\beta_o] + [\beta'_o]$, $c = \rho[\alpha] + [\beta_1] + [\beta_2] + \delta([\beta_o] + [\beta'_o])$ and $C = C_{m,\alpha,\beta_1,\beta_2,\beta_o,\beta'_o} > 0$ is a constant independent of $\mathcal{A}$ and x, y.

2. *For any $\alpha, \beta_1, \beta_2, \beta_o, \beta'_o \in \mathbb{N}_0^n$ satisfying*

$$m - \rho[\alpha] + [\beta_1] + [\beta_2] + \delta([\beta_o] + [\beta'_o]) < -Q,$$

the distribution $z \mapsto X_z^{\beta_1} \tilde{X}_z^{\beta_2} X_x^{\beta_o} X_y^{\beta'_o} \tilde{q}_\alpha(z)\kappa_{x,y}(z)$ is continuous on G for every $(x, y) \in G \times G$ and we have

$$\sup_{z \in G} \left| X_z^{\beta_1} \tilde{X}_z^{\beta_2} \left\{ X_x^{\beta_o} X_y^{\beta'_o} \tilde{q}_\alpha(z)\kappa_{x,y}(z) \right\} \right| \leq C \sup_{\pi \in \widehat{G}} \|\mathcal{A}(x, \pi)\|_{AS^m_{\rho,\delta}, [\alpha], [\beta_o]+[\beta'_o], [\beta_2]},$$

where $C = C_{m,\alpha,\beta_1,\beta_2,\beta_o,\beta'_o} > 0$ is a constant independent of $\mathcal{A}$ and x, y.

We now assume $\rho > 0$. Then the map $\kappa : (x, y, z) \mapsto \kappa_{x,y}(z)$ is smooth on $G \times G \times (G \setminus \{0\})$. Fixing a homogeneous quasi-norm $|\cdot|$ on G, we have the following more precise estimates:

at infinity: *For any $M \in \mathbb{R}$ and any $\alpha, \beta_1, \beta_2, \beta_o, \beta'_o \in \mathbb{N}_0^n$ there exist $C > 0$ and $a, b, c \in \mathbb{N}$ independent of $\mathcal{A}$ such that for all $x \in G$ and $z \in G$ satisfying $|z| \geq 1$, we have*

$$\left| X_z^{\beta_1} \tilde{X}_z^{\beta_2} (X_x^{\beta_o} X_y^{\beta'_o} \tilde{q}_\alpha(z)\kappa_{x,y}(z)) \right| \leq C \sup_{\pi \in \widehat{G}} \|\mathcal{A}(x, y, \pi)\|_{AS^m_{\rho,\delta}, a, b, c} |z|^{-M}.$$

at the origin: *For any $\alpha, \beta_1, \beta_2, \beta_o, \beta'_o \in \mathbb{N}_0^n$ with $Q+m+\delta([\beta_o]+[\beta'_o])-\rho[\alpha]+[\beta_1]+[\beta_2] \geq 0$ there exist a constant $C > 0$ and computable integers $a, b, c \in \mathbb{N}_0$ independent of $\mathcal{A}$ such that for all $x \in G$ and $z \in G\setminus\{0\}$, we have, if*

$$Q + m + \delta([\beta_o] + [\beta'_o]) - \rho[\alpha] + [\beta_1] + [\beta_2] > 0,$$

then

$$\left|X_z^{\beta_1}\tilde{X}_z^{\beta_2}(X_x^{\beta_o}X_y^{\beta'_o}\tilde{q}_\alpha(z)\kappa_{x,y}(z))\right|$$
$$\leq C\sup_{\pi\in\widehat{G}}\|\mathcal{A}(x,\pi)\|_{AS^m_{\rho,\delta},a,b,c}|z|^{-\frac{Q+m+\delta([\beta_o]+[\beta'_o])-\rho[\alpha]+[\beta_1]+[\beta_2]}{\rho}},$$

and if

$$Q+m+\delta([\beta_o]+[\beta'_o])-\rho[\alpha]+[\beta_1]+[\beta_2]=0,$$

then

$$\left|X_z^{\beta_1}\tilde{X}_z^{\beta_2}(X_x^{\beta_o}X_y^{\beta'_o}\tilde{q}_\alpha(z)\kappa_{x,y}(z))\right|\leq C\sup_{\pi\in\widehat{G}}\|\mathcal{A}(x,y,\pi)\|_{AS^m_{\rho,\delta},a,b,c}\ln|z|.$$

5.6.4 Link between symbols and amplitudes

Symbols can be viewed as amplitudes which do not depend on the second variable of the group. Then $S^m_{\rho,\delta}\subset AS^m_{\rho,\delta}$ and, by Remark 5.6.4, we have the inclusion

$$\Psi^m_{\rho,\delta}=\mathrm{Op}(S^m_{\rho,\delta})\subset \mathrm{AOp}(AS^m_{\rho,\delta}).$$

The next theorem shows the converse, namely, that the class of operators $\mathrm{AOp}(AS^m_{\rho,\delta})$ is included in $\Psi^m_{\rho,\delta}$. Therefore this will show that the amplitude quantization of $AS^m_{\rho,\delta}$ coincides with the symbol quantization of $S^m_{\rho,\delta}$.

Theorem 5.6.14. *Let $\mathcal{A}\in AS^m_{\rho,\delta}$ with $1\geq\rho\geq\delta\geq 0$, $\delta\neq 1$. Then $\mathrm{AOp}(\mathcal{A})$ is in $\Psi^m_{\rho,\delta}$, that is, there exists a (unique) symbol $\sigma\in S^m_{\rho,\delta}$ such that*

$$\mathrm{AOp}(\mathcal{A})=\mathrm{Op}(\sigma).$$

Furthermore, for any $M\in\mathbb{N}_0$, the map

$$\begin{cases} AS^m_{\rho,\delta} & \longrightarrow & S^{m-(\rho-\delta)(M+1)}_{\rho,\delta} \\ \mathcal{A} & \longmapsto & \sigma(x,\pi)-\sum_{[\alpha]\leq M}\Delta^\alpha X_y^\alpha\mathcal{A}(x,y,\pi)|_{y=x} \end{cases},$$

is continuous. If $\rho>\delta$, we have the asymptotic expansion

$$\sigma(x,\pi)\sim\sum_\alpha\Delta^\alpha X_y^\alpha\mathcal{A}(x,y,\pi)|_{y=x}.$$

The proof of Theorem 5.6.14 is in essence close to the proofs of product and adjoint of operators in $\cup_{m\in\mathbb{R}}\Psi^m_{\rho,\delta}$, see Theorems 5.5.12 and 5.5.3. As for these theorems, it is helpful to understand formally the steps of the rigorous proof.

From the amplitude quantization in (5.66), we see that if $\mathrm{AOp}(\mathcal{A})$ can be written as $\mathrm{Op}(\sigma)$, then, denoting by $\kappa_{\sigma,x}$ the symbol kernel and by $\kappa_{\mathcal{A},x,y}$ the amplitude kernel, we have

$$\mathrm{AOp}(\mathcal{A})(\phi)(x)=\int_G\phi(y)\kappa_{\mathcal{A},x,y}(y^{-1}x)dy=\int_G\phi(xz^{-1})\kappa_{\mathcal{A},x,xz^{-1}}(z)dz$$

whereas

$$\mathrm{Op}(\sigma)(\phi)(x) = \int_G \phi(y)\kappa_{\sigma,x}(y^{-1}x)dy = \int_G \phi(xz^{-1})\kappa_{\sigma,x}(z)dz.$$

Therefore, formally we must have

$$\kappa_{\mathcal{A},x,xz^{-1}}(z) = \kappa_{\sigma,x}(z) \quad \left(\text{or equivalently } \kappa_{\mathcal{A},x,y}(y^{-1}x) = \kappa_{\sigma,x}(y^{-1}x)\right).$$

Using the Taylor expansion in $y = xz^{-1}$ for $\kappa_{\mathcal{A},x,y}$ at x, we have (again formally)

$$\kappa_{\sigma,x}(z) = \kappa_{\mathcal{A},x,xz^{-1}}(z) \approx \sum_\alpha \tilde{q}_\alpha(z) X_y^\alpha \kappa_{\mathcal{A},x,y}(z)|_{y=x}. \tag{5.70}$$

Note that the group Fourier transform in z of each term in the sum above is

$$\begin{aligned}\mathcal{F}_{z\in G}\{\tilde{q}_\alpha(z)X_{y=x}^\alpha \kappa_{\mathcal{A},x,y}(z)\}(\pi) &= \Delta^\alpha X_{y=x}^\alpha \mathcal{F}_{z\in G}\{\kappa_{\mathcal{A},x,y}(z)\}(\pi)\\ &= \Delta^\alpha X_{y=x}^\alpha \mathcal{A}(x,y,\pi).\end{aligned}$$

Taking the group Fourier transform in z on both sides of (5.70), we obtain still formally that

$$\sigma(x,\pi) \approx \sum_\alpha \Delta^\alpha X_y^\alpha \mathcal{A}(x,y,\pi)|_{y=x}.$$

As in the proofs of Theorems 5.5.12 and 5.5.3, the crucial point is to control the remainder while using Taylor's expansion. The method is similar as in the proof of Theorem 5.5.12 and the adaptation is easy and left to the reader.

Note that Theorem 5.6.14 together with Proposition 5.6.6 give another proof of Theorem 5.5.12. This is not surprising given the similarity between the proof of Theorems 5.6.14 and 5.5.12.

5.7 Calderón-Vaillancourt theorem

In this section, we prove the analogue of the Calderón-Vaillancourt theorem, now in the setting of graded Lie groups. This extends the L^2-boundedness of operators in the class $\Psi^0_{1,0}$ given in Theorem 5.4.17 to the classes $\Psi^0_{\rho,\delta}$.

Theorem 5.7.1. *Let $T \in \Psi^0_{\rho,\delta}$ with $1 \geq \rho \geq \delta \geq 0$ and $\delta \neq 1$. Then T extends to a bounded operator on $L^2(G)$.*

Moreover, there exist a constant $C > 0$ and a seminorm $\|\cdot\|_{\Psi^0_{\rho,\delta},a,b,c}$ with computable integers $a, b, c \in \mathbb{N}_0$ independent of T such that

$$\forall \phi \in \mathcal{S}(G) \qquad \|T\phi\|_{L^2(G)} \leq C\|T\|_{\Psi^0_{\rho,\delta},a,b,c}\|\phi\|_{L^2(G)}.$$

Before showing Theorem 5.7.1, let us mention that together with the pseudo-differential calculus, it implies the following boundedness on Sobolev spaces L^2_s.

Corollary 5.7.2. *Let $T \in \Psi^m_{\rho,\delta}$ with $1 \geq \rho \geq \delta \geq 0$ and $\delta \neq 1$. Then for any $s \in \mathbb{R}$, the operator T extends to a continuous operator from $L^2_s(G)$ to $L^2_{s-m}(G)$:*

$$\forall \phi \in \mathcal{S}(G) \qquad \|T\phi\|_{L^2_{s-m}(G)} \leq C_{s,m,\rho,\delta} \|T\|_{\Psi^m_{\rho,\delta},a,b,c} \|\phi\|_{L^2_s(G)},$$

with some (computable) integers a, b, c depending on s, m, ρ, δ.

Proof of Corollary 5.7.2. Let $\mathcal{R}$ be a positive Rockland operator. By the composition theorem (e.g. Theorem 5.5.3), we have

$$(\mathrm{I} + \mathcal{R})^{\frac{-m+s}{\nu}} T (\mathrm{I} + \mathcal{R})^{-\frac{s}{\nu}} \in \Psi^0_{\rho,\delta}.$$

Therefore, by Theorem 5.7.1, we have

$$\begin{aligned} \|(\mathrm{I} + \mathcal{R})^{\frac{-m+s}{\nu}} T (\mathrm{I} + \mathcal{R})^{-\frac{s}{\nu}} \phi\|_{\mathscr{L}(L^2(G))} &\lesssim \|(\mathrm{I} + \mathcal{R})^{\frac{-m+s}{\nu}} T (\mathrm{I} + \mathcal{R})^{-\frac{s}{\nu}} \|_{\Psi^0_{\rho,\delta},a_1,b_1,c_1} \\ &\lesssim \|T\|_{\Psi^m_{\rho,\delta},a_2,b_2,c_2}, \end{aligned}$$

by Theorem 5.5.3. □

Remark 5.7.3. Combining the results obtained so far, for each (ρ, δ) with $1 \geq \rho \geq \delta \geq 0$ and $\delta \neq 1$, we have therefore obtained an operator calculus, in the sense that the set $\bigcup_{m \in \mathbb{R}} \Psi^m_{\rho,\delta}$ forms an algebra of operators, stable under taking the adjoint, and acting on the Sobolev spaces in such a way that the loss of derivatives in L^2 is controlled by the order of the operator.

Note that the L^2-boundedness in the case $(\rho, \delta) = (1, 0)$ was already proved by different methods, see Theorem 5.4.17 and its proof. With the same proof as in the corollary above, one obtains easily boundedness for L^p-Sobolev spaces in this case:

Corollary 5.7.4. *Let $T \in \Psi^m_{1,0}$. Then for any $s \in \mathbb{R}$ and $p \in (1, \infty)$ the operator T extends to a continuous operator from $L^p_s(G)$ to $L^p_{s-m}(G)$:*

$$\forall \phi \in \mathcal{S}(G) \qquad \|T\phi\|_{L^p_{s-m}(G)} \leq C_{s,m,\rho,\delta} \|T\|_{\Psi^m_{\rho,\delta},a,b,c} \|\phi\|_{L^p_s(G)},$$

with some (computable) integers a, b, c depending on s, m, ρ, δ.

Proof of Corollary 5.7.4. As above, $(\mathrm{I} + \mathcal{R})^{\frac{-m+s}{\nu}} T (\mathrm{I} + \mathcal{R})^{-\frac{s}{\nu}} \in \Psi^0$ therefore, by Corollary 5.4.20 we have

$$\begin{aligned} \|(\mathrm{I} + \mathcal{R})^{\frac{-m+s}{\nu}} T (\mathrm{I} + \mathcal{R})^{-\frac{s}{\nu}} \phi\|_{\mathscr{L}(L^p(G))} &\lesssim \|(\mathrm{I} + \mathcal{R})^{\frac{-m+s}{\nu}} T (\mathrm{I} + \mathcal{R})^{-\frac{s}{\nu}} \|_{\Psi^0,a_1,b_1,c_1} \\ &\lesssim \|T\|_{\Psi^0,a_2,b_2,c_2}, \end{aligned}$$

by Theorem 5.5.3. □

The rest of this section is devoted to the proof of the Calderón-Vaillancourt Theorem, that is, Theorem 5.7.1. In Section 5.7.2, we prove the result for $\rho = \delta = 0$.

The proof will rely on an analogue on G of the familiar decomposition of $\mathbb{R}^n$ into unit cubes presented in Section 5.7.1. The case $\rho = \delta \in (0,1)$ will be proved in Section 5.7.4 and its proof relies on the case $\rho = \delta = 0$ and on a bilinear estimate proved in Section 5.7.3. The case of $\rho = \delta \in [0,1)$ will then be proved and this will imply Theorem 5.7.1 thanks to the continuous inclusions between symbol classes (see (5.31)).

5.7.1 Analogue of the decomposition into unit cubes

In this section, we present an analogue of the dyadic cubes, more precisely we construct a useful covering of the general homogeneous Lie group G by unit balls and the corresponding partition of unity with a number of advantageous properties. The proof is an adaptation of [FS82, Lemma 7.14].

Lemma 5.7.5. *Let $|\cdot|$ be a fixed homogeneous quasi-norm on the homogeneous Lie group G. We denote by $C_o \geq 1$ a constant for the triangle inequality*

$$\forall x, y \in G \qquad |xy| \leq C_o(|x| + |y|). \tag{5.71}$$

Denoting by $B(x,R)$ the $|\cdot|$-ball centred at point x with radius R,

$$B(x,R) := \{y \in G \ : \ |x^{-1}y| < R\},$$

there exists a maximal family $\{B(x_i, \frac{1}{2C_o})\}_{i=1}^{\infty}$ of disjoint balls of radius $\frac{1}{2C_o}$, and we choose one such family. Then the following properties hold:

1. *The balls $\{B(x_i, 1)\}_{i=1}^{\infty}$ cover G.*
2. *For any $C \geq 1$, no point of G belongs to more than $\lceil (4C_o^2 C)^Q \rceil$ of the balls $\{B(x_i, C)\}_{i=1}^{\infty}$.*
3. *There exists a sequence of functions $\chi_i \in \mathcal{D}(G)$, $i \in \mathbb{N}$, such that each χ_i is supported in $B(x_i, 2)$ and satisfies $0 \leq \chi_i \leq 1$ while we have $\sum_{i=1}^{\infty} \chi_i = 1$. Moreover, for any $\beta \in \mathbb{N}_0^n$, $X^\beta \chi_i$ is uniformly bounded in $i \in \mathbb{N}$.*
4. *For any $p_1 > Q + 1$, we have*

$$\exists C_{p_1} > 0 \quad \forall i_o \in \mathbb{N} \qquad \sum_{i=1}^{\infty} (1 + |x_{i_o}^{-1} x_i|)^{-p_1} \leq C_{p_1} < \infty.$$

Remark 5.7.6. The conclusion of Part (4) is rough but will be sufficient for our purposes. We note, however, that if the quasi-norm in Lemma 5.7.5 is actually a norm, i.e. if the constant C_o in (5.71) is equal to one, $C_o = 1$, then the conclusion of Part (4) of Lemma 5.7.5 holds true for all $p_1 > Q$. This will be proved together with the lemma.

Proof of Lemma 5.7.5 and of Remark 5.7.6. If $x \in G$ then by maximality there exists i such that the distance from x to $B(x_i, \frac{1}{2C_o})$ is $< 1/(2C_o)$. Denoting by y a point in $\bar{B}(x_i, \frac{1}{2C_o})$ which realises the distance, we have

$$|x_i^{-1}x| \leq C_o(|x_i^{-1}y| + |y^{-1}x|) < C_o\left(\frac{1}{2C_o} + \frac{1}{2C_o}\right) = 1.$$

This proves Part (1).

If x is in all the balls $B(x_{i_\ell}, C)$, $\ell = 1, \ldots, \ell_o$, then

$$\forall y \in \cup_{\ell=1}^{\ell_o} B(x_{i_\ell}, C) \quad \exists \ell \in [1, \ell_o] \quad |x^{-1}y| \leq C_o(|x^{-1}x_{i_\ell}| + |x_{i_\ell}^{-1}y|) \leq C_o 2C.$$

This shows that $B(x, 2C_oC)$ contains $\cup_{\ell=1}^{\ell_o} B(x_{i_\ell}, C)$ and, therefore, it must contain the disjoint balls $\cup_{\ell=1}^{\ell_o} B(x_{i_\ell}, \frac{1}{2C_o})$. Taking the Haar measure and denoting $c_1 := |B(0,1)|$, we have

$$|\cup_{\ell=1}^{\ell_o} B(x_{i_\ell}, \frac{1}{2C_o})| = \ell_o c_1 \left(\frac{1}{2C_o}\right)^Q \leq |B(x, 2C_oC)| = (2C_oC)^Q \, c_1.$$

This proves Part (2).

Let us fix $\chi \in \mathcal{D}(G)$ satisfying $0 \leq \chi \leq 1$ with $\chi = 1$ on $B(0,1)$ and $\chi = 0$ on $B(0,2)$. The sum $\sum_{i'=1}^{\infty} \chi(x_{i'}^{-1} \cdot)$ is locally finite by Part (2); it is a smooth function with values between 1 and $\lceil (4C_o^2 \times 2)^Q \rceil$. We define

$$\chi_i(x) := \frac{\chi(x_i^{-1}x)}{\sum_{i'=1}^{\infty} \chi(x_{i'}^{-1}x)}.$$

This gives Part (3).

To prove Part (4), we fix a point x_{i_o} and observe that if $x \in G$ is in one of the balls $B(x_i, \frac{1}{2C_o})$ with $|x_{i_o}^{-1}x_i| \in [\ell, \ell+1)$ for some $\ell \in \mathbb{N}$, let us say $B(x_{i_1}, \frac{1}{2C_o})$, then

$$|x_{i_o}^{-1}x| \leq C_o(|x_{i_1}^{-1}x| + |x_{i_o}^{-1}x_{i_1}|) \leq C_o(\frac{1}{2C_o} + \ell + 1).$$

This yields the inclusion

$$\sqcup_{|x_{i_o}^{-1}x_i| \in [\ell, \ell+1)} B(x_i, \frac{1}{2C_o}) \ \subset \ B(x_{i_o}, C_o(\frac{1}{2C_o} + \ell + 1)).$$

The measure of the left hand side is $c_1(2C_o)^{-Q}\text{card}\{i : |x_{i_o}^{-1}x_i| \in [\ell, \ell+1)\}$ and the measure of the right hand side is $c_1(C_o(\frac{1}{2C_o} + \ell + 1))^Q$. Therefore,

$$\text{card}\{i : |x_{i_o}^{-1}x_i| \in [\ell, \ell+1)\} \leq c\ell^Q.$$

Now we decompose

$$\sum_{i=1}^{\infty}(1+|x_{i_o}^{-1}x_i|)^{-p_1} = \sum_{|x_i^{-1}x_{i_o}|<1}(1+|x_{i_o}^{-1}x_i|)^{-p_1} + \sum_{\ell=1}^{\infty}\sum_{|x_i^{-1}x_{i_o}|\in[\ell,\ell+1)}(1+|x_{i_o}^{-1}x_i|)^{-p_1}.$$

By Part (2) the first sum on the right hand side is $\leq \lceil (4C_o^2)^Q \rceil$ whereas from the observation just above, the second sum is $\leq \sum_{\ell=0}^{\infty}(1+\ell)^{-p_1}c'(1+\ell)^Q$. This last sum being convergent whenever $-p_1+Q<-1$, Part (4) is proved.

Let us finally prove Remark 5.7.6, that is, Part (4) of the lemma for $p_1>Q$ provided that $C_o=1$. This will follow by the same argument as above if we can show a refined estimate

$$\mathrm{card}\{i : |x_{i_o}^{-1}x_i| \in [\ell,\ell+1)\} \leq c\ell^{Q-1}.$$

We claim that this estimate holds true. Since $C_o=1$, we can estimate

$$|x_{i_o}^{-1}x| \geq |x_{i_o}^{-1}x_{i_1}| - |x_{i_1}^{-1}x| > \ell - \frac{1}{2C_o} = \ell - \frac{1}{2}.$$

We also have $C_o(\frac{1}{2C_o}+\ell+1)=\ell+\frac{3}{2}$. Consequently, we have the inclusion

$$\sqcup_{|x_{i_o}^{-1}x_i|\in[\ell,\ell+1)}B(x_i,\frac{1}{2C_o}) \subset B(x_{i_o},\ell+\frac{3}{2})\backslash B(x_{i_o},\ell-\frac{1}{2}),$$

with the measure on the right hand side being $c_1(\ell+\frac{3}{2})^Q - c_1(\ell-\frac{1}{2})^Q$. Therefore,

$$\mathrm{card}\{i : |x_{i_o}^{-1}x_i| \in [\ell,\ell+1)\} \leq c\ell^{Q-1},$$

so that the required claim is proved. □

5.7.2 Proof of the case $S^0_{0,0}$

This section is devoted to the proof of the following result which is a particular case of Theorem 5.7.1. We also give an explicit estimate on the number of derivatives and differences of the symbol needed for the L^2-boundedness.

Proposition 5.7.7. *Let $T \in \Psi^0_{0,0}$. Then T extends to a bounded operator on $L^2(G)$. Furthermore, if we fix a positive Rockland operator $\mathcal{R}$ (in order to define the semi-norms on $\Psi^m_{\rho,\delta}$) then*

$$\forall \phi \in \mathcal{S}(G) \qquad \|T\phi\|_{L^2(G)} \leq C\|T\|_{\Psi^0_{0,0},a,b,c}\|\phi\|_{L^2(G)},$$

where $C>0$ and $a,b,c\in\mathbb{N}_0$ are independent of T. In particular, this estimate holds with $a=rp_o$, $b=r\nu+\lceil\frac{Q}{2}\rceil$, $c=r\nu$, where ν is the degree of $\mathcal{R}$, $p_o/2$ is the smallest positive integer divisible by $\upsilon_1,\ldots,\upsilon_n$, and $r\in\mathbb{N}_0$ is the smallest integer such that $rp_o>Q+1$.

Throughout Section 5.7.2, we fix the homogeneous norm $|\cdot| = |\cdot|_{p_o}$ given by (3.21), where $p_o/2$ is the smallest positive integer divisible by $\upsilon_1, \ldots, \upsilon_n$. We fix a maximal family $\{B(x_i, \frac{1}{2C_o})\}_{i=1}^{\infty}$ of disjoint balls and a sequence of functions $(\chi_i)_{i=1}^{\infty}$ so that the properties of Lemma 5.7.5 hold. We also fix $\psi_0, \psi_1 \in \mathcal{D}(\mathbb{R})$ supported in $[-1,1]$ and $[1/2,2]$, respectively, such that $0 \le \psi_0, \psi_1 \le 1$ and

$$\forall \lambda \ge 0 \qquad \sum_{j=0}^{\infty} \psi_j(\lambda) = 1 \ \text{ with } \psi_j(\lambda) := \psi_1(2^{-(j-1)}\lambda), \ j \in \mathbb{N}.$$

Let us start the proof of Proposition 5.7.7. Let $\sigma \in S^0_{0,0}$.
For each $I = (i,j) \in \mathbb{N} \times \mathbb{N}_0$, we define

$$\sigma_I(x,\pi) := \chi_i(x)\sigma(x,\pi)\psi_j(\pi(\mathcal{R})).$$

We denote by T_I and κ_I the corresponding operator and kernel.

Roughly speaking, the parameters i and j correspond to localising in space and frequency, respectively. The localisation in space corresponds to the covering of G by the balls centred at the x_i's, while the localisation in frequency is determined by the spectral projection of $\mathcal{R}$ to the $L^2(G)$-eigenspaces corresponding to eigenvalues close to each 2^j.

It is not difficult to see that each T_I is bounded on $L^2(G)$:

Lemma 5.7.8. *Each operator T_I is bounded on $L^2(G)$.*

Since σ_I is localised both in space and in frequency, we may use one of the two localisations.

Proof of Lemma 5.7.8 using frequency localisation. Let $\alpha, \beta \in \mathbb{N}_0^n$. By the Leibniz formulae for difference operators (see Proposition 5.2.10) and for vector fields, we have

$$X_x^\beta \Delta^\alpha \sigma_I(x,\pi) = \sum_{\substack{[\beta_1]+[\beta_2]=[\beta] \\ [\alpha_1]+[\alpha_2]=[\alpha]}} X_x^{\beta_1}\chi_i(x)\; X_x^{\beta_2}\Delta^{\alpha_1}\sigma(x,\pi)\; \Delta^{\alpha_2}\psi_j(\pi(\mathcal{R})).$$

Therefore,

$$\begin{aligned}
&\|\pi(\mathrm{I}+\mathcal{R})^{\frac{[\alpha]+\gamma}{\nu}} X_x^\beta \Delta^\alpha \sigma_I(x,\pi)\pi(\mathrm{I}+\mathcal{R})^{-\frac{\gamma}{\nu}}\|_{\mathscr{L}(\mathcal{H}_\pi)} \\
&\quad \le C \sum_{\substack{[\beta_2]\le[\beta] \\ [\alpha_1]+[\alpha_2]=[\alpha]}} \|\pi(\mathrm{I}+\mathcal{R})^{\frac{[\alpha]+\gamma}{\nu}} X_x^{\beta_2}\Delta^{\alpha_1}\sigma(x,\pi)\; \Delta^{\alpha_2}\psi_j(\pi(\mathcal{R}))\pi(\mathrm{I}+\mathcal{R})^{-\frac{\gamma}{\nu}}\|_{\mathscr{L}(\mathcal{H}_\pi)} \\
&\quad \le C \sum_{\substack{[\beta_2]\le[\beta] \\ [\alpha_1]+[\alpha_2]=[\alpha]}} \|\pi(\mathrm{I}+\mathcal{R})^{\frac{[\alpha]+\gamma}{\nu}} X_x^{\beta_2}\Delta^{\alpha_1}\sigma(x,\pi)\pi(\mathrm{I}+\mathcal{R})^{-\frac{[\alpha_2]+\gamma}{\nu}}\|_{\mathscr{L}(\mathcal{H}_\pi)} \\
&\qquad\qquad \|\pi(\mathrm{I}+\mathcal{R})^{\frac{[\alpha_2]+\gamma}{\nu}} \Delta^{\alpha_2}\psi_j(\pi(\mathcal{R}))\pi(\mathrm{I}+\mathcal{R})^{-\frac{\gamma}{\nu}}\|_{\mathscr{L}(\mathcal{H}_\pi)}.
\end{aligned}$$

Therefore, by Lemma 5.4.7, we obtain

$$\|\sigma_I\|_{S^0_{1,0},a,b,c} \leq \|\sigma\|_{S^0_{0,0},a,b,c+a} 2^{ja/\nu}.$$

This shows that the operator T_I is in Ψ^0 and is therefore bounded on $L^2(G)$ by Theorem 5.4.17. □

Proof of Lemma 5.7.8 using space localisation. Another proof is to apply the following lemma since the symbol $\sigma_I(x,\pi)$ has compact support in x. □

Lemma 5.7.9. *Let $\sigma(x,\pi)$ be a symbol (in the sense of Definition 5.1.33) supported in $x \in S$, and assume that S is compact. Then the operator norm of the associated operator on $L^2(G)$ is*

$$\|\mathrm{Op}(\sigma)\|_{\mathscr{L}(L^2(G))} \leq C|S|^{1/2} \sup_{\substack{x\in G \\ [\beta]\leq\lceil\frac{Q}{2}\rceil}} \|X_x^\beta \sigma(x,\pi)\|_{L^\infty(\widehat{G})}.$$

Proof of Lemma 5.7.9. Let $T = \mathrm{Op}(\sigma)$ and let κ_x be the associated kernel. We have by the Sobolev inequality in Theorem 4.4.25,

$$\begin{aligned} |T\phi(x)|^2 &= |\phi * \kappa_x(x)|^2 \leq \sup_{x_o\in G} |\phi * \kappa_{x_o}(x)|^2 \\ &\leq C \sum_{[\beta]\leq\lceil\frac{Q}{2}\rceil} \left\|\phi * X^\beta_{x_o}\kappa_{x_o}(x)\right\|^2_{L^2(dx_o)}. \end{aligned}$$

Hence

$$\begin{aligned} \|T\phi\|^2_{L^2(G)} &\leq C \sum_{[\beta]\leq\lceil\frac{Q}{2}\rceil} \int_G\int_G |\phi * X^\beta_{x_o}\kappa_{x_o}(x)|^2 dx_o dx \\ &\leq C \sum_{[\beta]\leq\lceil\frac{Q}{2}\rceil} \int_G \|\phi * X^\beta_{x_o}\kappa_{x_o}\|^2_{L^2(dx)} dx_o \\ &\leq C|S| \sup_{x_o\in G,[\beta]\leq\lceil\frac{Q}{2}\rceil} \left\|\phi * X^\beta_{x_o}\kappa_{x_o}(x)\right\|^2_{L^2(dx)}. \end{aligned}$$

Now by Plancherel's Theorem,

$$\left\|\phi * X^\beta_{x_o}\kappa_{x_o}(x)\right\|_{L^2(dx)} \leq \|\phi\|_{L^2(dx)} \left\|X^\beta_{x_o}\sigma(x_o,\pi)\right\|_{L^\infty(\widehat{G})}.$$

This implies that the L^2-operator norm of T is

$$\leq C|S|^{1/2} \sup_{x_o\in G,[\beta]\leq\lceil\frac{Q}{2}\rceil} \|X^\beta_{x_o}\sigma(x_o,\pi)\|_{L^\infty(\widehat{G})},$$

and concludes the proof of Lemma 5.7.9. □

Let us go back to the proof of Proposition 5.7.7. The approach is to apply the following version of Cotlar's lemma:

Lemma 5.7.10 (Cotlar's lemma here). *Suppose that $r \in \mathbb{N}_0$ is such that $rp_o > Q+1$ and that there exists $A_r > 0$ satisfying for all $(I, I') \in \mathbb{N} \times \mathbb{N}_0$:*

$$\max\left(\|T_I T_{I'}^*\|_{\mathscr{L}(L^2(G))}, \|T_I^* T_{I'}\|_{\mathscr{L}(L^2(G))}\right) \leq A_r 2^{-|j-j'|r}(1 + |x_{i'}^{-1} x_i|)^{-rp_o}.$$

Then $T = \mathrm{Op}(\sigma)$ is L^2-bounded with operator norm $\leq C\sqrt{A_r}$.

Lemma 5.7.10 can be easily shown, adapting for instance the proof given in [Ste93, ch. VII §2] using Part (4) of Lemma 5.7.5. Indeed, the numbering of the sequence of operators to which the Cotlar-Stein lemma (see Theorem A.5.2) is applied is not important, and the condition $rp_o > Q+1$ is motivated by Lemma 5.7.5, Part (4). This is left to the reader.

Lemma 5.7.11 which follows gives the operator norm for $T_I T_{I'}^*$ and $T_I^* T_{I'}$. Combining Lemmata 5.7.10 and 5.7.11 gives the proof of Proposition 5.7.7.

Lemma 5.7.11. *1. For any $r \in \mathbb{N}_0$, the operator norm of $T_I T_{I'}^*$ on $L^2(G)$ is*

$$\|T_I T_{I'}^*\|_{\mathscr{L}(L^2(G))} \leq C_r 1_{|j-j'|\leq 1}(1 + |x_{i'}^{-1} x_i|)^{-rp_o} \|\sigma\|^2_{S^0_{0,0}, rp_o, \lceil \frac{Q}{2} \rceil, 0}.$$

2. For any $r \in \mathbb{N}_0$, the operator norm of $T_I^* T_{I'}$ on $L^2(G)$ is

$$\|T_I^* T_{I'}\|_{\mathscr{L}(L^2(G))} \leq C_r 1_{|x_{i'}^{-1} x_i| \leq 4C_o} 2^{-|j-j'|r} \|\sigma\|^2_{S^0_{0,0}, 0, r\nu + \lceil \frac{Q}{2} \rceil, r\nu}.$$

In the proof of Lemma 5.7.11, we will also use the symbols σ_i, $i \in \mathbb{N}$, given by

$$\sigma_i(x, \pi) := \chi_i(x)\, \sigma(x, \pi),$$

and the corresponding operators $T_i = \mathrm{Op}(\sigma_i)$ and kernels κ_i. We observe that σ_i is compactly supported in x, therefore by Lemma 5.7.9, the operator T_i is bounded on $L^2(G)$.

Proof of Lemma 5.7.11 Part (1). We have (see the end of Lemma 5.5.4)

$$T_I = \mathrm{Op}(\sigma_I) = T_i\, \psi_j(\mathcal{R}),$$

thus

$$T_I T_{I'}^* = T_i \psi_j(\mathcal{R}) \psi_{j'}(\mathcal{R}) T_{i'}^*.$$

Since $\psi_j(\mathcal{R})\psi_{j'}(\mathcal{R}) = (\psi_j \psi_{j'})(\mathcal{R})$, this is 0 if $|j - j'| > 1$. Let us assume $|j - j'| \leq 1$. We set

$$T_{i'j'j} := T_{i'} \circ (\psi_j \psi_{j'})(\mathcal{R}) = \mathrm{Op}\left(\sigma_{i'} \circ (\psi_j \psi_{j'})(\pi(\mathcal{R}))\right),$$

see again the end of Lemma 5.5.4. Therefore $T_I T_{I'}^* = T_i T_{i'j'j}^*$, and we have by the Sobolev inequality in Theorem 4.4.25,

$$\begin{aligned}
|T_I T_{I'}^* \phi(x)| &= \left| \int_G T_{i'j'j}^* \phi(z)\ \kappa_{ix}(z^{-1}x) dz \right| \\
&\leq \sup_{x_o} \left| \int_G T_{i'j'j}^* \phi(z)\ \kappa_{ix_o}(z^{-1}x) dz \right| 1_{x \in B(x_i,2)} \\
&\leq C \sum_{[\beta] \leq \lceil \frac{Q}{2} \rceil} \left\| X_{x_o}^\beta \int_G T_{i'j'j}^* \phi(z) \kappa_{ix_o}(z^{-1}x) dz \right\|_{L^2(dx_o)} 1_{x \in B(x_i,2)}.
\end{aligned}$$

Hence,

$$\|T_I T_{I'}^* \phi\|_{L^2} \leq C \sum_{[\beta] \leq \lceil \frac{Q}{2} \rceil} \left\| \int_G T_{i'j'j}^* \phi(z) X_{x_o}^\beta \kappa_{ix_o}(z^{-1}x) dz\ 1_{x \in B(x_i,2)} \right\|_{L^2(dx_o dx)}.$$

The idea of the proof is to use a quantity which will help the space localisation; so we introduce this quantity $1 + |z^{-1}x|^{rp_o}$ and its inverse, where the integer $r \in \mathbb{N}$ is to be chosen suitably. Notice that for the inverse we have

$$(1 + |z^{-1}x|^{rp_o})^{-1} \leq C_r (1 + |z^{-1}x|)^{-rp_o} \leq C_r (1 + |x_{i'}^{-1} x_i|)^{-rp_o},$$

for any $z \in \text{supp}\chi_{i'}$ and $x \in B(x_i, 2)$. Therefore, we obtain

$$\begin{aligned}
&\left\| \int_G T_{i'j'j}^* \phi(z) X_{x_o}^\beta \kappa_{ix_o}(z^{-1}x) dz\ 1_{x \in B(x_i,2)} \right\|_{L^2(dx_o dx)} \\
&= \left\| \int_G T_{i'j'j}^* \phi(z) \frac{1 + |z^{-1}x|^{rp_o}}{1 + |z^{-1}x|^{rp_o}} X_{x_o}^\beta \kappa_{ix_o}(z^{-1}x) dz\ 1_{x \in B(x_i,2)} \right\|_{L^2(dx_o, dx)} \\
&\leq C (1 + |x_{i'}^{-1} x_i|)^{-rp_o} \left\| T_{i'j'j}^* \phi(z_1) \right\|_{L^2(dz_1)} \\
&\qquad \left\| (1 + |z_2^{-1} x|^{rp_o}) X_{x_o}^\beta \kappa_{ix_o}(z_2^{-1} x)\ 1_{x \in B(x_i,2)} \right\|_{L^2(dz_2, dx_o, dx)}
\end{aligned}$$

by the observation just above and the Cauchy-Schwartz inequality. The last term can be estimated as

$$\begin{aligned}
&\left\| (1 + |z_2^{-1} x|^{rp_o}) X_{x_o}^\beta \kappa_{ix_o}(z_2^{-1} x)\ 1_{x \in B(x_i,2)} \right\|_{L^2(dz_2, dx_o, dx)} \\
&\leq |B(x_i, 2)| \sup_{x_o \in G} \left\| (1 + |z'|^{rp_o}) X_{x_o}^\beta \kappa_{ix_o}(z') \right\|_{L^2(dz')} \\
&\leq C \sup_{x_o \in G} \sum_{[\alpha]=0}^{rp_o} \left\| X_{x_o}^\beta \Delta^\alpha \sigma_i(x_o, \pi) \right\|_{L^\infty(\widehat{G})}
\end{aligned}$$

by the Plancherel theorem and Theorem 5.2.22, since $|z'|^{rp_o}$ can be written as a linear combination of $\tilde{q}_\alpha(z)$, $[\alpha] = rp_o$. Combining the estimates above, we have obtained

$$\|T_I T_{I'}^* \phi\|_{L^2} \le C(1+|x_{i'}^{-1}x_i|)^{-rp_o} \left\|T_{i'j'j}^*\phi\right\|_{L^2} \sup_{\substack{x_o\in G\\ [\beta']\le\lceil\frac{Q}{2}\rceil,[\alpha]\le rp_o}} \left\|\Delta^\alpha X_{x_o}^{\beta'}\sigma(x_o,\pi)\right\|_{L^\infty(\widehat{G})}.$$

The supremum is equal to $\|\sigma\|_{S^0_{0,0},rp_o,\lceil\frac{Q}{2}\rceil,0}$. So we now want to study the operator norm of $T^*_{i'j'j}$, which is equal to the operator norm of $T_{i'j'j}$. Since the symbol of $T_{i'j'j}$ is localised in space we may apply Lemma 5.7.9 and obtain

$$\begin{aligned}
\|T^*_{i'j'j}\|_{\mathscr{L}(L^2(G))} &= \|T_{i'j'j}\|_{\mathscr{L}(L^2(G))} = \|\mathrm{Op}\left(\sigma_i\left(\psi_j\psi_{j'}\right)(\pi(\mathcal{R}))\right)\|_{\mathscr{L}(L^2(G))}\\
&\le C|B(x_i,2)|^{1/2} \sup_{\substack{x\in G\\ [\beta]\le\lceil Q/2\rceil}} \|X_x^\beta\left\{\chi_i(x)\sigma(x,\pi)\left(\psi_j\psi_{j'}\right)(\pi(\mathcal{R}))\right\}\|_{L^\infty(\widehat{G})}\\
&\le C \sup_{\substack{x\in G,\,\pi\in\widehat{G}\\ [\beta]\le\lceil Q/2\rceil}} \sum_{[\beta_1]+[\beta_2]=[\beta]} |X^{\beta_1}\chi_i(x)|\, \|X_x^{\beta_2}\sigma(x,\pi)\|_{\mathscr{L}(\mathcal{H}_\pi)} \|\left(\psi_j\psi_{j'}\right)(\pi(\mathcal{R}))\|_{\mathscr{L}(\mathcal{H}_\pi)}\\
&\le C \sup_{\substack{x\in G,\,\pi\in\widehat{G}\\ [\beta_2]\le\lceil Q/2\rceil}} \|X_x^{\beta_2}\sigma(x,\pi)\|_{\mathscr{L}(\mathcal{H}_\pi)} = C\|\sigma\|_{S^0_{0,0},0,\lceil Q/2\rceil,0},
\end{aligned}$$

since the $X^{\beta_2}\chi_i$'s are uniformly bounded on G and over i.

Thus, we have obtained

$$\|T_I T_{I'}^*\phi\|_{L^2} \le C(1+|x_{i'}^{-1}x_i|)^{-rp_o}\|\sigma\|_{S^0_{0,0},0,\lceil Q/2\rceil,0}\,\|\phi\|_{L^2}\,\|\sigma\|_{S^0_{0,0},rp_o,\lceil\frac{Q}{2}\rceil,0},$$

and this concludes the proof of the first part of Lemma 5.7.11. □

Proof of Lemma 5.7.11 Part (2). Recall that each $\kappa_{Ix}(y)$ is supported, with respect to x, in the ball $B(x_i,2)$. We compute easily that the kernel of $T_I^*T_{I'}$ is

$$\kappa_{I*I'}(x,w) = \int_G \kappa_{I'xz^{-1}}(wz^{-1})\kappa^*_{Ixz^{-1}}(z)dz.$$

Therefore, $\kappa_{I*I'}$ is identically 0 if there is no z such that $xz^{-1}\in B(x_i,2)\cap B(x_{i'},2)$. So if $|x_{i'}^{-1}x_i| > 4C_o$ (which implies $B(x_i,2)\cap B(x_{i'},2)=\emptyset$) then $T_I^*T_{I'}=0$. So we may assume $|x_{i'}^{-1}x_i|\le 4C_o$.

The idea of the proof is to use a quantity which will help the frequency localisation; so we introduce this quantity $(\mathrm{I}+\mathcal{R})^r$ and its inverse, where the integer $r\in\mathbb{N}$ is to be chosen suitably. We can write

$$T_I^*T_{I'} = T_I^*T_{i'}\psi_{j'}(\mathcal{R}) = T_I^*T_{i'}(\mathrm{I}+\mathcal{R})^r\ (\mathrm{I}+\mathcal{R})^{-r}\psi_{j'}(\mathcal{R}).$$

By the functional calculus (see Corollary 4.1.16),

$$\|(\mathrm{I}+\mathcal{R})^{-r}\psi_{j'}(\mathcal{R})\|_{\mathscr{L}(L^2(G))} = \sup_{\lambda\ge0}(1+\lambda)^{-r}\psi_{j'}(\lambda)\le C_r 2^{-j'r}.$$

Thus we need to study $T_I^* T_{i'}(\mathrm{I}+\mathcal{R})^r$. We see that its kernel is

$$\begin{aligned}\kappa_x(w) &= \int_G (\mathrm{I}+\tilde{\mathcal{R}})^r \kappa_{i' xz^{-1}}(wz^{-1}) \kappa^*_{I xz^{-1}}(z) dz \\ &= \int_G (\mathrm{I}+\tilde{\mathcal{R}})^r \kappa_{i' xw^{-1}z}(z) \kappa^*_{I xw^{-1}z}(z^{-1}w) dz.\end{aligned}$$

We introduce $(\mathrm{I}+\mathcal{R})^r(\mathrm{I}+\mathcal{R})^{-r}$ on the first term of the integrand acting on the variable of $\kappa_{i'xw^{-1}z}$, and then integrate by parts to obtain

$$\begin{aligned}\kappa_x(w) &= \overline{\sum_{[\beta_1]+[\beta_2]+[\beta_3]=r\nu}} \int_G X^{\beta_1}_{z_1=z} (\mathrm{I}+\mathcal{R})^{-r} (\mathrm{I}+\tilde{\mathcal{R}})^r \kappa_{i' xw^{-1}z_1}(z) \\ &\qquad X^{\beta_2}_{z_2=z} X^{\beta_3}_{z_3=z} \kappa^*_{I xw^{-1}z_2}(z_3^{-1}w) dz \\ &= \overline{\sum_{[\beta_1]+[\beta_2]+[\beta_3]=r\nu}} \int_G X^{\beta_1}_{z_1=xw^{-1}z} (\mathrm{I}+\mathcal{R})^{-r} (\mathrm{I}+\tilde{\mathcal{R}})^r \kappa_{i' z_1}(z) \\ &\qquad X^{\beta_2}_{z_2=xw^{-1}z} (X^{\beta_3}\kappa_{I z_2})^*(z^{-1}w) dz.\end{aligned}$$

Re-interpreting this in terms of operators, we obtain

$$\begin{aligned}T_I^* T_{i'}(\mathrm{I}+\mathcal{R})^r = \overline{\sum_{[\beta_1]+[\beta_2]+[\beta_3]=r\nu}} &\mathrm{Op}\left(\pi(X^{\beta_3}) X_x^{\beta_2} \sigma_I(x,\pi)\right)^* \\ &\mathrm{Op}\left(\pi(\mathrm{I}+\mathcal{R})^{-r} X_x^{\beta_1} \sigma_{i'}(x,\pi) \pi(\mathrm{I}+\mathcal{R})^r\right).\end{aligned}$$

By Lemma 5.7.9,

$$\begin{aligned}&\|\mathrm{Op}\left(\pi(\mathrm{I}+\mathcal{R})^{-r} X_x^{\beta_1} \sigma_{i'}(x,\pi)\pi(\mathrm{I}+\mathcal{R})^r\right)\|_{\mathscr{L}(L^2(G))} \\ &\quad \le C \sup_{\substack{x\in G \\ [\beta]\le \lceil \frac{Q}{2}\rceil}} \|\pi(\mathrm{I}+\mathcal{R})^{-r} X_x^{\beta} X_x^{\beta_1} \sigma_{i'}(x,\pi)\pi(\mathrm{I}+\mathcal{R})^r\|_{L^\infty(\widehat{G})} \\ &\quad \le \|\sigma\|_{S^0_{0,0},0,[\beta_1]+\lceil \frac{Q}{2}\rceil, r\nu},\end{aligned}$$

and

$$\begin{aligned}&\|\mathrm{Op}\left(\pi(X^{\beta_3}) X_x^{\beta_2} \sigma_I(x,\pi)\right)\|_{\mathscr{L}(L^2(G))} \\ &\quad \le \sup_{[\beta]\le \lceil \frac{Q}{2}\rceil} \|\pi(X^{\beta_3}) X_x^{\beta} X_x^{\beta_2} \sigma_i(x,\pi) \psi_j(\pi(\mathcal{R}))\|_{L^\infty(\widehat{G})} \\ &\quad \le \sup_{[\beta]\le \lceil \frac{Q}{2}\rceil} \|\pi(X^{\beta_3})\pi(\mathrm{I}+\mathcal{R})^{-\frac{[\beta_3]}{\nu}}\|_{L^\infty(\widehat{G})} \times \\ &\qquad \times \|\pi(\mathrm{I}+\mathcal{R})^{\frac{[\beta_3]}{\nu}} X_x^{\beta+\beta_2} \sigma_i(x,\pi)\pi(\mathrm{I}+\mathcal{R})^{-\frac{[\beta_3]}{\nu}}\|_{L^\infty(\widehat{G})} \times \\ &\qquad \times \|\pi(\mathrm{I}+\mathcal{R})^{\frac{[\beta_3]}{\nu}} \psi_j(\pi(\mathcal{R}))\|_{L^\infty(\widehat{G})} \\ &\quad \le C_\beta 2^{j\frac{[\beta_3]}{\nu}} \|\sigma\|_{S^0_{0,0},0,[\beta_2]+\lceil \frac{Q}{2}\rceil,[\beta_3]},\end{aligned}$$

by Lemma 5.4.7. Hence we have obtained

$$\begin{aligned}\|T_I^* T_{I'}\|_{\mathscr{L}(L^2(G))} &\leq C_r 2^{-j'r} \sum_{[\beta_1]+[\beta_2]+[\beta_3]=r\nu} \|\sigma\|^2_{S^0_{0,0},0,r\nu+\lceil\frac{Q}{2}\rceil,r\nu} 2^{j\frac{[\beta_3]}{\nu}} \\ &\leq C_r 2^{(j-j')r} \|\sigma\|^2_{S^0_{0,0},0,r\nu+\lceil\frac{Q}{2}\rceil,r\nu}.\end{aligned}$$

This shows Part 2 of Lemma 5.7.11 up to the fact that we should have $-|j-j'|$ instead of $(j-j')$ but this can be deduced easily by reversing the rôle of I and I', and using $\|T\|_{\mathscr{L}(L^2(G))} = \|T^*\|_{\mathscr{L}(L^2(G))}$. □

This concludes the proof of Lemma 5.7.11. Therefore, by Lemma 5.7.10, Proposition 5.7.7 is also proved.

5.7.3 A bilinear estimate

In this section, we prove a bilinear estimate which will be the major ingredient in the proof of the L^2-boundedness for operators of orders 0 in the case $\rho = \delta \in (0,1)$ in Section 5.7.4.

Note that if $f, g \in \mathcal{S}(G)$ and if $\gamma \in \mathbb{N}_0$ then the Leibniz properties together with the properties of the Sobolev spaces (cf. Theorem 4.4.28, especially the Sobolev embeddings in Part (5)) imply

$$\begin{aligned}\|(\mathrm{I}+\mathcal{R})^\gamma (fg)\|_{L^2(G)} &\lesssim \sum_{[\beta_1]+[\beta_2]\leq\nu\gamma} \|X^{\alpha_1} f\, X^{\alpha_2} g\|_{L^2(G)} \\ &\lesssim \sum_{[\beta_1]+[\beta_2]\leq\nu\gamma} \|X^{\alpha_1} f\|_{L^\infty(G)} \|X^{\alpha_2} g\|_{L^2(G)} \\ &\lesssim \sum_{[\beta_1]+[\beta_2]\leq\nu\gamma} \|X^{\alpha_1} f\|_{H^s(G)} \|X^{\alpha_2} g\|_{L^2(G)} \\ &\lesssim \|f\|_{H^{s+\nu\gamma}(G)} \|g\|_{H^{\nu\gamma}(G)},\end{aligned}$$

where $s > Q/2$. As usual, $\mathcal{R}$ is a positive Rockland operator of homogeneous degree ν; we denote by E its spectral decomposition, see Corollary 4.1.16. Consequently, if f, g are localised in the spectrum of $\mathcal{R}$ in the sense that $f = E(I_i)f$, $g = E(I_j)g$, where I_i, I_j are the dyadic intervals given via

$$I_j := (2^{j-2}, 2^j), \quad j \in \mathbb{N}, \quad \text{and} \quad I_0 := [0,1), \tag{5.72}$$

we obtain easily

$$\|(\mathrm{I}+\mathcal{R})^\gamma (fg)\|_{L^2(G)} \lesssim \|f\|_{L^2(G)} \|g\|_{L^2(G)} 2^{(\gamma+\frac{s}{\nu})\max(i,j)}. \tag{5.73}$$

Our aim in this section is to prove a similar result but for $\gamma \ll 0$:

Proposition 5.7.12. *Let $\mathcal{R}$ be a positive Rockland operator of homogeneous degree ν. As usual, we denote by E its spectral decomposition. There exists a constant $C > 0$ such that for any $\gamma \in \mathbb{R}$ with $\gamma + Q/(2\nu) < 0$, for any $i, j \in \mathbb{N}_0$ with $|i - j| > 3$, we have*

$$\forall f, g \in L^2(G) \qquad f = E(I_i)f \quad \text{and} \quad g = E(I_j)g$$
$$\Longrightarrow \|(\mathrm{I} + \mathcal{R})^\gamma (fg)\|_{L^2(G)} \leq C \|f\|_{L^2} \|g\|_{L^2} 2^{(\gamma + \frac{Q}{2\nu}) \max(i,j)}.$$

The intervals I_i, I_j were defined via (5.72). The proof of Proposition 5.7.12 relies on the following lemma:

Lemma 5.7.13. *Let $\mathcal{R}$ be a positive Rockland operator. As in Corollary 4.1.16, for any strongly continuous unitary representation π_1 on G, E_{π_1} denotes the spectral decomposition of $\pi_1(\mathcal{R})$. There exists a 'gap' constant $a \in \mathbb{N}$ such that for any $i, j, k \in \mathbb{N}_0$ with $k < j - a$ and $i \leq j - 4$, we have*

$$\forall \tau, \pi \in \widehat{G} \qquad E_{\tau \otimes \pi}(I_i)\big(E_\tau(I_j) \otimes E_\pi(I_k)\big) = 0.$$

and

$$\forall \tau, \pi \in \widehat{G} \qquad \big(E_\tau(I_j) \otimes E_\pi(I_k)\big) E_{\tau \otimes \pi}(I_i) = 0.$$

Proof of Lemma 5.7.13. We keep the notation of the statement. We also set

$$\mathcal{H}_{\pi_1, j} := E_{\pi_1}(I_j), \qquad j \in \mathbb{N}_0,$$

for any strongly continuous unitary representation π_1 on G. We can write $\mathcal{R}$ as a linear combination

$$\mathcal{R} = \sum_{[\alpha] = \nu} c_\alpha X^\alpha,$$

for some complex coefficients c_α. For any strongly continuous unitary representation π_1, we have

$$\pi_1(\mathcal{R}) = \sum_{[\alpha] = \nu} c_\alpha \pi_1(X)^\alpha.$$

Let $\tau, \pi \in \widehat{G}$. We consider the strongly continuous unitary representation $\pi_1 = \tau \otimes \pi$. For any $X \in \mathfrak{g}$, its infinitesimal representation is given via $\pi_1(X) = X_{x=0}\{\pi_1(x)\}$, see Section 1.7. Consequently, we have for any $u \in \mathcal{H}_\tau, v \in \mathcal{H}_\pi$,

$$\begin{aligned}
\pi_1(X)(u, v) &= X_{x=0} \pi_1(x)(u, v) \\
&= X_{x=0} \tau(x) u \otimes \pi(x) v \\
&= \tau(X) u \otimes v + u \otimes \pi(X) v.
\end{aligned}$$

In other words,

$$(\tau \otimes \pi)(X) = \tau(X) \otimes \mathrm{I}_{\mathcal{H}_\pi} + \mathrm{I}_{\mathcal{H}_\tau} \otimes \pi(X).$$

We obtain iteratively

$$(\tau\otimes\pi)(X)^\alpha = \tau(X)^\alpha\otimes \mathrm{I}_{\mathcal{H}_\pi} + \mathrm{I}_{\mathcal{H}_\tau}\otimes\pi(X)^\alpha + \overline{\sum_{\substack{[\beta_1]+[\beta_2]=[\alpha]\\ 0<[\beta_1],[\beta_2]<[\alpha]}}} \tau(X)^{\beta_1}\otimes\pi(X)^{\beta_2},$$

where $\overline{\sum}$ denotes a linear combination which depends only on $\alpha\in\mathbb{N}_0^n$ and on the structure of G but not on $\tau,\pi\in\widehat{G}$. This easily implies

$$\begin{aligned}(\tau\otimes\pi)(\mathcal{R}) &= \sum_{[\alpha]=\nu} c_\alpha(\tau\otimes\pi)(X)^\alpha\\ &= \tau(\mathcal{R})\otimes \mathrm{I}_{\mathcal{H}_\pi} + \mathrm{I}_{\mathcal{H}_\tau}\otimes\pi(\mathcal{R}) + \overline{\sum_{\substack{[\beta_1]+[\beta_2]=\nu\\ 0<[\beta_1],[\beta_2]<\nu}}} \tau(X)^{\beta_1}\otimes\pi(X)^{\beta_2},\end{aligned}$$

where $\overline{\sum}$ denotes a linear combination which depends only on $\mathcal{R}$ and on the structure of G but not on π,τ. Hence there exists a constant $C>0$ independent of π,τ such that for any $u\in\mathcal{H}_\tau$, $v\in\mathcal{H}_\pi$, we have

$$\begin{aligned}\|(\tau\otimes\pi)(\mathcal{R})(u\otimes v)\|_{\mathcal{H}_{\tau\otimes\pi}} \geq\ & \|\tau(\mathcal{R})u\|_{\mathcal{H}_\tau}\|v\|_{\mathcal{H}_\pi} - \|u\|_{\mathcal{H}_\tau}\|\pi(\mathcal{R})v\|_{\mathcal{H}_\pi}\\ & -C\sum_{\substack{[\beta_1]+[\beta_2]=\nu\\ 0<[\beta_1],[\beta_2]<\nu}} \|\tau(X)^{\beta_1}u\|_{\mathcal{H}_\tau}\|\pi(X)^{\beta_2}v\|_{\mathcal{H}_\pi}.\end{aligned}$$

If $u\in\mathcal{H}_{\tau,j}$ then from the properties of the functional calculus of $\tau(\mathcal{R})$, we have

$$\|\tau(\mathcal{R})u\|_{\mathcal{H}_\tau}\in\|u\|_{\mathcal{H}_\tau}I_j.$$

Furthermore, the properties of the functional calculus of $\mathcal{R}$ and $\tau(\mathcal{R})$ yield

$$\|\tau(X)^{\beta_1}u\|_{\mathcal{H}_\tau}\leq\|\tau(X)^{\beta_1}E_\tau(I_j)\|_{\mathscr{L}(\mathcal{H}_\tau)}\|u\|_{\mathcal{H}_\tau},$$

and, as $X^{\beta_1}\mathcal{R}^{-\frac{[\beta_1]}{\nu}}$ is bounded on $L^2(G)$ by Theorem 4.4.16, we have

$$\begin{aligned}\|\tau(X)^{\beta_1}E_\tau(I_j)\|_{\mathscr{L}(\mathcal{H}_\tau)} &\leq \|X^{\beta_1}E(I_j)\|_{\mathscr{L}(L^2(G))}\\ &\leq \|X^{\beta_1}\mathcal{R}^{-\frac{[\beta_1]}{\nu}}\|_{\mathscr{L}(L^2(G))}\|\mathcal{R}^{\frac{[\beta_1]}{\nu}}E(I_j)\|_{\mathscr{L}(L^2(G))}\\ &\lesssim 2^{j\frac{[\beta_1]}{\nu}}.\end{aligned}$$

We have similar inequalities for $v\in\mathcal{H}_{\pi,k}$. For any unit vectors $u\in\mathcal{H}_{\tau,j}$ and $v\in\mathcal{H}_{\pi,k}$ with $j,k\in\mathbb{N}$, we then have

$$\|(\tau\otimes\pi)(\mathcal{R})(u\otimes v)\|_{\mathcal{H}_{\tau\otimes\pi}}\geq 2^{j-2}-2^k-C_1\sum_{\substack{[\beta_1]+[\beta_2]=\nu\\ 0<[\beta_1],[\beta_2]<\nu}} 2^{\frac{j[\beta_1]+k[\beta_2]}{\nu}},$$

where the constant C_1 depends only on $\mathcal{R}$ and on the structure of G. We notice that

$$\sum_{\substack{[\beta_1]+[\beta_2]=\nu \\ 0<[\beta_1],[\beta_2]<\nu}} 2^{\frac{j[\beta_1]+k[\beta_2]}{\nu}} = 2^j \sum_{\substack{[\beta_1]+[\beta_2]=\nu \\ 0<[\beta_1],[\beta_2]<\nu}} 2^{\frac{[\beta_2]}{\nu}(k-j)} \leq 2^j C' 2^{-a v_1},$$

if $k-j \leq -a$. Here C' is a constant which depends on the structure of G and on ν. We choose $a \in \mathbb{N}$ the smallest integer such that

$$CC' 2^{-a v_1 + 2} < 1/2 \quad \text{and} \quad 2^{-a+3} < 1/2.$$

Note that a depends only on the structure of G and on $\mathcal{R}$. When $k-j \leq -a$, we have obtained

$$\begin{aligned} \|(\tau \otimes \pi)(\mathcal{R})(u \otimes v)\|_{\mathcal{H}_{\tau\otimes\pi}} &\geq 2^{j-2} - 2^k - C 2^j C' 2^{-a v_1} \\ &= 2^{j-2}(1 - CC' 2^{-a v_1 + 2}) - 2^k \\ &> 2^{j-3} - 2^{j-a} > 2^{j-4}. \end{aligned}$$

This implies that $u \otimes v$ can not be in $\mathcal{H}_{\tau\otimes\mathcal{R},\pi}$ for $i \in \mathbb{N}_0$ such that $2^i \leq 2^{j-4}$. This shows the first equality of the statement when $i, j, k \in \mathbb{N}$. The case of $k = 0$ or $i = 0$ requires to modify slightly some constants above and is left to the reader. This shows the first equality of the statement and the second follows by taking the adjoint. This concludes the proof of Lemma 5.7.13. □

Proof of Proposition 5.7.12. We keep the notation of Proposition 5.7.12 and Lemma 5.7.13. We notice that it suffices to prove the statement for large enough $\max(i, j)$ and that the rôles of i and j are symmetric. Hence we may assume that $i \leq j - 4$ and that $j \geq a$ where a is the 'gap' constant of Lemma 5.7.13

Let $f, g \in L^2(G)$ such that $f = E(I_i)f$ and $g = E(I_j)g$. The inverse formula for g yields

$$(\mathrm{I} + \mathcal{R})^\gamma (fg)(x) = \int_{\widehat{G}} \mathrm{Tr}\big(\pi(g)(\mathrm{I} + \mathcal{R})_x^\gamma \{f(x)\pi(x)\}\big) d\mu(\pi).$$

We also have $\pi(g) = E_\pi(I_j)\pi(g)$. By the Cauchy-Schwartz inequality and the Plancherel formula, we obtain

$$|(\mathrm{I} + \mathcal{R})^\gamma (fg)(x)|^2 \leq \|g\|_{L^2(G)}^2 \int_{\widehat{G}} \|E_\pi(I_j)(\mathrm{I} + \mathcal{R})_x^\gamma \{f(x)\pi(x)\}\|_{\mathtt{HS}}^2 d\mu(\pi).$$

Integrating on both side over $x \in G$, we have

$$\|(\mathrm{I} + \mathcal{R})^\gamma (fg)\|_{L^2(G)}^2 \leq \|g\|_{L^2}^2 \int_{\widehat{G}} \int_G \|E_\pi(I_j)(\mathrm{I} + \mathcal{R})_x^\gamma \{f(x)\pi(x)\}\|_{\mathtt{HS}}^2 dx d\mu(\pi).$$

For each $\pi \in \widehat{G}$, we fix an orthonormal basis of $\mathcal{H}_\pi$, so that we can write the Hilbert-Schmidt norm as the square of the coefficients of a (possibly infinite dimensional) matrix. The Plancherel formula then yields

$$\begin{aligned}
&\int_G \|E_\pi(I_j)(\mathrm{I}+\mathcal{R})_x^\gamma\{f(x)\pi(x)\}\|_{\mathtt{HS}}^2 dx \\
&\quad = \sum_{kl} \int_G |[E_\pi(I_j)(\mathrm{I}+\mathcal{R})_x^\gamma\{f(x)\pi(x)\}]_{kl}|^2 dx \\
&\quad = \sum_{kl} \int_{\widehat{G}} \|\mathcal{F}\left[E_\pi(I_j)(\mathrm{I}+\mathcal{R})^\gamma f\pi\right]_{kl}(\tau)\|_{\mathtt{HS}(\mathcal{H}_\tau)}^2 d\mu(\tau),
\end{aligned}$$

where

$$\begin{aligned}
&\mathcal{F}\left[E_\pi(I_j)(\mathrm{I}+\mathcal{R})^\gamma f\pi\right]_{kl}(\tau) \\
&\quad = \int_G (\mathrm{I}+\mathcal{R})_x^\gamma \{f(x)[E_\pi(I_j)\pi(x)]_{kl}\}\ \tau(x)^* dx \\
&\quad = \tau(\mathrm{I}+\mathcal{R})^\gamma \int_G f(x)[E_\pi(I_j)\pi(x)]_{kl}\ \tau(x)^* dx \\
&\quad = \left[E_\pi(I_j) \otimes \tau(\mathrm{I}+\mathcal{R})^\gamma \int_G f(x)(\pi\otimes\tau^*)(x)dx\right]_{kl,\cdot}.
\end{aligned}$$

Here the notation $[\cdot]_{kl,\cdot}$ means considering the (kl)-coefficients in $\mathcal{H}_\pi$ in the tensor product over $\mathcal{H}_\pi \otimes \mathcal{H}_\tau$. We recognise

$$\int_G f(x)(\pi\otimes\tau^*)(x)dx = (\pi^*\otimes\tau)(f)$$

thus

$$\begin{aligned}
&\sum_{kl} \|\mathcal{F}[E_\pi(I_j)(\mathrm{I}+\mathcal{R})^\gamma f\pi]_{kl}(\tau)\|_{\mathtt{HS}(\mathcal{H}_\tau)}^2 \\
&\qquad = \|\left(E_\pi(I_j)\otimes\tau(\mathrm{I}+\mathcal{R})^\gamma\right)\left((\pi^*\otimes\tau)(f)\right)\|_{\mathtt{HS}(\mathcal{H}_\pi\otimes\mathcal{H}_\tau)}^2.
\end{aligned}$$

So far, we have obtained

$$\begin{aligned}
&\int_{\widehat{G}}\int_G \|E_\pi(I_j)(\mathrm{I}+\mathcal{R})_x^\gamma\{f(x)\pi(x)\}\|_{\mathtt{HS}}^2 dx d\mu(\pi) \\
&\quad = \int_{\widehat{G}}\int_{\widehat{G}} \|\left(E_\pi(I_j)\otimes\tau(\mathrm{I}+\mathcal{R})^\gamma\right)\left((\pi^*\otimes\tau)(f)\right)\|_{\mathtt{HS}(\mathcal{H}_\pi\otimes\mathcal{H}_\tau)}^2 d\mu(\tau)d\mu(\pi) \\
&\quad = \|\|\left(E_\pi(I_j)\otimes\tau(\mathrm{I}+\mathcal{R})^\gamma\right)\left((\pi^*\otimes\tau)(f)\right)\|_{\mathtt{HS}(\mathcal{H}_\pi\otimes\mathcal{H}_\tau)}\|_{L^2(d\mu(\tau),d\mu(\pi))}^2.
\end{aligned}$$

We fix a dyadic decomposition, that is, we fix $\psi_0, \psi_1 \in \mathcal{D}(\mathbb{R})$ supported in $(-1,1)$ and $(1/2,2)$, respectively, valued in $[0,1]$ and such that

$$\forall\lambda\geq 0 \quad \sum_{k=0}^{\infty}\psi_k(\lambda) = 1 \quad \text{with } \psi_k(\lambda) = \psi_1(2^{-(k-1)}\lambda) \text{ if } k\in\mathbb{N}.$$

The series $\sum_k \psi_k(\tau(\mathcal{R}))$ converges to $\mathrm{I}_{\mathcal{H}_\tau}$ in the strong operator topology and we can apply the following general property:

$$\begin{aligned}\|(B\otimes C)A\|_{\mathtt{HS}(\mathcal{H}_\pi\otimes\mathcal{H}_\tau)}\\ \leq \sum_{k=0}^{\infty}\|E_\tau(I_k)C\|_{\mathscr{L}(\mathcal{H}_\tau)}\|(B\otimes\psi_k(\tau(\mathcal{R})))A\|_{\mathtt{HS}(\mathcal{H}_\pi\otimes\mathcal{H}_\tau)},\end{aligned}$$

to $B=E_\pi(I_j)$, $C=\tau(\mathrm{I}+\mathcal{R})^\gamma$, and

$$A=(\pi^*\otimes\tau)(f).$$

We keep momentarily this notation for A and C. As $\|E_\tau(I_k)C\|_{\mathscr{L}(\mathcal{H}_\tau)}\lesssim 2^{\gamma k}$, we have obtained

$$\begin{aligned}\|\|\left(E_\pi(I_j)\otimes\tau(\mathrm{I}+\mathcal{R})^\gamma\right)A\|_{\mathtt{HS}(\mathcal{H}_\pi\otimes\mathcal{H}_\tau)}\|_{L^2(d\mu(\tau),d\mu(\pi))}\\ \lesssim\sum_{k=0}^{\infty}2^{\gamma k}\|\|\left(E_\pi(I_j)\otimes\psi_k(\tau(\mathcal{R}))\right)A\|_{\mathtt{HS}(\mathcal{H}_\pi\otimes\mathcal{H}_\tau)}\|_{L^2(d\mu(\tau),d\mu(\pi))}.\end{aligned}$$

Now

$$A=((\pi^*\otimes\tau)(f))=E_{\pi^*\otimes\tau}(I_i)\left((\pi^*\otimes\tau)(f)\right),$$

thus we can apply Lemma 5.7.13 and the sum over k above is in fact from $k\geq j-a$. We claim that

$$\|\|\left(E_\pi(I_j)\otimes\psi_k(\tau(\mathcal{R}))\right)A\|_{\mathtt{HS}(\mathcal{H}_\pi\otimes\mathcal{H}_\tau)}\|_{L^2(d\mu(\tau),d\mu(\pi))}\lesssim\|f\|_{L^2(G)}2^{k\frac{Q}{2\nu}}. \tag{5.74}$$

Collecting the equalities and estimates above, (5.74) would then imply

$$\|(\mathrm{I}+\mathcal{R})^\gamma(fg)\|^2_{L^2(G)}\lesssim\|g\|^2_{L^2}\|f\|^2_{L^2(G)}\sum_{k=j-a}^{\infty}2^{k(\gamma+\frac{Q}{2\nu})},$$

and would conclude the proof of Proposition 5.7.12.

Hence it just remains to prove (5.74). Natural properties of tensor product and functional calculus yield

$$\begin{aligned}\|\left(E_\pi(I_j)\otimes\psi_k(\tau(\mathcal{R}))\right)A\|_{\mathtt{HS}(\mathcal{H}_\pi\otimes\mathcal{H}_\tau)}\\ \leq\|E_\pi(I_j)\|_{\mathscr{L}(\mathcal{H}_\pi)}\|\left(\mathrm{I}_{\mathcal{H}_\pi}\otimes\psi_k(\tau(\mathcal{R}))\right)A\|_{\mathtt{HS}(\mathcal{H}_\pi\otimes\mathcal{H}_\tau)}\\ \leq\|\left(\mathrm{I}_{\mathcal{H}_\pi}\otimes\psi_k(\tau(\mathcal{R}))\right)A\|_{\mathtt{HS}(\mathcal{H}_\pi\otimes\mathcal{H}_\tau)}.\end{aligned}$$

We notice that

$$\left(\mathrm{I}_{\mathcal{H}_\pi}\otimes\psi_k(\tau(\mathcal{R}))\right)A=\int_G f(x)(\pi\otimes\psi_k(\tau(\mathcal{R}))\tau^*)(x)dx,$$

and introducing an orthonormal basis on $\mathcal{H}_\tau$,

$$[(\mathrm{I}_{\mathcal{H}_\pi} \otimes \psi_k(\tau(\mathcal{R}))) A]_{\cdot,l'k'} = \int_G f(x) \left[\psi_k(\tau(\mathcal{R}))\right]_{l'k'} \pi(x) dx$$
$$= \mathcal{F}\left[f\psi_k(\tau(\mathcal{R}))\right]_{l'k'} (\pi^*) = \mathcal{F}\{\left[f\psi_k(\tau(\mathcal{R}))\right]_{l'k'} (\cdot^{-1})\}(\pi).$$

Therefore we have

$$\|\|\, (\mathrm{I}_{\mathcal{H}_\pi} \otimes \psi_k(\tau(\mathcal{R}))) A\|_{\mathrm{HS}(\mathcal{H}_\pi\otimes\mathcal{H}_\tau)}\|^2_{L^2(d\mu(\tau),d\mu(\pi))}$$
$$= \int_{\widehat{G}} \sum_{k'l'} \int_{\widehat{G}} \|\mathcal{F}\left[f\psi_k(\tau(\mathcal{R}))\right]_{l'k'} (\pi^*)\|^2_{\mathrm{HS}(\mathcal{H}_\pi)} d\mu(\pi) d\mu(\tau)$$
$$= \int_{\widehat{G}} \sum_{k'l'} \|\left[f\psi_k(\tau(\mathcal{R}))\right]_{l'k'} (\cdot^{-1})\|^2_{L^2(G)}\, d\mu(\tau),$$

having applied the Plancherel formula in π. Simple manipulations yield

$$\sum_{k'l'} \|\left[f\psi_k(\tau(\mathcal{R}))\right]_{l'k'} (\cdot^{-1})\|^2_{L^2(G)} = \sum_{k'l'} \|\left[f\psi_k(\tau(\mathcal{R}))\right]_{l'k'} \|^2_{L^2(G)}$$
$$= \sum_{k'l'} \int_G |f(x) \left[\psi_k(\tau(\mathcal{R}))\right]_{l'k'} |^2 dx$$
$$= \int_G |f(x)|^2 dx \sum_{k'l'} | \left[\psi_k(\tau(\mathcal{R}))\right]_{l'k'} |^2$$
$$= \|f\|^2_{L^2(G)} \|\psi_k(\tau(\mathcal{R}))\|^2_{\mathrm{HS}(\mathcal{H}_\tau)}.$$

Integrating over $\tau \in \widehat{G}$, we can apply the Plancherel formula and obtain

$$\int_{\widehat{G}} \sum_{k'l'} \|\left[f\psi_k(\tau(\mathcal{R}))\right]_{l'k'} (\cdot^{-1})\|^2_{L^2(G)}\, d\mu(\tau) = \|f\|^2_{L^2(G)} \|\psi_k(\mathcal{R})\delta_0\|^2_{L^2(G)}.$$

Using the properties of dilations, we have for any $k \in \mathbb{N}$:

$$\|\psi_k(\mathcal{R})\delta_0\|_{L^2(G)} = 2^{\frac{Q}{2}\frac{k-1}{\nu}} \|\psi_1(\mathcal{R})\delta_0\|_{L^2(G)}.$$

Collecting the equalities and inequalities above yields that the left-hand side of (5.74) is

$$\|\|\, (E_\pi(I_j) \otimes \psi_k(\tau(\mathcal{R}))) A\|_{\mathrm{HS}(\mathcal{H}_\pi\otimes\mathcal{H}_\tau)}\|_{L^2(d\mu(\tau),d\mu(\pi))}$$
$$\leq \|f\|_{L^2(G)} 2^{\frac{Q}{2}\frac{k-1}{\nu}} \|\psi_1(\mathcal{R})\delta_0\|_{L^2(G)}.$$

By Hulanicki's theorem, see Corollary 4.5.2, $\|\psi_1(\mathcal{R})\delta_0\|_{L^2(G)}$ is a finite constant. This shows (5.74) and concludes the proof of Proposition 5.7.12. □

5.7.4 Proof of the case $S^0_{\rho,\rho}$

In this section, we prove the L^2-boundedness of operators in $\Psi^0_{\rho,\rho}$ with $\rho \in (0,1)$:

Proposition 5.7.14. *Let $\sigma \in S^0_{\rho,\rho}$ with $\rho \in (0,1)$. Then $\text{Op}(\sigma)$ is bounded on $L^2(G)$ and the operator norm is, up to a constant, less than a seminorm of $\sigma \in S^0_{\rho,\rho}$; the parameters of the seminorm depend on ρ but not on σ and could be computed explicitly.*

The rest of this section is devoted to the proof of Proposition 5.7.14. The strategy is broadly similar to the one in [Ste93, ch VII §2.5] for the Euclidean case. Technically, this means using analogous rescaling arguments but also replacing certain integrations by parts on the (Euclidean) Fourier side with the bilinear estimate obtained in Proposition 5.7.12.

Strategy of the proof

We fix a dyadic decomposition, that is, we fix $\psi_0, \psi_1 \in \mathcal{D}(\mathbb{R})$ supported in $(-1,1)$ and $(1/2,2)$, respectively, valued in $[0,1]$ and such that

$$\forall \lambda \geq 0 \quad \sum_{j=0}^{\infty} \psi_j(\lambda) = 1 \quad \text{with } \psi_j(\lambda) = \psi_1(2^{-(j-1)}\lambda) \text{ if } j \in \mathbb{N}.$$

Let $\sigma \in S^0_{\rho,\rho}$. We define

$$\sigma_j(x,\pi) := \sigma(x,\pi)\psi_j(\pi(\mathcal{R})) \quad \text{and} \quad T_j := \text{Op}(\sigma_j) = T\psi_j(\mathcal{R}),$$

where $T = \text{Op}(\sigma)$.

It is clear that $T_j T_i^* = T(\psi_j\psi_i)(\mathcal{R})T^*$ is zero if $|j-i| > 1$ and the strategy of the proof is to apply the crude version of the Cotlar-Stein Lemma, see Proposition A.5.3. We will first prove that the operator norms of the T_j's are uniformly bounded in j by a $S^0_{\rho,\rho}$-seminorm, see Lemma 5.7.15. Then we will show that there exist a constant $C > 0$ and a $S^0_{\rho,\rho}$-seminorm such that

$$\sum_{|i-j|>3} \|T_j^* T_i\|_{\mathscr{L}(L^2(G))} \leq C\|\sigma\|^2_{S^0_{\rho,\rho},a,b,c}. \tag{5.75}$$

These two claims together with Proposition A.5.3 and Remark A.5.4 imply that the series $\sum_j T_j \in \mathscr{L}(L^2(G))$ converges in the strong operator topology of $\mathscr{L}(L^2(G))$ and that the operator norm of the sum is $\lesssim \|\sigma\|_{S^0_{\rho,\rho},a,b,c}$. As $\text{Op}(\sigma) = \sum_j T_j$ in the strong operator topology, this will conclude the proof of Proposition 5.7.14.

Step 1

Let us show that the operator norms of the T_j's are uniformly bounded with respect to j:

Lemma 5.7.15. *The operator* $T_j = \mathrm{Op}(\sigma_j)$ *is bounded on* $L^2(G)$ *with operator norm* $\leq C\|\sigma\|_{S^0_{\rho,\rho},a,b,c}$ *with* a, b, c *as in Proposition 5.7.7.*

The proof of Lemma 5.7.15 uses the following result which is of interest on its own. In particular, it describes the action of the dilations on $\widehat{G}$.

Lemma 5.7.16. *Let* σ *be a symbol with kernel* κ_x *and operator* $T = \mathrm{Op}(\sigma)$. *Let* $r > 0$. *We define the operator*

$$T_r : \mathcal{S}(G) \ni \phi \longmapsto (T\phi(r\,\cdot))\,(r^{-1}\cdot).$$

Then (with operator norm possibly infinite)

$$\|T\|_{\mathscr{L}(L^2(G))} = \|T_r\|_{\mathscr{L}(L^2(G))}.$$

Furthermore, the symbol of T_r *is*

$$\sigma_r := \mathrm{Op}^{-1}(T_r) \quad \text{given by} \quad \sigma_r(x,\pi) := \sigma\left(r^{-1}x, \pi^{(r)}\right),$$

where the representation $\pi^{(r)}$ *is defined by*

$$\pi^{(r)}(y) := \pi(ry).$$

The kernel of σ_r *is* $r^{-Q}\kappa_{r^{-1}x}(r^{-1}\cdot)$. *Moreover, we have*

$$\begin{aligned}
\mathcal{F}_G(\kappa)(\pi^{(r)}) &= \mathcal{F}_G\left(r^{-Q}\kappa(r^{-1}\cdot)\right)(\pi),\\
\Delta^\alpha\left\{\mathcal{F}_G(\kappa)(\pi^{(r)})\right\} &= r^{[\alpha]}\left\{\Delta^\alpha\mathcal{F}_G(\kappa)\right\}(\pi^{(r)}),\\
f(\pi^{(r)}(\mathcal{R})) &= f(r^\nu\pi(\mathcal{R})),
\end{aligned}$$

for any $\alpha \in \mathbb{N}_0^n$, *any positive Rockland operator* $\mathcal{R}$ *of homogeneous degree* ν, *and any reasonable functions* f *and* κ *(for instance* f *measurable bounded and* κ *in some* $\mathcal{K}_{a,b}$*).*

Proof of Lemma 5.7.16. We keep the notation of the statement. The property $\|T\|_{\mathscr{L}(L^2(G))} = \|T_r\|_{\mathscr{L}(L^2(G))}$ follows easily from $\|\phi(r\cdot)\|_2 = r^{-Q/2}\|\phi\|_2$. We compute

$$\begin{aligned}
(T\phi(r\,\cdot))\,(r^{-1}x) &= \int_G \phi(ry)\;\kappa_{r^{-1}x}(y^{-1}r^{-1}x)dy\\
&= \int_G \phi(z)\;\kappa_{r^{-1}x}(r^{-1}z^{-1}r^{-1}x)r^{-Q}dz\\
&= \phi * \left(r^{-Q}\kappa_{r^{-1}x}(r^{-1}\cdot)\right)(x).
\end{aligned}$$

Therefore, the kernel of the operator T_r is $r^{-Q}\kappa_{r^{-1}x}(r^{-1}\cdot)$. The computation of its symbol follows from

$$\begin{aligned}
\mathcal{F}_G\left(r^{-Q}\kappa(r^{-1}\cdot)\right)(\pi) &= \int_G r^{-Q}\kappa(r^{-1}x)\pi(x)^*dx\\
&= \int_G \kappa(y)\pi(ry)^*dx \;=\; \mathcal{F}_G(\kappa)(\pi^{(r)}).
\end{aligned}$$

The difference operator applied to the above expression is

$$\begin{aligned}\Delta^\alpha \left\{\mathcal{F}_G(\kappa)(\pi^{(r)})\right\} &= \Delta^\alpha \left\{\mathcal{F}_G\left(r^{-Q}\kappa(r^{-1}\cdot)\right)(\pi)\right\} \\ &= \mathcal{F}_G\left(\tilde{q}_\alpha(\cdot)\, r^{-Q}\kappa(r^{-1}\cdot)\right)(\pi) \\ &= r^{[\alpha]}\left\{\mathcal{F}_G\left(r^{-Q}(\tilde{q}_\alpha\kappa)(r^{-1}\cdot)\right)(\pi)\right\} \\ &= r^{[\alpha]}\left\{\Delta^\alpha\mathcal{F}_G(\kappa)\right\}(\pi^{(r)}).\end{aligned}$$

The kernels of the operators $f(\mathcal{R})$ and $f(r^\nu\mathcal{R})$ are respectively $f(\mathcal{R})\delta_o$ and $r^{-Q}f(\mathcal{R})\delta_o(r^{-1}\cdot)$ (see (4.3) in Corollary 4.1.16, and Example 3.1.20 for the homogeneity of δ_o). Since the group Fourier transform of the former is $f(\pi(\mathcal{R}))$, the group Fourier transform of the latter is $f(r^\nu\pi(\mathcal{R})) = f(\pi^{(r)}(\mathcal{R}))$. □

We can now show Lemma 5.7.15 using the rescaling arguments (together with the lemma above) and the case $\rho = \delta = 0$.

Proof of Lemma 5.7.15. Using the Leibniz formula in Proposition 5.2.10, we first estimate

$$\begin{aligned}&\|\pi(\mathrm{I}+\mathcal{R})^{\frac{\gamma}{\nu}} X_x^{\beta_o}\Delta^{\alpha_o}\sigma_j(x,\pi)\pi(\mathrm{I}+\mathcal{R})^{-\frac{\gamma}{\nu}}\|_{\mathscr{L}(\mathcal{H}_\pi)} \\ &\leq C_{\alpha_o}\sum_{[\alpha_1]+[\alpha_2]=[\alpha_o]} \|\pi(\mathrm{I}+\mathcal{R})^{\frac{\gamma}{\nu}} X_x^{\beta_o}\Delta^{\alpha_1}\sigma(x,\pi)\pi(\mathrm{I}+\mathcal{R})^{\frac{\rho([\alpha_1]-[\beta_o])-\gamma}{\nu}}\|_{\mathscr{L}(\mathcal{H}_\pi)} \\ &qquad \quad \|\pi(\mathrm{I}+\mathcal{R})^{-\frac{\rho([\alpha_1]-[\beta_o])-\gamma}{\nu}}\Delta^{\alpha_2}\psi_j(\pi(\mathcal{R}))\pi(\mathrm{I}+\mathcal{R})^{-\frac{\gamma}{\nu}}\|_{\mathscr{L}(\mathcal{H}_\pi)} \\ &\leq C_{\alpha_o}\|\sigma\|_{S^0_{\rho,\rho},[\alpha_o],[\beta_o],|\gamma|}\sum_{[\alpha_1]+[\alpha_2]=[\alpha_o]} 2^{-j\frac{\nu}{\rho}\frac{[\alpha_2]+\rho([\alpha_1]-[\beta_o])}{\nu}} \\ &\leq C_{\alpha_o}\|\sigma\|_{S^0_{\rho,\rho},[\alpha_o],[\beta_o],|\gamma|}2^{-j([\alpha_o]-[\beta_o])},\end{aligned} \tag{5.76}$$

by Lemma 5.4.7.

For each $j\in\mathbb{N}_0$, we define the symbol σ_j' given by setting

$$\sigma_j'(x,\pi) := \sigma_j\left(2^{-j\rho}x, \pi^{(2^{j\rho})}\right).$$

By Lemma 5.7.16, the corresponding operator $T_j' := \mathrm{Op}(\sigma_j')$ satisfies

$$(T_j'\phi)(x) = \left(T_j\phi(2^{j\rho}\cdot)\right)(2^{-j\rho}x).$$

Lemma 5.7.16 and Proposition 5.7.7 imply that

$$\|T_j\|_{\mathscr{L}(L^2(G))} = \|T_j'\|_{\mathscr{L}(L^2(G))} \leq C\|\sigma_j'\|_{S^0_{0,0},a,b,c}, \tag{5.77}$$

with a,b,c as in Proposition 5.7.7. So we are led to compute $\|\sigma_j'\|_{S^0_{0,0},a,b,c}$. By Lemma 5.7.16, we have

$$\begin{aligned}&X_x^{\beta_o}\Delta^{\alpha_o}\sigma_j'(x,\pi) = 2^{-j\rho[\beta_o]}2^{j\rho[\alpha_o]}X_{x_o=2^{-j\rho}x}^{\beta_o}\Delta^{\alpha_o}_{\pi_o=\pi^{(2^{j\rho})}}\sigma_j(x_o,\pi_o) \\ &= 2^{j\rho([\alpha_o]-[\beta_o])}\pi(\mathrm{I}+2^{j\rho}\mathcal{R})^{-\frac{\gamma}{\nu}} \\ &\left(\pi_o(\mathrm{I}+\mathcal{R})^{\frac{\gamma}{\nu}}X_{x_o=2^{-j\rho}x}^{\beta_o}\Delta^{\alpha_o}\sigma_j(x_o,\pi_o)\pi_o(\mathrm{I}+\mathcal{R})^{-\frac{\gamma}{\nu}}\right)_{\pi_o=\pi^{(2^{j\rho})}}\pi(\mathrm{I}+2^{j\rho}\mathcal{R})^{\frac{\gamma}{\nu}},\end{aligned}$$

so that

$$\begin{aligned}
&\|\pi(\mathrm{I}+\mathcal{R})^{\frac{\gamma}{\nu}} X_x^{\beta_o}\Delta^{\alpha_o}\sigma_j'(x,\pi)\pi(\mathrm{I}+\mathcal{R})^{-\frac{\gamma}{\nu}}\|_{\mathscr{L}(\mathcal{H}_\pi)}\\
&\quad\le 2^{j\rho([\alpha_o]-[\beta_o])}\|\pi(\mathrm{I}+\mathcal{R})^{\frac{\gamma}{\nu}}\pi(\mathrm{I}+2^{j\rho}\mathcal{R})^{-\frac{\gamma}{\nu}}\|_{\mathscr{L}(\mathcal{H}_\pi)}\\
&\qquad\|\left(\pi_o(\mathrm{I}+\mathcal{R})^{\frac{\gamma}{\nu}} X_{x_o=2^{-j\rho}x}^{\beta_o}\Delta^{\alpha_o}\sigma_j(x_o,\pi_o)\pi_o(\mathrm{I}+\mathcal{R})^{-\frac{\gamma}{\nu}}\right)_{\pi_o=\pi^{(2^{j\rho})}}\|_{\mathscr{L}(\mathcal{H}_\pi)}\\
&\qquad\|\pi(\mathrm{I}+2^{j\rho}\mathcal{R})^{\frac{\gamma}{\nu}}\pi(\mathrm{I}+\mathcal{R})^{-\frac{\gamma}{\nu}}\|_{\mathscr{L}(\mathcal{H}_\pi)}.
\end{aligned}$$

By the functional calculus (Corollary 4.1.16),

$$\begin{aligned}
\|\pi(\mathrm{I}+\mathcal{R})^{\frac{\gamma}{\nu}}\pi(\mathrm{I}+2^{j\rho}\mathcal{R})^{-\frac{\gamma}{\nu}}\|_{\mathscr{L}(\mathcal{H}_\pi)} &\le \sup_{\lambda\ge 0}\left(\frac{1+\lambda}{1+2^{j\rho}\lambda}\right)^{\frac{\gamma}{\nu}} \le C2^{-j\rho\frac{\gamma}{\nu}},\\
\|\pi(\mathrm{I}+2^{j\rho}\mathcal{R})^{\frac{\gamma}{\nu}}\pi(\mathrm{I}+\mathcal{R})^{-\frac{\gamma}{\nu}}\|_{\mathscr{L}(\mathcal{H}_\pi)} &\le \sup_{\lambda\ge 0}\left(\frac{1+2^{j\rho}\lambda}{1+\lambda}\right)^{\gamma\nu} \le C2^{j\rho\frac{\gamma}{\nu}},
\end{aligned}$$

for any $j\in\mathbb{N}_0$. Thus, we have obtained

$$\begin{aligned}
&\|\pi(\mathrm{I}+\mathcal{R})^{\frac{\gamma}{\nu}} X_x^{\beta_o}\Delta^{\alpha_o}\sigma_j'(x,\pi)\pi(\mathrm{I}+\mathcal{R})^{-\frac{\gamma}{\nu}}\|_{\mathscr{L}(\mathcal{H}_\pi)}\\
&\le C2^{j\rho([\alpha_o]-[\beta_o])}\sup_{x_o\in G,\,\pi_o\in\widehat{G}}\|\pi_o(\mathrm{I}+\mathcal{R})^{\frac{\gamma}{\nu}} X_{x_o}^{\beta_o}\Delta^{\alpha_o}\sigma_j(x_o,\pi_o)\pi_o(\mathrm{I}+\mathcal{R})^{-\frac{\gamma}{\nu}}\|_{\mathscr{L}(\mathcal{H}_\pi)}\\
&\le C\|\sigma\|_{S^0_{\rho,\rho},[\alpha_o],[\beta_o],|\gamma|},
\end{aligned}$$

because of (5.76). Taking the supremum over $\pi\in\widehat{G}$, $x\in G$, $[\alpha_o]\le a$, $[\beta_o]\le b$ and $|\gamma|\le c$ yields

$$\|\sigma_j'\|_{S^0_{0,0},a,b,c}\le C\|\sigma\|_{S^0_{\rho,\rho},a,b,c}.$$

With (5.77), we conclude that $\|T_j\|_{\mathscr{L}(L^2(G))}\le C\|\sigma\|_{S^0_{\rho,\rho},a,b,c}$. □

Step 2

Now let us prove Claim (5.75). This relies on the bilinear estimate obtained in Proposition 5.7.12.

Proof of Claim (5.75). For each $i\in\mathbb{N}_0$, we denote by $\kappa_{i,x}$ the kernel associated with σ_i. Then one computes easily the integral kernel $K_{ji}(x,y)$ of the operator $T_j^*T_i$, that is,

$$(T_j^*T_i)f(x)=\int_G K_{ji}(x,y)f(y)dy,\quad f\in\mathcal{S}(G),$$

with

$$K_{ji}(x,y)=\int_G\bar{\kappa}_{j,z}(x^{-1}z)\kappa_{i,z}(y^{-1}z)dz.$$

By Schur's lemma [Ste93, §2.4.1], we have

$$\begin{aligned}\|T_j^*T_i\|_{\mathscr{L}(L^2(G))} &\leq \max\left(\sup_{x\in G}\int_G |K_{ji}(x,y)|dy, \sup_{y\in G}\int_G |K_{ji}(x,y)|dx\right), \\ &\lesssim \|T_j^*T_i\|_{\Psi^2_{\rho,\rho},a,b,c} + \max_{|y^{-1}x|\leq 1}|K_{ji}(x,y)|,\end{aligned}$$

since the estimates at infinity for the kernels of a pseudo-differential operator obtained in Theorem 5.4.1 for $\rho \neq 0$ yield

$$|K_{ji}(x,y)| \lesssim \|T_j^*T_i\|_{\Psi^2_{\rho,\rho},a_1,b_1,c_1}|y^{-1}x|^{-N}$$

for any $N \in \mathbb{N}_0$. (We have assumed that a quasi-norm $|\cdot|$ has been fixed on G.) The properties of composition and of taking the adjoint of pseudo-differential operators (see Theorems 5.5.3 and 5.5.12) together with Lemma 5.4.7 yield

$$\|T_j^*T_i\|_{\Psi^2_{\rho,\rho},a_1,b_1,c_1} \lesssim \|\sigma_j\|_{S^1_{\rho,\rho},a_2,b_2,c_2}\|\sigma_i\|_{S^1_{\rho,\rho},a_3,b_3,c_3} \lesssim \|\sigma\|^2_{S^0_{\rho,\rho},a_4,b_4,c_4}2^{-\frac{i+j}{\nu}}.$$

We now analyse $\max_{|y^{-1}x|\leq 1}|K_{ji}(x,y)|$. So let $x,y\in G$ with $|y^{-1}x|\leq 1$. We fix a function $\chi\in\mathcal{D}(G)$ which is a smooth version of the indicatrix function of the ball $B(0,10)=\{z\in G: |x^{-1}z|<10\}$ about 0 with radius 10, that is, we assume that $\chi\equiv 1$ on $B(0,10)$ and $\chi\equiv 0$ on $B(0,11)$. Let us assume that the quasi-norm is in fact a norm, that is, it satisfies the triangle inequality 'with constant 1' (although we could give a proof without this restriction, it simplifies the choice of constants and therefore avoids dwelling on unimportant technical points). We can always decompose

$$\begin{aligned}K_{ji}(x,y) &= \int_{z\in G}\bar{\kappa}_{j,z}(x^{-1}z)\kappa_{i,z}(y^{-1}z)\left(\chi(x^{-1}z)+(1-\chi(x^{-1}z))\right)dz \\ &= I_1+I_2.\end{aligned}$$

We first estimate the second integral via

$$|I_2| \lesssim \|\sigma_j\|_{S^1_{\rho,\rho},a_5,b_5,c_5}\|\sigma_i\|_{S^1_{\rho,\rho},a_6,b_6,c_6}\int_{|x^{-1}z|>10}|x^{-1}z|^{-N_1}|y^{-1}z|^{-N_1}dz.$$

having used the estimates at infinity for the kernels of a pseudo-differential operator obtained in Theorem 5.4.1 for $\rho\neq 0$. As $|y^{-1}x|\leq 1$, the last integral is just a finite constant if we choose $N_1 = Q+1$ for instance. We estimate the $S^1_{\rho,\rho}$-seminorms with Lemma 5.4.7 and we obtain then

$$|I_2| \lesssim \|\sigma\|^2_{S^0_{\rho,\rho},a_7,b_7,c_7}2^{-\frac{i+j}{\nu}}.$$

We now estimate the integral I_1:

$$I_1 = \int_G \bar{\kappa}_{j,z}(x^{-1}z)\kappa_{i,z}(y^{-1}z)\chi(x^{-1}z)dz.$$

It is of the form $\int_G f(z,z)dz$ for a given function f on $G \times G$. Simple formal manipulations yield for any $N \in \mathbb{N}_0$

$$\begin{aligned}\int_G f(z,z)dz &= \int_G (\mathrm{I}+\mathcal{R})^N_{z_2=z}(\mathrm{I}+\mathcal{R})^{-N}_{z_2} f(z,z_2)dz \\ &= \int_G (\mathrm{I}+\bar{\mathcal{R}})^N_{z_1=z}(\mathrm{I}+\mathcal{R})^{-N}_{z_2=z} f(z_1,z_2)dz,\end{aligned}$$

having used integration by parts or equivalently $\mathcal{R}^t = \bar{\mathcal{R}}$, since $\mathcal{R}$ is essentially self-adjoint. Hence, we obtain formally in our case

$$I_1 = \int_G (\mathrm{I}+\bar{\mathcal{R}})^N_{z_1=z}(\mathrm{I}+\mathcal{R})^{-N}_{z_2=z}\left\{\bar{\kappa}_{j,z_1}(x^{-1}z_2)\kappa_{i,z_1}(y^{-1}z_2)\chi(x^{-1}z_1)\right\}dz,$$

where $N \in \mathbb{N}_0$ is to be fixed later. Note that the expression in z_1 is supported in $B(x_1, 11)$, hence so is the integrand in z. This produces the following estimate

$$|I_1| \leq \int_{|x^{-1}z_2|\leq 11} S(z_2)dz_2$$

where $S(z_2)$ is the supremum

$$\begin{aligned}S(z_2) &= \sup_{z_1\in G}\left|(\mathrm{I}+\bar{\mathcal{R}})^N_{z_1}(\mathrm{I}+\mathcal{R})^{-N}_{z_2}\left\{\bar{\kappa}_{j,z_1}(x^{-1}z_2)\kappa_{i,z_1}(y^{-1}z_2)\chi(x^{-1}z_1)\right\}\right| \\ &\lesssim \left\|(\mathrm{I}+\bar{\mathcal{R}})^{N+\frac{s_0}{\nu}}_{z_1}(\mathrm{I}+\mathcal{R})^{-N}_{z_2}\bar{\kappa}_{j,z_1}(x^{-1}z_2)\kappa_{i,z_1}(y^{-1}z_2)\chi(x^{-1}z_1)\right\|_{L^2(dz_1)} \\ &\lesssim \sum_{\substack{[\beta_{01}]+[\beta_{02}]\\ \leq \nu N+s_0}}\left\|(\mathrm{I}+\mathcal{R})^{-N}_{z_2}\{X^{\beta_{01}}_{z_1}\bar{\kappa}_{j,z_1}(x^{-1}z_2)\; X^{\beta_{02}}_{z_1}\kappa_{i,z_1}(y^{-1}z_2)\}\right\|_{L^2(B(x,11),dz_1)},\end{aligned}$$

by the properties of the Sobolev spaces, see Theorem 4.4.28, especially the Sobolev embedding in Part (5). Here $s_0 \in \nu\mathbb{N}$ denotes the smallest integer multiple of ν such that $\frac{s_0}{\nu} > Q/2$. By the Cauchy-Schwartz inequality, as $B(x, 11)$ has finite volume independent of x, we obtain

$$\begin{aligned}|I_1| &\lesssim \sum_{\substack{[\beta_{01}]+[\beta_{02}]\\ \leq \nu N+s_0}}\left\|(\mathrm{I}+\mathcal{R})^{-N}_{z_2}\{X^{\beta_{01}}_{z_1}\bar{\kappa}_{j,z_1}(x^{-1}z_2)\; X^{\beta_{02}}_{z_1}\kappa_{i,z_1}(y^{-1}z_2)\}\right\|_{L^2(B(x,11)^2,dz_1dz_2)} \\ &\lesssim \sup_{\substack{z_1\in B(x,11)\\ [\beta_{01}]+[\beta_{02}]\leq \nu N+s_0}}\left\|(\mathrm{I}+\mathcal{R})^{-N}_{z_2}\{X^{\beta_{01}}_{z_1}\bar{\kappa}_{j,z_1}(x^{-1}z_2)\; X^{\beta_{02}}_{z_1}\kappa_{i,z_1}(y^{-1}z_2)\}\right\|_{L^2(dz_2)}.\end{aligned}$$

Choosing $N > \frac{Q}{2\nu}$, we can apply Proposition 5.7.12 to the L^2-norm above, so that

$$\begin{aligned}&\left\|(\mathrm{I}+\mathcal{R})^{-N}_{z_2}\{X^{\beta_{01}}_{z_1}\bar{\kappa}_{j,z_1}(x^{-1}z_2)\; X^{\beta_{02}}_{z_1}\kappa_{i,z_1}(y^{-1}z_2)\}\right\|_{L^2(dz_2)} \\ &\quad\lesssim \left\|X^{\beta_{01}}_{z_1}\bar{\kappa}_{j,z_1}(z_2)\right\|_{L^2(dz_2)}\left\|X^{\beta_{02}}_{z_1}\kappa_{i,z_1}(z_2)\right\|_{L^2(dz_2)} 2^{(-N+\frac{Q}{2\nu})\max(i,j)}.\end{aligned}$$

By Corollary 5.4.3, we have

$$\left\|X_{z_1}^{\beta_{01}}\bar{\kappa}_{j,z_1}(z_2)\right\|_{L^2(dz_2)} \lesssim \|X_x^{\beta_{01}}\sigma_j\|_{S^{m'}_{\rho,\rho},a_7,b_7,c_7},$$

where m' is a number such that $m' < -Q/2$, for instance $m' := -1-Q/2$. By Lemma 5.4.7, we have (with $\rho=\delta$)

$$\|X_x^{\beta_{01}}\sigma_j\|_{S^{m'}_{\rho,\rho},a_7,b_7,c_7} \lesssim \|\sigma\|_{S^0_{\rho,\rho},a_8,b_8,c_8} 2^{-j\frac{m'-\delta[\beta_{01}]}{\nu}}.$$

We have similar estimates for $\left\|X_{z_1}^{\beta_{02}}\kappa_{i,z_1}(z_2)\right\|_{L^2(dz_2)}$, thus

$$\begin{aligned}
&\max_{\substack{[\beta_{01}]+[\beta_{02}]\\ \leq \nu N+s_0}} \left\|X_{z_1}^{\beta_{01}}\bar{\kappa}_{j,z_1}(z_2)\right\|_{L^2(dz_2)} \left\|X_{z_1}^{\beta_{02}}\kappa_{i,z_1}(z_2)\right\|_{L^2(dz_2)}\\
&\qquad \lesssim \|\sigma\|^2_{S^0_{\rho,\rho},a_9,b_9,c_9} \max_{\substack{[\beta_{01}]+[\beta_{02}]\\ \leq \nu N+s_0}} 2^{-j\frac{m'-\delta[\beta_{01}]}{\nu}} 2^{-i\frac{m'-\delta[\beta_{02}]}{\nu}}\\
&\qquad \lesssim \|\sigma\|^2_{S^0_{\rho,\rho},a_9,b_9,c_9} 2^{\max(i,j)(-2m'+\delta(N+s_0))}.
\end{aligned}$$

The estimates above show that the first formal manipulations on I_1 are justified and we obtain

$$|I_1| \lesssim \|\sigma\|^2_{S^0_{\rho,\rho},a_9,b_9,c_9} 2^{\max(i,j)(-(1-\delta)N-2m'+s_0+\frac{Q}{2\nu})}.$$

Consequently, we have

$$\max_{|y^{-1}x|\leq 1} |K_{ji}(x,y)| \lesssim \|\sigma\|^2_{S^0_{\rho,\rho},a,b,c}\left(2^{-\frac{i+j}{\nu}} + 2^{\max(i,j)(-(1-\delta)N-2m'+s_0+\frac{Q}{2\nu})}\right),$$

thus

$$\|T_j^* T_i\|_{\mathscr{L}(L^2(G))} \lesssim \|\sigma\|^2_{S^0_{\rho,\rho},a,b,c}\left(2^{-\frac{i+j}{\nu}} + 2^{\max(i,j)(-(1-\delta)N-2m'+s_0+\frac{Q}{2\nu})}\right).$$

As $\delta=\rho\in(0,1)$, we can choose N such that $-(1-\delta)N-2m'+s_0+\frac{Q}{2\nu}<-1$. Summing over $i>j+3$ and using the symmetry of the rôle played by i and j yield (5.75). □

Hence we have shown Proposition 5.7.14 and this concludes the proof of Theorem 5.7.1.

5.8 Parametrices, ellipticity and hypoellipticity

In this section, we obtain statements regarding ellipticity and hypoellipticity which are similar to the compact case presented in Section 2.2.3 where the Laplacian has the role of the positive Rockland operator. However, on nilpotent Lie groups, since $\widehat{G}$ is not discrete and the representations are often not (and can be almost never) finite dimensional, the precise hypotheses become more technical to present.

5.8.1 Ellipticity

Roughly speaking, we define the ellipticity by requiring that the symbol is invertible for 'high frequencies'. These 'high frequencies' are determined with respect to the spectral projection E of a positive Rockland operator $\mathcal{R}$, and its group Fourier transform E_π, see Corollary 4.1.16.

We will use the following shorthand notation:

$$\mathcal{H}^\infty_{\pi,\Lambda} := E_\pi(\Lambda, +\infty)\mathcal{H}^\infty_\pi. \tag{5.78}$$

Since $E_\pi(\Lambda,\infty) = \mathcal{F}_G(1_{(\Lambda,\infty)}(\mathcal{R})\delta_0)$ yields a symbol acting on smooth vectors (see Examples 5.1.27 and 5.1.38), $\mathcal{H}^\infty_{\pi,\Lambda}$ is a subspace of $\mathcal{H}^\infty_\pi$.

We can now define our notion of ellipticity:

Definition 5.8.1. Let $\mathcal{R}$ be a positive Rockland operator of homogeneous degree ν. Let σ be a symbol given by fields of operators acting on smooth vectors, i.e. $\sigma(x,\cdot) = \{\sigma(x,\cdot) : \mathcal{H}^\infty_\pi \to \mathcal{H}^\infty_\pi, \pi \in \widehat{G}\}$ is in some $L^\infty_{a,b}(\widehat{G})$ for each $x \in G$.

The symbol σ is said to be *elliptic* with respect to $\mathcal{R}$ of *elliptic order* m_o if there is $\Lambda \in \mathbb{R}$ such that for any $\gamma \in \mathbb{R}$, $x \in G$, μ-almost all $\pi \in \widehat{G}$, and any $u \in \mathcal{H}^\infty_{\pi,\Lambda}$ we have

$$\forall\gamma \in \mathbb{R} \qquad \|\pi(\mathrm{I}+\mathcal{R})^{\frac{\gamma}{\nu}}\sigma(x,\pi)u\|_{\mathcal{H}_\pi} \geq C_\gamma\|\pi(\mathrm{I}+\mathcal{R})^{\frac{\gamma}{\nu}}\pi(\mathrm{I}+\mathcal{R})^{\frac{m_o}{\nu}}u\|_{\mathcal{H}_\pi}. \tag{5.79}$$

with $C_\gamma = C_{\sigma,\mathcal{R},m_o,\Lambda,\gamma}$ independent of $(x,\pi) \in G \times \widehat{G}$ and $u \in \mathcal{H}^\infty_{\pi,\Lambda}$.

We will say that the symbol σ or the corresponding operator $\mathrm{Op}(\sigma)$ is $(\mathcal{R},\Lambda,m_o)$-*elliptic*, or elliptic of elliptic order m_o, or just elliptic.

The notation $\mathcal{H}^\infty_{\pi,\Lambda}$ was defined in (5.78). As $\mathcal{H}^\infty_{\pi,\Lambda}$ is a subspace of $\mathcal{H}^\infty_\pi$ and since $\pi(\mathrm{I}+\mathcal{R})^{\frac{\gamma}{\nu}}$ and $\sigma(x,\cdot)$ are fields of operators acting on smooth vectors, the expression in the norm of the left-hand side of (5.79) makes sense.

In our elliptic condition in Definition 5.8.1, σ is a symbol in the sense of Definition 5.1.33 which is given by fields of operators acting on smooth vectors. It will be natural to consider symbols in the classes $S^m_{\rho,\delta}$ to construct parametrices, see Proposition 5.8.5 and Theorem 5.8.7.

Our definition of ellipticity requires a property of 'x-uniform partial injectivity'. Of course, we note that $\pi(\mathrm{I}+\mathcal{R})^{\frac{\gamma}{\nu}}\pi(\mathrm{I}+\mathcal{R})^{\frac{m_o}{\nu}} = \pi(\mathrm{I}+\mathcal{R})^{\frac{\gamma+m_o}{\nu}}$.

Naturally, we will see shortly in Corollary 5.8.4 that it suffices to check (5.79) for a sequence of real numbers $\{\gamma_\ell,\ \ell \in \mathbb{Z}\}$ which tends to $\pm\infty$ as $\ell \to \pm\infty$.

Our first examples of elliptic operators are provided by positive Rockland operators:

Proposition 5.8.2. *Let $\mathcal{R}$ be a positive Rockland operator of homogeneous degree ν. Then we have the following properties.*

1. *The operator $(\mathrm{I}+\mathcal{R})^{\frac{m_o}{\nu}}$, for any $m_o \in \mathbb{R}$, is elliptic with respect to $\mathcal{R}$ of elliptic order m_o.*

2. *If f_1 and f_2 are complex-valued (smooth) functions on G such that*

$$\inf_{x\in G,\lambda\geq\Lambda} \frac{|f_1(x)+f_2(x)\lambda|}{1+\lambda} > 0 \quad \text{for some } \Lambda\geq 0,$$

then the differential operator $f_1(x)+f_2(x)\mathcal{R}$ is $(\mathcal{R},\Lambda,\nu)$-elliptic.

3. *The operator $E(\Lambda,\infty)\mathcal{R}$, for any $\Lambda>0$, is $(\mathcal{R},\Lambda,\nu)$-elliptic.*

   More generally, if f is a complex-valued function on G such that $\inf_G|f| > 0$, then $f(x)E(\Lambda,\infty)\mathcal{R}$ is $(\mathcal{R},\Lambda,\nu)$-elliptic.

4. *Let $\psi\in C^\infty(\mathbb{R})$ be such that*

$$\psi_{|(-\infty,\Lambda_1]}=0 \quad \text{and} \quad \psi_{|[\Lambda_2,\infty)}=1,$$

for some real numbers Λ_1,Λ_2 satisfying $0<\Lambda_1<\Lambda_2$, Then the operator $\psi(\mathcal{R})\mathcal{R}$ is $(\mathcal{R},\Lambda_2,\nu)$-elliptic.

   More generally, if f is a complex-valued function on G such that $\inf_G|f| > 0$, then $f(x)\psi(\mathcal{R})\mathcal{R}$ is $(\mathcal{R},\Lambda_2,\nu)$-elliptic.

Proof. The symbols involved in the statement are multipliers in $\mathcal{R}$. By Example 5.1.27 and Corollary 5.1.30, the corresponding symbols are symbols in the sense of Definition 5.1.33 which are given by fields of operators acting on smooth vectors. Hence it remains just to check the condition in (5.79).

Part (1) is easy to check using the functional calculus of $\pi(\mathcal{R})$.

Let us prove Part (2). Let Λ, f_1, f_2, and m be as in the statement. The properties of the functional calculus for $\pi(\mathcal{R})$ yield that, for each $x\in G$ fixed and $u\in\mathcal{H}^\infty_{\pi,\Lambda}$ we have

$$\pi(\mathrm{I}+\mathcal{R})^{\frac{\gamma}{\nu}}\pi(\mathrm{I}+\mathcal{R})u=\phi_x(\pi(\mathcal{R}))\pi(\mathrm{I}+\mathcal{R})^{\frac{\gamma}{\nu}}(f_1(x)+f_2(x)\pi(\mathcal{R}))u,$$

where $\phi_x\in L^\infty[0,\infty)$ is given by

$$\phi_x(\lambda)=\frac{1+\lambda}{f_1(x)+f_2(x)\lambda}1_{\lambda\geq\Lambda}.$$

Our assumption implies that ϕ_x is bounded on $[0,\infty)$ with

$$C:=\sup_{x\in G}\|\phi_x\|_\infty=\left(\inf_{x\in G,\lambda\geq\Lambda}\frac{|f_1(x)+f_2(x)\lambda|}{1+\lambda}\right)^{-1}<\infty.$$

The property of the functional calculus for $\pi(\mathcal{R})$ yields

$$\forall x\in G \quad \|\phi_x(\pi(\mathcal{R}))\|_{\mathscr{L}(\mathcal{H}_\pi)}\leq C.$$

Thus we have

$$\begin{aligned}\|\pi(\mathrm{I}+\mathcal{R})^{\frac{\gamma}{\nu}}\pi(\mathrm{I}+\mathcal{R})u\|_{\mathcal{H}_\pi} &= \|\phi_x(\pi(\mathcal{R}))\pi(\mathrm{I}+\mathcal{R})^{\frac{\gamma}{\nu}}(f_1(x)+f_2(x)\pi(\mathcal{R}))u\|_{\mathcal{H}_\pi}\\ &\leq C\|\pi(\mathrm{I}+\mathcal{R})^{\frac{\gamma}{\nu}}(f_1(x)+f_2(x)\pi(\mathcal{R}))u\|_{\mathcal{H}_\pi}.\end{aligned}$$

This proves Part (2).

Let us prove Part (3). The properties of the functional calculus for $\pi(\mathcal{R})$ yield

$$\pi(\mathrm{I}+\mathcal{R})u = \phi(\pi(\mathcal{R}))E_\pi(\Lambda,\infty)\pi(\mathcal{R})u,$$

where $\phi \in L^\infty[0,\infty)$ is given by

$$\phi(\lambda) = \frac{1+\lambda}{\lambda}1_{(\Lambda,\infty)}(\lambda).$$

Moreover,

$$\begin{aligned}\|\pi(\mathrm{I}+\mathcal{R})^{1+\frac{\gamma}{\nu}}u\|_{\mathcal{H}_\pi} &= \|\phi(\pi(\mathcal{R}))\pi(\mathrm{I}+\mathcal{R})^{\frac{\gamma}{\nu}}E_\pi(\Lambda,\infty)\pi(\mathcal{R})u\|_{\mathcal{H}_\pi}\\ &\leq \|\phi\|_\infty\|\pi(\mathrm{I}+\mathcal{R})^{\frac{\gamma}{\nu}}E_\pi(\Lambda,\infty)\pi(\mathcal{R})u\|_{\mathcal{H}_\pi}.\end{aligned}$$

Since $C=\|\phi\|_\infty^{-1}$ is a finite positive constant, we have obtained

$$C\|\pi(\mathrm{I}+\mathcal{R})^{1+\frac{\gamma}{\nu}}u\|_{\mathcal{H}_\pi} \leq \|\pi(\mathrm{I}+\mathcal{R})^{\frac{\gamma}{\nu}}E_\pi(\Lambda,\infty)\pi(\mathcal{R})u\|_{\mathcal{H}_\pi}.$$

This shows that $E(\Lambda,\infty)\mathcal{R}$, is elliptic.

If f is as in the statement, we proceed as above, replacing ϕ by

$$\phi_x(\lambda) = \frac{1+\lambda}{f(x)\lambda}1_{(\Lambda,\infty)}(\lambda),$$

and C such that C^{-1} is equal to the right-hand side of the estimate

$$\|\phi_x\|_\infty \leq \frac{1}{\inf_G|f|}\sup_{\lambda\geq\Lambda}\frac{1+\lambda}{\lambda} := C^{-1}.$$

This shows Part (3).

For Part (4), we proceed as in Part (3) replacing $1_{(\Lambda,\infty)}$ by $\psi(\lambda)$ and Λ by Λ_2. □

The next lemma is technical. It states that we can construct a partial inverse of an elliptic symbol. The analogue for scalar-valued symbols would be obvious: if $|a(x,\xi)|$ does not vanish for $|\xi|>\Lambda$ then we can consider $1_{|\xi|>\Lambda}1/a(x,\xi)$. However, in the context of operator-valued symbols, we need to proceed with caution.

Lemma 5.8.3. *Let σ be a symbol $(\mathcal{R}, \Lambda, m_o)$-elliptic as in Definition 5.8.1.*

For any $v \in \mathcal{H}_\pi^\infty$, if there is a vector $u \in \mathcal{H}_{\pi,\Lambda}^\infty$ such that $\sigma(x,\pi)u = v$ then this u is necessarily unique. In this sense $\sigma(x,\pi)$ is invertible on $\mathcal{H}_{\pi,\Lambda}^\infty$ and we can set

$$E_\pi(\Lambda,\infty)\sigma(x,\pi)^{-1}(v) := \begin{cases} u & \text{if } v = \sigma(x,\pi)u, \; u \in \mathcal{H}_{\pi,\Lambda}^\infty, \\ 0 & \text{if } \mathcal{H}_\pi^\infty \ni v \perp \sigma(x,\pi)\mathcal{H}_{\pi,\Lambda}^\infty. \end{cases} \tag{5.80}$$

This yields the symbol (in the sense of Definition 5.1.33) given by fields of operators acting on smooth vectors

$$\{E_\pi(\Lambda,\infty)\sigma(x,\pi)^{-1} : \mathcal{H}_\pi^\infty \to \mathcal{H}_\pi^\infty, (x,\pi) \in G \times \widehat{G}\}. \tag{5.81}$$

Furthermore, for every γ,

$$\|E_\pi(\Lambda,\infty)\sigma(x,\pi)^{-1}\|_{L^\infty_{\gamma,\gamma+m_o}(\widehat{G})} \leq C_\gamma^{-1}, \tag{5.82}$$

where C_γ is the constant appearing in (5.79) of Definition 5.8.1.

If σ is continuous in the sense of Definition 5.1.34, then the symbol in (5.81) is continuous in the sense of Definition 5.1.34. If σ is smooth, then the symbol in (5.81) is continuous and depends smoothly on $x \in G$ in the sense of Remark 1.8.16.

Proof. Recall that $E_\pi(\Lambda,\infty) = \mathcal{F}_G(1_{(\Lambda,\infty)}(\mathcal{R})\delta_0)$ yields a symbol acting on smooth vectors, see Examples 5.1.27 and 5.1.38.

If $v = \sigma(x,\pi)u$ where $u \in \mathcal{H}_{\pi,\Lambda}^\infty$, then, using (5.79), we have

$$\|\pi(\mathrm{I}+\mathcal{R})^{\frac{m_o+\gamma}{\nu}}u\|_{\mathcal{H}_\pi} \leq C_\gamma^{-1}\|\pi(\mathrm{I}+\mathcal{R})^{\frac{\gamma}{\nu}}\sigma(x,\pi)u\|_{\mathcal{H}_\pi} = C_\gamma^{-1}\|\pi(\mathrm{I}+\mathcal{R})^{\frac{\gamma}{\nu}}v\|_{\mathcal{H}_\pi}.$$

It is now easy to check $\{E_\pi(\Lambda,\infty)\sigma(x,\pi)^{-1}, (x,\pi) \in G \times \widehat{G}\}$ is a symbol in the sense of Definition 5.1.33 and that the estimates in (5.82) hold.

If σ is continuous, then one checks easily that the map

$$G \ni x \mapsto E_\pi(\Lambda,\infty)\sigma(x,\pi)^{-1} \in L^\infty_{\gamma,\gamma+m_o}(\widehat{G})$$

is continuous. Consequently $\{E_\pi(\Lambda,\infty)\sigma(x,\pi)^{-1}, (x,\pi) \in G \times \widehat{G}\}$ is continuous.

If σ is smooth, then $\{E_\pi(\Lambda,\infty)\sigma(x,\pi)^{-1}, (x,\pi) \in G \times \widehat{G}\}$ depends smoothly in $x \in G$, see Remark 1.8.16. □

Corollary 5.8.4. *Let $\mathcal{R}$ be a positive Rockland operator of homogeneous degree ν. The symbol σ satisfies (5.79) for each $\gamma \in \mathbb{R}$ if and only if σ satisfies (5.79) for a sequence of real numbers $\{\gamma_\ell, \ell \in \mathbb{Z}\}$ which tends to $\pm\infty$ as $\ell \to \pm\infty$.*

We may choose the constants C_γ such that $\max_{|\gamma|\leq c} C_\gamma$ in (5.79) is finite for any $c \geq 0$.

Proof. From the proof of Lemma 5.8.3, we see that σ satisfies (5.79) for γ if and only if

$$\sup_{x\in G} \|E_\pi(\Lambda,\infty)\sigma(x,\pi)^{-1}\|_{L^\infty_{\gamma,\gamma+m_o}(\widehat{G})} < \infty$$

is finite. The conclusion follows from Corollary 4.4.10. □

The next statement says that if a symbol in some $S^m_{\rho,\delta}$ is elliptic and if the elliptic order is equal to the order m of the symbol, then we can define a symbol in $S^{-m}_{\rho,\delta}$ using the operator $E_\pi(\Lambda,\infty)\sigma(x,\pi)^{-1}$ defined via (5.80). This will be the main ingredient in the construction of a parametrix, see the proof of Theorem 5.8.7.

Proposition 5.8.5. *Assume $1\geq\rho\geq\delta\geq 0$. Let $\sigma\in S^m_{\rho,\delta}$ be a symbol which is $(\mathcal{R},\Lambda,m)$-elliptic with respect to a positive Rockland operator $\mathcal{R}$. If $\psi\in C^\infty(\mathbb{R})$ is such that*

$$\psi_{|(-\infty,\Lambda_1]}=0 \quad \text{and} \quad \psi_{|[\Lambda_2,\infty)}=1,$$

for some real numbers Λ_1,Λ_2 satisfying $\Lambda<\Lambda_1<\Lambda_2$, then the symbol

$$\{\psi(\pi(\mathcal{R}))\sigma^{-1}(x,\pi)\ ,\ (x,\pi)\in G\times\widehat{G}\},$$

given by

$$\psi(\pi(\mathcal{R}))\sigma^{-1}(x,\pi) := \psi(\pi(\mathcal{R}))E_\pi(\Lambda_1,\infty)\sigma(x,\pi)^{-1},$$

is in $S^{-m}_{\rho,\delta}$. Moreover, for any $a_o,b_o\in\mathbb{N}_0$, we have

$$\begin{aligned}\|\psi(\pi(\mathcal{R}))\sigma^{-1}(x,\pi)\|_{S^{-m}_{\rho,\delta},a_o,b_o,0} \\ \leq C \sum_{\substack{a_1',a_2'\leq a_o\\ b_1',b_2'\leq b_o}} \max_{|\gamma|\leq\rho a_o+\delta b_o} C_{\gamma,\sigma,\Lambda_1}^{a_1'+b_1'+1}\|\sigma(x,\pi)\|^{a_2'+b_2'}_{S^m_{\rho,\delta},a_o,b_o,|m|},\end{aligned}$$

where $C>0$ is a positive constant depending on a_o,b_o,ψ, and where the constant $C_{\gamma,\sigma,\Lambda_1}$ was given in (5.79).

The following lemma is helpful in the proof of Proposition 5.8.5. Indeed, in the case of $\mathbb{R}^n$, if a cut-off function $\psi(\xi)$ on the Fourier side is constant for $|\xi|>\Lambda$ (Λ large enough), then its derivatives are $\partial_\xi^\alpha\psi(\xi)=0$ if $|\xi|>\Lambda$. In our case, we can not say anything in general. If we use $\psi(\pi(\mathcal{R}))$ as 'a cut-off in frequency' with ψ as in Proposition 5.8.5 for example, it is not true in general that its (Δ^α-)derivatives will vanish on $E_\pi(\Lambda,\infty)$ or will be of the form $\psi_1(\pi(\mathcal{R}))$. However, we can show that these derivatives are smoothing:

Lemma 5.8.6. *Let $\psi\in C^\infty(\mathbb{R})$ satisfy $\psi_{|[\Lambda,+\infty)}=1$ for some $\Lambda\in\mathbb{R}$. Then for any $\alpha\in\mathbb{N}_0^n\backslash\{0\}$, the symbol given by $\Delta^\alpha\psi(\pi(\mathcal{R}))$ is smoothing, i.e. is in $S^{-\infty}$.*

Proof of Lemma 5.8.6. Let $\alpha \in \mathbb{N}_0^n \backslash \{0\}$. Then $\Delta^\alpha \mathrm{I} = 0$ by Example 5.2.8. Therefore

$$\Delta^\alpha \psi(\pi(\mathcal{R})) = -\Delta^\alpha (1-\psi)(\pi(\mathcal{R})).$$

As $1-\psi$ is a smooth function such that $\text{supp}(1-\psi)\cap[0,\infty)$ is compact, the symbol $(1-\psi)(\pi(\mathcal{R}))$ is smoothing. Hence so is $\Delta^\alpha(1-\psi)(\pi(\mathcal{R}))$ and $\Delta^\alpha\psi(\pi(\mathcal{R}))$. □

Proof of Proposition 5.8.5. Recall that by the Leibniz formula (Proposition 5.2.10), we have

$$\Delta^{\alpha_o}(\sigma_1\sigma_2) = \sum_{[\alpha_1]+[\alpha_2]=[\alpha_o]} c_{\alpha_1,\alpha_2} \Delta^{\alpha_1}\sigma_1 \, \Delta^{\alpha_2}\sigma_2,$$

with

$$c_{\alpha_1,0} = \begin{cases} 1 & \text{if } \alpha_1 = \alpha_o \\ 0 & \text{otherwise} \end{cases}, \quad c_{0,\alpha_2} = \begin{cases} 1 & \text{if } \alpha_2 = \alpha_o \\ 0 & \text{otherwise} \end{cases}.$$

It is also easy to see that

$$X^{\beta_o}(f_1 f_2) = \sum_{[\beta_1]+[\beta_2]=[\beta_o]} c'_{\beta_1,\beta_2} X^{\beta_1} f_1 \, X^{\beta_2} f_2,$$

with

$$c'_{\beta_1,0} = \begin{cases} 1 & \text{if } \beta_1 = \alpha_o \\ 0 & \text{otherwise} \end{cases}, \quad c'_{0,\beta_2} = \begin{cases} 1 & \text{if } \beta_2 = \beta_o \\ 0 & \text{otherwise} \end{cases}.$$

Let $\sigma = \sigma(x,\pi) \in S^m_{\rho,\delta}$ and $\psi \in C^\infty(\mathbb{R})$ as in the statement. By Lemma 5.8.3, the continuous symbol

$$\{E_\pi(\Lambda,\infty)\sigma(x,\pi)^{-1} : \mathcal{H}_\pi^\infty \to \mathcal{H}_\pi^\infty, (x,\pi) \in G \times \widehat{G}\},$$

depends smoothly on $x \in G$. Hence so does the continuous symbol σ_o defined via

$$\sigma_o(x,\pi) := \psi(\pi(\mathcal{R}))\sigma^{-1}(x,\pi).$$

Since $\psi(\pi(\mathcal{R}))$ commutes with powers of $\pi(\mathrm{I}+\mathcal{R})$ and

$$\|\psi(\pi(\mathcal{R}))\|_{\mathscr{L}(\mathcal{H}_\pi)} \le \|\psi\|_\infty,$$

we have

$$\begin{aligned}
&\|\pi(\mathrm{I}+\mathcal{R})^{\frac{m}{\nu}}\sigma_o(x,\pi)\|_{\mathscr{L}(\mathcal{H}_\pi)} \\
&\quad \le \|\psi\|_\infty \|\pi(\mathrm{I}+\mathcal{R})^{\frac{m}{\nu}} \left\{E_\pi(\Lambda,\infty)\sigma(x,\pi)^{-1}\right\}\|_{\mathscr{L}(\mathcal{H}_\pi)} \\
&\quad = \|\psi\|_\infty C_0^{-1},
\end{aligned}$$

where by Lemma 5.8.3, C_0 is the finite constant intervening in the ellipticity condition for $\gamma = 0$ in (5.79). More generally, in this proof, C_γ denotes the constant depending on γ in (5.79), see also Corollary 5.8.4.

By Proposition 5.3.4, $\psi(\pi(\mathcal{R})) \in S^0$. We also see that

$$\psi(\pi(\mathcal{R})) = \sigma_o(x,\pi)\sigma(x,\pi). \tag{5.83}$$

Hence for any left-invariant vector field X we have

$$\begin{aligned} 0 &= X_x\psi(\pi(\mathcal{R})) \\ &= X_x\sigma_o(x,\pi)\ \sigma(x,\pi) + \sigma_o(x,\pi)\ X_x\sigma(x,\pi). \end{aligned}$$

Thus

$$X_x\sigma_o(x,\pi)\sigma(x,\pi) = -\sigma_o(x,\pi)\ X_x\sigma(x,\pi),$$

and since $\sigma(x,\pi)$ is invertible on $E_\pi(\Lambda_1,\infty)\mathcal{H}_\pi^\infty$,

$$X_x\sigma_o(x,\pi) = -\sigma_o(x,\pi)\ \{X_x\sigma(x,\pi)\}\ \ E(\Lambda_1,\infty)\sigma^{-1}(x,\pi).$$

Assuming that X is homogeneous of degree d, we can take the operator norm and estimate

$$\begin{aligned} &\|\pi(\mathrm{I}+\mathcal{R})^{\frac{m-\delta d}{\nu}} X_x\sigma_o(x,\pi)\|_{\mathscr{L}(\mathcal{H}_\pi)} \\ &\quad\le \|\pi(\mathrm{I}+\mathcal{R})^{\frac{m-\delta d}{\nu}}\sigma_o(x,\pi)\pi(\mathrm{I}+\mathcal{R})^{\frac{\delta d}{\nu}}\|_{\mathscr{L}(\mathcal{H}_\pi)} \\ &\qquad \|\pi(\mathrm{I}+\mathcal{R})^{-\frac{\delta d}{\nu}} X_x\sigma(x,\pi)\pi(\mathrm{I}+\mathcal{R})^{-\frac{m}{\nu}}\|_{\mathscr{L}(\mathcal{H}_\pi)} \\ &\qquad \|\pi(\mathrm{I}+\mathcal{R})^{\frac{m}{\nu}}\ \{E_\pi(\Lambda_1,\infty)\sigma(x,\pi)^{-1}\}\ \|_{\mathscr{L}(\mathcal{H}_\pi)} \\ &\quad\le \|\psi\|_\infty C_{-\delta d}^{-1}C_0^{-1}\|\sigma(x,\pi)\|_{S^m_{\rho,\delta},0,d,|-m|}. \end{aligned}$$

Recursively on $d=[\beta_o]$, we can show similar properties for $X_x^{\beta_o}\ \{\psi(\pi(\mathcal{R}))\sigma(x,\pi)^{-1}\}$, and obtain

$$\begin{aligned} &\|\psi(\pi(\mathcal{R}))\sigma(x,\pi)^{-1}\|_{S^{-m}_{\rho,\delta},0,b_o,0} \\ &\quad\le C_{b_o,\|\psi\|_\infty} \sum_{b_1',b_2'\le b_o} \max_{|\gamma|\le\delta b_o} C_\gamma^{-(b_1'+1)}\|\sigma(x,\pi)\|^{b_2'}_{S^m_{\rho,\delta},0,b_o,|m|}. \end{aligned}$$

We can proceed in a parallel way for difference operators. Indeed, for any $\alpha_o \in \mathbb{N}_0^n$ with $|\alpha_o| = 1$, we apply Δ^{α_o} to both sides of (5.83) and obtain

$$\Delta^{\alpha_o}\{\psi(\pi(\mathcal{R}))\} = \Delta^{\alpha_o}\sigma_o(x,\pi)\ \sigma(x,\pi) + \sigma_o(x,\pi)\ \Delta^{\alpha_o}\{\sigma(x,\pi)\},$$

thus

$$\begin{aligned} \Delta^{\alpha_o}\sigma_o(x,\pi) &= \Delta^{\alpha_o}\{\psi(\pi(\mathcal{R}))\}E(\Lambda_1,\infty)\sigma^{-1}(x,\pi) \\ &\quad -\sigma_o(x,\pi)\ \{\Delta^{\alpha_o}\sigma(x,\pi)\}\ E(\Lambda_1,\infty)\ \sigma^{-1}(x,\pi). \end{aligned}$$

Then

$$\|\pi(\mathrm{I}+\mathcal{R})^{\frac{\rho[\alpha_o]+m}{\nu}}\Delta^{\alpha_o}\sigma_o(x,\pi)\|_{\mathscr{L}(\mathcal{H}_\pi)} \le N_1 + N_2,$$

with

$$\begin{aligned} N_1 &= \|\pi(\mathrm{I}+\mathcal{R})^{\frac{\rho[\alpha_o]+m}{\nu}} \Delta^{\alpha_o}\{\psi(\pi(\mathcal{R}))\}E(\Lambda_1,\infty)\sigma^{-1}(x,\pi)\|_{\mathscr{L}(\mathcal{H}_\pi)}, \\ N_2 &= \|\pi(\mathrm{I}+\mathcal{R})^{\frac{\rho[\alpha_o]+m}{\nu}} \sigma_o(x,\pi)\ \{\Delta^{\alpha_o}\sigma(x,\pi)\}\ E(\Lambda_1,\infty)\ \sigma^{-1}(x,\pi)\|_{\mathscr{L}(\mathcal{H}_\pi)}. \end{aligned}$$

For the first norm, we see that

$$\begin{aligned} N_1 &\leq \|\pi(\mathrm{I}+\mathcal{R})^{\frac{\rho[\alpha_o]+m}{\nu}} \Delta^{\alpha_o}\{\psi(\pi(\mathcal{R}))\}\pi(\mathrm{I}+\mathcal{R})^{-\frac{m}{\nu}}\|_{\mathscr{L}(\mathcal{H}_\pi)} \\ &\qquad \|\pi(\mathrm{I}+\mathcal{R})^{\frac{m}{\nu}} E(\Lambda_1,\infty)\sigma^{-1}(x,\pi)\|_{\mathscr{L}(\mathcal{H}_\pi)} \\ &\leq C_\psi C_0^{-1}, \end{aligned}$$

since $\Delta^{\alpha_o}\{\psi(\pi(\mathcal{R}))\} \in S^{-\infty}$ by Lemma 5.8.6. For the second norm, we see that

$$\begin{aligned} N_2 &\leq \|\pi(\mathrm{I}+\mathcal{R})^{\frac{\rho[\alpha_o]+m}{\nu}} \sigma_o(x,\pi)\pi(\mathrm{I}+\mathcal{R})^{-\frac{\rho[\alpha_o]}{\nu}}\|_{\mathscr{L}(\mathcal{H}_\pi)} \\ &\qquad \|\pi(\mathrm{I}+\mathcal{R})^{\frac{\rho[\alpha_o]}{\nu}} \Delta^{\alpha_o}\sigma(x,\pi)\pi(\mathrm{I}+\mathcal{R})^{-\frac{m}{\nu}}\|_{\mathscr{L}(\mathcal{H}_\pi)} \\ &\qquad \|\pi(\mathrm{I}+\mathcal{R})^{\frac{m}{\nu}}\ E(\Lambda_1,\infty)\ \sigma^{-1}(x,\pi)\|_{\mathscr{L}(\mathcal{H}_\pi)} \\ &\leq \|\psi\|_\infty C_{\rho[\alpha_o]}^{-1} C_0^{-1}\|\sigma\|_{S^m_{\rho,\delta},[\alpha_o],0,|m|}. \end{aligned}$$

Recursively on $[\alpha_o]$, we can show similar properties for $\Delta^{\alpha_o}\left\{\psi(\pi(\mathcal{R}))\sigma(x,\pi)^{-1}\right\}$, and obtain

$$\begin{aligned} &\|\sigma_o(x,\pi)\|_{S^{-m}_{\rho,\delta},a_o,0,0} \\ &\leq C_{a_o,\psi} \sum_{a_1',a_2'\leq a_o} \max_{|\gamma|\leq \rho a_o} C_\gamma^{-(a_1'+1)} \|\sigma(x,\pi)\|^{a_2'}_{S^m_{\rho,\delta},a_o,0,|m|}. \end{aligned}$$

More generally, we have

$$\begin{aligned} X_x^{\beta_o}\Delta^{\alpha_o}\left\{\psi(\pi(\mathcal{R}))\right\} &= \sum_{\substack{[\alpha_1]+[\alpha_2]=[\alpha_o] \\ [\beta_1]+[\beta_2]=[\beta_o]}} c'_{\beta_1,\beta_2}c_{\alpha_1,\alpha_2}\ X_x^{\beta_1}\Delta^{\alpha_1}\sigma_o(x,\pi) \\ &\qquad\qquad X_x^{\beta_2}\Delta^{\alpha_2}\sigma(x,\pi). \end{aligned}$$

Because of the very first remark of this proof, we obtain $X^{\beta_o}\Delta^{\alpha_o}\sigma_o$ in terms of $X^{\beta'}\Delta^{\alpha'}\sigma_o$ with $[\beta'] < [\beta_o]$ and $[\alpha'] < [\alpha_o]$ and of some derivatives of $\psi(\pi(\mathcal{R}))$ and σ. If we assume that we can control all the seminorms $\|\sigma_o\|_{S^{-m}_{\rho,\delta},a,b,c}$ with $a < [\alpha_o]$, $b < [\beta_o]$ and any $c \in \mathbb{R}$, then we can proceed as above introducing powers of $\mathrm{I}+\mathcal{R}$ to obtain the estimate for the seminorms of $\psi(\pi(\mathcal{R}))\sigma(x,\pi)^{-1}$. Recursively this shows Proposition 5.8.5. □

5.8.2 Parametrix

In the next theorem, we show that our notion of ellipticity implies the construction of a parametrix.

Theorem 5.8.7. *Let $\sigma \in S^m_{\rho,\delta}$ be elliptic of elliptic order m with $1 \geq \rho > \delta \geq 0$. We can construct a left parametrix $B \in \Psi^{-m}_{\rho,\delta}$ for the operator $A = \text{Op}(\sigma)$, that is, there exists $B \in \Psi^{-m}_{\rho,\delta}$ such that*

$$BA - \text{I} \in \Psi^{-\infty}.$$

Comparing with two-sided parametrices in the case of compact Lie groups (Theorem 2.2.17), this parametrix is one-sided. It was also the case in [CGGP92].

Proof. We can adapt the proof in [Tay81, §0.4] to our setting. Let $\psi \in C^\infty(\mathbb{R})$ be such that $\psi_{|(-\infty,\Lambda_1]} = 0$ and $\psi_{|[\Lambda_2,\infty)} = 1$ for some $\Lambda_1, \Lambda_2 \in \mathbb{R}$ with $\Lambda < \Lambda_1 < \Lambda_2$. By Proposition 5.8.5,

$$\psi(\pi(\mathcal{R}))\sigma^{-1}(x,\pi) \in S^{-m}_{\rho,\delta}.$$

Since $\psi(\pi(\mathcal{R})) = \psi(\pi(\mathcal{R}))\sigma^{-1}(x,\pi)\sigma(x,\pi)$, by Corollary 5.5.8,

$$\text{Op}\big(\psi(\pi(\mathcal{R}))\sigma^{-1}(x,\pi)\big)\ A = \psi(\mathcal{R}) \ \text{mod}\Psi^{-(\rho-\delta)}_{\rho,\delta}\ ;$$

now $\psi(\mathcal{R}) = \text{I} - (1-\psi)(\mathcal{R})$ and $(1-\psi) \in \mathcal{D}([0,\infty))$ so $(1-\psi)(\mathcal{R}) \in \Psi^{-\infty}$. This shows

$$\text{Op}\big(\psi(\pi(\mathcal{R}))\sigma^{-1}(x,\pi)\big)\ A \quad = \text{I} \ \text{mod}\Psi^{-(\rho-\delta)}_{\rho,\delta}.$$

So we have

$$\text{Op}\big(\psi(\pi(\mathcal{R}))\sigma^{-1}(x,\pi)\big)\ A \quad = \text{I} - U \quad \text{with} \quad U \in \Psi^{-(\rho-\delta)}_{\rho,\delta}.$$

By Theorem 5.5.1, there exists $T \in \Psi^0_{\rho,\delta}$ such that

$$T \sim \text{I} + U + U^2 + \ldots + U^j + \ldots$$

By Theorem 5.5.3,

$$B := T\ \text{Op}\big(\psi(\pi(\mathcal{R}))\sigma^{-1}\big)\ \in \Psi^{-m}_{\rho,\delta}.$$

Therefore, we obtain

$$BA = T(\text{I} - U) = \text{I} \ \text{mod}\Psi^{-\infty},$$

completing the proof. □

It is not difficult to construct the following examples of elliptic operators satisfying Theorem 5.8.7 out of any Rockland operator. Indeed, combining Proposition 5.3.4 or Corollary 5.3.8 together with Proposition 5.8.2 yield

Example 5.8.8. Let $\mathcal{R}$ be a positive Rockland operator of homogeneous degree ν.

1. For any $m \in \mathbb{R}$, the operator $(\text{I} + \mathcal{R})^{\frac{m}{\nu}} \in \Psi^m$ is elliptic with respect to $\mathcal{R}$ of elliptic order m.

2. If f_1 and f_2 are complex-valued smooth functions on G such that
$$\inf_{x\in G,\lambda\geq\Lambda}\frac{|f_1(x)+f_2(x)\lambda|}{1+\lambda}>0 \quad \text{for some } \Lambda\geq 0,$$
and such that $X^{\alpha_1}f_1$, $X^{\alpha_2}f_2$ are bounded for each $\alpha_1,\alpha_2\in\mathbb{N}_0^n$, then the differential operator
$$f_1(x)+f_2(x)\mathcal{R}\in\Psi^\nu$$
is $(\mathcal{R},\Lambda,\nu)$-elliptic.

3. Let $\psi\in C^\infty(\mathbb{R})$ be such that
$$\psi_{|(-\infty,\Lambda_1]}=0 \quad\text{and}\quad \psi_{|[\Lambda_2,\infty)}=1,$$
for some real numbers Λ_1,Λ_2 satisfying $0<\Lambda_1<\Lambda_2$, Then the operator $\psi(\mathcal{R})\mathcal{R}\in\Psi^\nu$ is $(\mathcal{R},\Lambda_2,\nu)$-elliptic.

   More generally, if f is a smooth complex-valued function on G such that $\inf_G|f|>0$ and that $X^\alpha f$ is bounded on G for every $\alpha\in\mathbb{N}_0^n$, then
$$f(x)\psi(\mathcal{R})\mathcal{R}\in\Psi^\nu$$
is elliptic with respect to $\mathcal{R}$ of elliptic order ν.

   Hence all the operators in Example 5.8.8 admit a left parametrix.

We will see other concrete examples of elliptic differential operators on the Heisenberg group in Section 6.6.1, see Example 6.6.2.

In fact we can prove the existence of left parametrices for symbols which are elliptic with an elliptic order lower than their order. Indeed, we can modify the hypothesis of the ellipticity in Section 5.8.1 to obtain the analogue of Hörmander's theorem about hypoellipticity involving lower order terms, similar to Theorem 2.2.18 in the compact case.

Theorem 5.8.9. *Let $\sigma\in S^m_{\rho,\delta}$ with $1\geq\rho>\delta\geq 0$. We assume that σ is elliptic with respect to a positive Rockland operator $\mathcal{R}$ in the sense of Definition 5.8.1, and that its elliptic order is $m_o\leq m$.*

We also assume that the following hypothesis on the lower order terms holds: there is $\Lambda\in\mathbb{R}$ such that for any $\gamma\in\mathbb{R}$, $x\in G$, μ-almost all $\pi\in\widehat{G}$, and any $u\in\mathcal{H}^\infty_{\pi,\Lambda}$, we have
$$\|\pi(\mathrm{I}+\mathcal{R})^{\frac{\rho[\alpha]-\delta[\beta]+\gamma}{\nu}}\left\{\Delta^\alpha X^\beta\sigma(x,\pi)\right\}\pi(\mathrm{I}+\mathcal{R})^{-\frac{\gamma}{\nu}}u\|_{\mathcal{H}_\pi} \leq C'_{\alpha,\beta,\gamma}\|\sigma(x,\pi)u\|_{\mathcal{H}_\pi}, \quad (5.84)$$
with $C'_{\alpha,\beta,\gamma}=C'_{\alpha,\beta,\gamma,\sigma,\mathcal{R},m_o,\Lambda,\gamma}$ independent of $(x,\pi)\in G\times\widehat{G}$ and $u\in\mathcal{H}^\infty_{\pi,\Lambda}$.

Then we can construct a left parametrix $B\in\Psi^{-m_o}_{\rho,\delta}$ for the operator $A=\mathrm{Op}(\sigma)$, that is, there exists $B\in\Psi^{-m_o}_{\rho,\delta}$ such that
$$BA-\mathrm{I}\in\Psi^{-\infty}.$$

Proceeding as in Corollary 5.8.4, we can show easily that it suffices to assume (5.79) and (5.84) for a countable sequence γ which goes to $+\infty$ and $-\infty$.

Proof. Let $\psi \in C^\infty(\mathbb{R})$ be such that $\psi_{|(-\infty,\Lambda_1]} = 0$ and $\psi_{|[\Lambda_2,\infty)} = 1$ for some $\Lambda_1, \Lambda_2 \in \mathbb{R}$ with $\Lambda < \Lambda_1 < \Lambda_2$. Proceeding as in the proof of Proposition 5.8.5, we see that

$$\sigma_o(x,\pi) := \psi(\pi(\mathcal{R}))\sigma^{-1}(x,\pi) \in S^{-m_o}_{\rho,\delta},$$

with similar estimates for the seminorms of σ_o and σ.

With similar ideas, using (5.84), we claim that, for any multi-index $\beta_o \in \mathbb{N}_0^n$, we have

$$X^{\beta_o}\sigma(x,\pi)\ \sigma_o(x,\pi) \in S^{\delta[\beta_o]}_{\rho,\delta}.$$

Indeed, from the proof of Proposition 5.8.5, we know that

$$X\sigma_o = -\sigma_o\ X\sigma\ E(\Lambda,\infty)\sigma^{-1},$$

hence

$$\begin{aligned} X\left(X^{\beta_o}\sigma(x,\pi)\ \sigma_o(x,\pi)\right) &= XX^{\beta_o}\sigma(x,\pi)\ \sigma_o(x,\pi) + X^{\beta_o}\sigma(x,\pi)\ X\sigma_o(x,\pi) \\ &= XX^{\beta_o}\sigma(x,\pi)\ \sigma_o(x,\pi) - X^{\beta_o}\sigma(x,\pi)\ \sigma_o\ X\sigma\ E(\Lambda,\infty)\sigma^{-1}, \end{aligned}$$

and we can use the hypothesis (5.84) on each term to control the $S^m_{\rho,\delta}$-seminorms of the expression on the right-hand side. For the difference operators, from the proof of Proposition 5.8.5, we know with $|\alpha_o| = 1$, that

$$\Delta^{\alpha_o}\sigma_o = \Delta^{\alpha_o}\psi(\pi(\mathcal{R}))\ E(\Lambda,\infty)\sigma^{-1} - \sigma_o\ \Delta^{\alpha_o}\sigma\ E(\Lambda,\infty)\sigma^{-1}.$$

Hence

$$\begin{aligned} &\Delta^{\alpha_o}\left\{X^{\beta_o}\sigma(x,\pi)\ \sigma_o(x,\pi)\right\} \\ &\quad = X^{\beta_o}\Delta^{\alpha_o}\sigma(x,\pi)\ \sigma_o(x,\pi) + X^{\beta_o}\sigma(x,\pi)\ \Delta^{\alpha_o}\sigma_o(x,\pi) \\ &\quad = X^{\beta_o}\Delta^{\alpha_o}\sigma(x,\pi)\ \sigma_o(x,\pi) - X^{\beta_o}\sigma(x,\pi)\ \sigma_o\ \Delta^{\alpha_o}\sigma\ E(\Lambda,\infty)\sigma^{-1} \\ &\qquad + X^{\beta_o}\sigma(x,\pi)\ \Delta^{\alpha_o}\psi(\pi(\mathcal{R}))\ \psi_o(\pi(\mathcal{R}))\sigma^{-1}, \end{aligned}$$

where $\psi_o \in C^\infty(\mathbb{R})$ is a fixed smooth function such that $\psi_{o|[\Lambda_1,\infty)} = 1$ and $\psi_{o|(-\infty,\Lambda_1/2)} = 0$. While we can use the hypothesis (5.84) on the first two terms, we use Lemma 5.8.6 for the last term which is then smoothing. Proceeding recursively as in the proof of Proposition 5.8.5, we obtain the estimates for the sum on the right-hand side.

We now define recursively

$$\sigma_n(x,\pi) := \left(\sum_{0<[\alpha]\leq n} \Delta^\alpha\sigma_{n-[\alpha]}X^\alpha\sigma\right)\sigma_o, \quad n = 1, 2, \ldots$$

It is easy to check that each symbol $\sigma_n(x,\pi)$ is in $S^{-m_o-n(\rho-\delta)}_{\rho,\delta}$ and that as in the compact case,

$$\mathrm{Op}(\sigma_o)\mathrm{Op}(\sigma) - I - \mathrm{Op}(\sigma_1)\mathrm{Op}(\sigma) - \ldots - \mathrm{Op}(\sigma_n)\mathrm{Op}(\sigma) \in \Psi^{m-m_0-n}_{\rho,\delta}.$$

Therefore, the operator $B \in \Psi^{-m_o}_{\rho,\delta}$ whose symbol is given by the asymptotic sum $\sigma_o - \sum_{j=1}^{\infty} \sigma_j$ is a left parametrix for $A = \mathrm{Op}(\sigma)$. □

We will see a concrete example of hypoelliptic differential operators on the Heisenberg group in Section 6.6.2, see Example 6.6.4.

We now note the following generalisation of Proposition 5.8.5 that we have already used in the proof of Theorem 5.8.9.

Proposition 5.8.10. *Assume $1 \geq \rho \geq \delta \geq 0$. Let $\sigma \in S^m_{\rho,\delta}$ be a symbol which is $(\mathcal{R}, \Lambda, m_o)$-elliptic with respect to a positive Rockland operator $\mathcal{R}$. If $\psi \in C^\infty(\mathbb{R})$ is such that*

$$\psi_{|(-\infty,\Lambda_1]} = 0 \quad \text{and} \quad \psi_{|[\Lambda_2,\infty)} = 1,$$

for some real numbers Λ_1, Λ_2 satisfying $\Lambda < \Lambda_1 < \Lambda_2$, then the symbol

$$\{\psi(\pi(\mathcal{R}))\sigma^{-1}(x,\pi)\ ,\ (x,\pi) \in G \times \widehat{G}\},$$

given by

$$\psi(\pi(\mathcal{R}))\sigma(x,\pi)^{-1} := \psi(\pi(\mathcal{R}))E_\pi(\Lambda_1,\infty)\sigma^{-1}(x,\pi),$$

is in $S^{-m_o}_{\rho,\delta}$. Moreover, for any $a_o, b_o \in \mathbb{N}_0$, we have

$$\begin{aligned}
&\|\psi(\pi(\mathcal{R}))\sigma^{-1}(x,\pi)\|_{S^{-m_o}_{\rho,\delta}, a_o, b_o, 0} \\
&\leq C \sum_{\substack{a_1', a_2' \leq a_o \\ b_1', b_2' \leq b_o}} \max_{|\gamma| \leq \rho a_o + \delta b_o} C^{a_1'+b_1'+1}_{\gamma,\sigma,\Lambda_1} \|\sigma(x,\pi)\|^{a_2'+b_2'}_{S^m_{\rho,\delta}, a_o, b_o, |m|},
\end{aligned}$$

where $C > 0$ is a positive constant depending on a_o, b_o, ψ, and where the constant $C_{\gamma,\sigma,\Lambda_1}$ was given in (5.79).

Here the elliptic order m_o and the symbol order m are different but the same results holds: one can construct a symbol $\psi(\pi(\mathcal{R}))\sigma^{-1}(x,\pi) \in S^{-m_o}_{\rho,\delta}$. The proof is easily obtained by generalising the proof of Proposition 5.8.5.

We now show that Theorem 5.8.7 has a partial inverse.

Proposition 5.8.11. *Suppose that the operator $A = \mathrm{Op}(\sigma) \in \Psi^m_{\rho,\delta}$, with $1 \geq \rho > \delta \geq 0$, admits a left parametrix $B \in \Psi^{-m}_{\rho,\delta}$, i.e. $BA - \mathrm{I} \in \Psi^{-\infty}$. Then σ is elliptic of order m, that is, there exist a positive Rockland operator $\mathcal{R}$ of homogeneous degree ν, and $\Lambda \in \mathbb{R}$ such that for any $\gamma \in \mathbb{R}$, $x \in G$, μ-almost all $\pi \in \widehat{G}$, and any $u \in \mathcal{H}^\infty_{\pi,\Lambda}$ we have*

$$\|\pi(\mathrm{I}+\mathcal{R})^{\frac{\gamma}{\nu}}\sigma(x,\pi)u\|_{\mathcal{H}_\pi} \geq C_\gamma \|\pi(\mathrm{I}+\mathcal{R})^{\frac{\gamma}{\nu}}\pi(\mathrm{I}+\mathcal{R})^{\frac{m}{\nu}}u\|_{\mathcal{H}_\pi}.$$

Moreover, if this property holds for one positive Rockland operator then it holds for any Rockland operator.

Proof. Let A and B be as in the statement. Let σ and τ be their respective symbols. Then the symbol

$$\begin{aligned} \varepsilon &:= \tau\sigma - \mathrm{I} \\ &= (\tau\sigma - \mathrm{Op}^{-1}(BA)) - (\mathrm{I} - \mathrm{Op}^{-1}(BA)), \end{aligned}$$

is in $S^{-(\rho-\delta)}_{\rho,\delta}$, and we can write

$$\pi(\mathrm{I}+\mathcal{R})^{\frac{m+\gamma}{\nu}}\tau\sigma = \pi(\mathrm{I}+\mathcal{R})^{\frac{m+\gamma}{\nu}} + \epsilon_0\pi(\mathrm{I}+\mathcal{R})^{-\frac{\rho-\delta}{\nu}}\pi(\mathrm{I}+\mathcal{R})^{\frac{m+\gamma}{\nu}},$$

where

$$\varepsilon_0 := \pi(\mathrm{I}+\mathcal{R})^{\frac{m+\gamma}{\nu}}\varepsilon\pi(\mathrm{I}+\mathcal{R})^{\frac{\rho-\delta}{\nu}-\frac{m+\gamma}{\nu}} \in S^0_{\rho,\delta}.$$

For any $u \in \mathcal{H}^\infty_\pi$, $(x,\pi) \in G \times \widehat{G}$, we thus have

$$\begin{aligned} &\|\pi(\mathrm{I}+\mathcal{R})^{\frac{m+\gamma}{\nu}}\tau(x,\pi)\sigma(x,\pi)u\|_{\mathcal{H}_\pi} \\ &\quad = \|\left(\pi(\mathrm{I}+\mathcal{R})^{\frac{m+\gamma}{\nu}} + \epsilon_0(x,\pi)\pi(\mathrm{I}+\mathcal{R})^{-\frac{\rho-\delta}{\nu}}\pi(\mathrm{I}+\mathcal{R})^{\frac{m+\gamma}{\nu}}\right)u\|_{\mathcal{H}_\pi}. \end{aligned}$$

We can bound the left hand side by

$$\begin{aligned} &\|\pi(\mathrm{I}+\mathcal{R})^{\frac{m+\gamma}{\nu}}\tau(x,\pi)\sigma(x,\pi)u\|_{\mathcal{H}_\pi} \\ &\quad \leq \|\pi(\mathrm{I}+\mathcal{R})^{\frac{m+\gamma}{\nu}}\tau(x,\pi)\pi(\mathrm{I}+\mathcal{R})^{-\frac{\gamma}{\nu}}\|_{\mathscr{L}(\mathcal{H}_\pi)}\|\pi(\mathrm{I}+\mathcal{R})^{\frac{\gamma}{\nu}}\sigma(x,\pi)u\|_{\mathcal{H}_\pi} \\ &\quad \leq \|\tau\|_{S^{-m}_{0,0,|\gamma|}}\|\pi(\mathrm{I}+\mathcal{R})^{\frac{\gamma}{\nu}}\sigma(x,\pi)u\|_{\mathcal{H}_\pi}, \end{aligned}$$

and the right hand side below by

$$\begin{aligned} &\|\left(\pi(\mathrm{I}+\mathcal{R})^{\frac{m+\gamma}{\nu}} + \epsilon_0(x,\pi)\pi(\mathrm{I}+\mathcal{R})^{-\frac{\rho-\delta}{\nu}}\pi(\mathrm{I}+\mathcal{R})^{\frac{m+\gamma}{\nu}}\right)u\|_{\mathcal{H}_\pi} \\ &\quad \geq \|\pi(\mathrm{I}+\mathcal{R})^{\frac{m+\gamma}{\nu}}u\|_{\mathcal{H}_\pi} - \|\epsilon_0(x,\pi)\pi(\mathrm{I}+\mathcal{R})^{-\frac{\rho-\delta}{\nu}}\pi(\mathrm{I}+\mathcal{R})^{\frac{m+\gamma}{\nu}}u\|_{\mathcal{H}_\pi} \\ &\quad \geq \|\pi(\mathrm{I}+\mathcal{R})^{\frac{m+\gamma}{\nu}}u\|_{\mathcal{H}_\pi} \\ &\qquad -\|\epsilon_0(x,\pi)\|_{\mathscr{L}(\mathcal{H}_\pi)}\|\pi(\mathrm{I}+\mathcal{R})^{-\frac{\rho-\delta}{\nu}}\pi(\mathrm{I}+\mathcal{R})^{\frac{m+\gamma}{\nu}}u\|_{\mathcal{H}_\pi}. \end{aligned}$$

Hence if $u \in E(\Lambda,\infty)\mathcal{H}^\infty_\pi$ where $\Lambda \geq 0$ then

$$\begin{aligned} &\|\tau\|_{S^{-m}_{0,0,|\gamma|}}\|\pi(\mathrm{I}+\mathcal{R})^{\frac{\gamma}{\nu}}\sigma(x,\pi)u\|_{\mathcal{H}_\pi} \\ &\quad \geq \|\pi(\mathrm{I}+\mathcal{R})^{\frac{m+\gamma}{\nu}}u\|_{\mathcal{H}_\pi} \\ &\qquad -\|\epsilon_0(x,\pi)\|_{\mathscr{L}(\mathcal{H}_\pi)}(1+\Lambda)^{-\frac{\rho-\delta}{\nu}}\|\pi(\mathrm{I}+\mathcal{R})^{\frac{m+\gamma}{\nu}}u\|_{\mathcal{H}_\pi}. \end{aligned}$$

Clearly $\tau \not\equiv 0$ and $\|\tau\|_{S^{-m}_{0,0,|\gamma|}} \neq 0$. Furthermore

$$\|\epsilon_0(x,\pi)\|_{\mathscr{L}(\mathcal{H}_\pi)} \leq \|\epsilon_0\|_{S^0_{\rho,\delta},0,0,0} < \infty,$$

hence we can choose $\Lambda \geq 0$ such that

$$\|\epsilon_0(x,\pi)\|_{\mathscr{L}(\mathcal{H}_\pi)}(1+\Lambda)^{-\frac{\rho-\delta}{\nu}} \leq \|\epsilon_0\|_{S^0_{\rho,\delta},0,0,0}(1+\Lambda)^{-\frac{\rho-\delta}{\nu}} \leq \frac{1}{2},$$

in view of $\rho > \delta$. We have therefore obtained for $u \in E(\Lambda,\infty)\mathcal{H}_\pi^\infty$ with the chosen Λ, that

$$\|\pi(\mathrm{I}+\mathcal{R})^{\frac{\gamma}{\nu}}\sigma(x,\pi)u\|_{\mathcal{H}_\pi} \geq \frac{1}{2\|\tau\|_{S^{-m}_{0,0,|\gamma|}}}\|\pi(\mathrm{I}+\mathcal{R})^{\frac{m+\gamma}{\nu}}u\|_{\mathcal{H}_\pi},$$

which is the required statement. □

5.8.3 Subelliptic estimates and hypoellipticity

The existence of a parametrix yields subelliptic estimates:

Corollary 5.8.12. *Let $m \in \mathbb{R}$ and $1 \geq \rho > \delta \geq 0$. If $A \in \Psi^m_{\rho,\delta}$ is elliptic of order m, then A satisfies the following subelliptic estimates*

$$\forall s \in \mathbb{R} \quad \forall N \in \mathbb{R} \quad \exists C > 0 \quad \forall f \in \mathcal{S}(G) \qquad \|f\|_{L^2_{s+m}} \leq C\Big(\|Af\|_{L^2_s} + \|f\|_{L^2_{-N}}\Big).$$

If $A \in \Psi^m_{\rho,\delta}$ is elliptic of order m_o and satisfies the hypotheses of Theorem 5.8.9, then A satisfies the subelliptic estimates

$$\forall s \in \mathbb{R} \quad \forall N \in \mathbb{R} \quad \exists C > 0 \quad \forall f \in \mathcal{S}(G) \qquad \|f\|_{L^2_{s+m_o}} \leq C\Big(\|Af\|_{L^2_s} + \|f\|_{L^2_{-N}}\Big).$$

In the case $(\rho,\delta) = (1,0)$, assume that $A \in \Psi^m$ is either elliptic of order $m_0 = m$ or is elliptic of some order m_0 and satisfies the hypotheses of Theorem 5.8.9. Then A satisfies the subelliptic estimates

$$\forall s \in \mathbb{R} \quad \forall N \in \mathbb{R} \quad \forall p \in (1,\infty) \quad \exists C > 0 \quad \forall f \in \mathcal{S}(G)$$
$$\|f\|_{L^p_{s+m_o}} \leq C\Big(\|Af\|_{L^p_s} + \|f\|_{L^p_{-N}}\Big).$$

In the estimates above, $\|\cdot\|_{L^p_s}$ denotes any (fixed) Sobolev norm, for example obtained from a (fixed) positive Rockland operator.

Proof. By Theorem 5.8.7 or Theorem 5.8.9, A admits a left parametrix B, i.e. $BA - \mathrm{I} = R \in \Psi^{-\infty}$. By using the boundedness on Sobolev spaces from Corollary 5.7.2, we get

$$\|f\|_{L^2_{s+m_o}} \leq \|BAf\|_{L^2_{s+m_o}} + \|Rf\|_{L^2_{s+m_o}} \leq C(\|Af\|_{L^2_s} + \|f\|_{L^2_{-N}}).$$

In the case $(\rho,\delta) = (1,0)$, the last statement follows from Corollary 5.7.4 with Sobolev L^p-boundedness instead. □

Local hypoelliptic properties

Our construction of parametrices implies the following local property:

Proposition 5.8.13. *Let $A \in \Psi^m_{\rho,\delta}$ with $m \in \mathbb{R}$, $1 \geq \rho > \delta \geq 0$. We assume that the operator A is elliptic of order m_0 and that*

- *either $m = m_0$,*
- *or $m > m_0$ and in this case A satisfies the hypotheses of Theorem 5.8.9.*

Then the singular support of any $f \in \mathcal{S}'(G)$ is contained the singular support of Af,

$$\operatorname{sing\,supp} f \subset \operatorname{sing\,supp} Af,$$

that is, if Af coincides with a smooth function on any open subset of G, then f is also smooth there.

Consequently, if A is a differential operator, then it is hypoelliptic.

The notion of hypoellipticity for a differential operator with smooth coefficients is explained in Appendix A.1.

Proposition 5.8.13 follows easily from the following property:

Lemma 5.8.14. *Let $A \in \Psi^m_{\rho,\delta}$ with $m \in \mathbb{R}$, $1 \geq \rho > \delta \geq 0$. We assume that there exists an open set Ω such that the symbol of A satisfies the elliptic condition in (5.79) for any $x \in \Omega$ only. We also assume that*

- *either $m = m_0$,*
- *or $m > m_0$ and in this case A satisfies the hypotheses of Theorem 5.8.9 with $x \in \Omega$.*

If $f \in \mathcal{S}'(G)$ and if Ω' is an open subset of Ω where Af is smooth, i.e. $Af \in C^\infty(\Omega')$, then $f \in C^\infty(\Omega')$.

The proof requires to revisit the construction of parametrices 'to make it local'.

Proof of Lemma 5.8.14. We keep the hypotheses and notation of the statement. As the properties are essentially local, we may assume that the open subsets Ω, Ω' are open bounded and that there exists an open subset Ω_1 such that $\bar{\Omega}' \subset \Omega_1$ and $\bar{\Omega}_1 \subset \Omega$. Let $\chi \in \mathcal{D}(G)$ be such that $\chi \equiv 1$ on Ω' and $\chi \equiv 0$ outside Ω_1. The symbol of the operator $A' := \chi(x)A$ is given via $\chi(x)\sigma(x,\pi)$. An easy modification of the proof of Proposition 5.8.5 implies that the symbol given by

$$\chi(x)\psi(\pi(\mathcal{R}))\sigma(x,\pi)^{-1}$$

is in $S^{-m_0}_{\rho,\delta}$ (here ψ is a function as in Proposition 5.8.5). Adapting the proof of Theorem 5.8.7 or Theorem 5.8.9, we construct an operator $B \in \Psi^{-m_0}_{\rho,\delta}$ such that $BA' = \chi(x) + R$ with $R \in \Psi^{-\infty}$.

Let $\chi_1 \in \mathcal{D}(G)$ be such that $\chi_1 \equiv 1$ on Ω_1 and $\chi_1 \equiv 0$ outside Ω. Let $f \in \mathcal{S}'(G)$. As A admits a singular integral representation, see Lemma 5.4.15 and its proof, the function $x \mapsto \chi(x)\ A\{(1-\chi_1)f\}(x)$ is smooth and compactly supported. Let us assume that Af is smooth on Ω'. Since we have for any $x \in G$

$$A'\{\chi_1 f\}(x) = \chi(x)\ Af(x) - \chi(x)\ A\{(1-\chi_1)f\}(x),$$

the function $A'\{\chi_1 f\}$ is necessarily smooth and compactly supported on G, i.e. $A'\{\chi_1 f\} \in \mathcal{D}(G)$. Applying B, we have $BA'\{\chi_1 f\} \in \mathcal{S}(G)$ by Theorem 5.2.15. By Corollary 5.5.13. $R\{\chi_1 f\} \in \mathcal{S}(G)$ since the distribution $\chi_1 f \in \mathcal{E}'(G)$ has compact support. Hence $\chi_1 f = BA'\{\chi_1 f\} - R\{\chi_1 f\}$ must be in $\mathcal{S}(G)$. This shows that f is smooth on Ω'. □

Global hypoelliptic-type properties

Our construction of parametrix is global. Hence we also obtain the following global property:

Proposition 5.8.15. *Let $A \in \Psi^m_{\rho,\delta}$ with $m \in \mathbb{R}$, $1 \geq \rho > \delta \geq 0$. We assume that the operator A is elliptic of order m_0 and that*

- *either $m = m_0$,*
- *or $m > m_0$ and in this case A satisfies the hypotheses of Theorem 5.8.9.*

If $f \in \mathcal{S}'(G)$ and $Af \in \mathcal{S}(G)$, then f is smooth and all its left-derivatives (hence also right-derivatives and abelian derivatives) have polynomial growth. More precisely, for any multi-index $\beta \in \mathbb{N}_0^n$, there exists a constant $C > 0$, an integer $M \in \mathbb{N}_0$ and seminorms $\|\cdot\|_{\mathcal{S}'(G),N_1}$, $\|\cdot\|_{\mathcal{S}(G),N_2}$ such that for any $f \in \mathcal{S}'(G)$ with $Af \in \mathcal{S}(G)$, we have

$$|X^\beta f(x)| \leq C\left((1+|x|)^M \|f\|_{\mathcal{S}'(G),N_1} + \|Af\|_{\mathcal{S}(G),N_2}\right), \qquad x \in G.$$

Proof. We keep the hypotheses and notation of the statement. By Theorem 5.8.7 or Theorem 5.8.9, A admits a left parametrix B, i.e. $BA - \mathrm{I} \in \Psi^{-\infty}$. By Corollary 5.4.10, $(BA - \mathrm{I})f$ is smooth with polynomial growth. As $Af \in \mathcal{S}(G)$, $B(Af) \in \mathcal{S}(G)$ by Theorem 5.2.15. Thus

$$f = -(BA - \mathrm{I})f + B(Af)$$

is smooth with polynomial growth. The estimate follows easily from the ones in Corollary 5.4.10 and Theorem 5.2.15. □

Examples

Hence we have obtained hypoellipticity and subelliptic estimates for the operators in Examples 5.8.8.

Corollary 5.8.16. *Let $\mathcal{R}$ be a positive Rockland operator of homogeneous degree ν and let $p \in (1,\infty)$.*

1. *If f_1 and f_2 are complex-valued smooth functions on G such that*
$$\inf_{x\in G, \lambda\geq\Lambda} \frac{|f_1(x)+f_2(x)\lambda|}{1+\lambda} > 0 \quad \textit{for some } \Lambda \geq 0,$$
and such that $X^{\alpha_1}f_1$, $X^{\alpha_2}f_2$ are bounded for each $\alpha_1,\alpha_2 \in \mathbb{N}_0^n$, then the differential operator
$$f_1(x)+f_2(x)\mathcal{R}$$
satisfies the following subelliptic estimates
$$\forall p \in (1,\infty) \quad \forall s \in \mathbb{R} \quad \forall N \in \mathbb{R} \quad \exists C>0 \quad \forall \varphi \in \mathcal{S}(G)$$
$$\|\varphi\|_{L^p_{s+\nu}} \leq C\Big(\|(f_1+f_2\mathcal{R})\varphi\|_{L^p_s} + \|\varphi\|_{L^p_{-N}}\Big),$$
and is (locally) hypoelliptic. It is also globally hypoelliptic in the sense of Proposition 5.8.15.
2. *Let $\psi \in C^\infty(\mathbb{R})$ be such that*
$$\psi_{|(-\infty,\Lambda_1]} = 0 \quad \textit{and} \quad \psi_{|[\Lambda_2,\infty)} = 1,$$
for some real numbers Λ_1, Λ_2 satisfying $0 < \Lambda_1 < \Lambda_2$. Let also f_1 be a smooth complex-valued function on G such that
$$\inf_G |f_1| > 0$$
and that $X^\alpha f_1$ is bounded on G for each $\alpha \in \mathbb{N}_0^n$. Then the operator
$$f_1(x)\psi(\mathcal{R})\mathcal{R} \in \Psi^\nu$$
satisfies the following subelliptic estimates
$$\forall p \in (1,\infty) \quad \forall s \in \mathbb{R} \quad \exists C>0 \quad \forall N \in \mathbb{R} \quad \forall \varphi \in \mathcal{S}(G)$$
$$\|\varphi\|_{L^p_{s+\nu}} \leq C\Big(\|f_1\psi(\mathcal{R})\mathcal{R}\varphi\|_{L^p_s} + \|\varphi\|_{L^p_{-N}}\Big),$$
and is (locally) hypoelliptic. It is also globally hypoelliptic in the sense of Proposition 5.8.15.

6

Differential Operators in Heisenberg Group

The Heisenberg group was introduced in Example 1.6.4. It was our primal example of a stratified Lie group, see Section 3.1.1. Due to the importance of the Heisenberg group and of its many realisations, we start this chapter by sketching various descriptions of the Heisenberg group. We also describe its dual via the well known Schrödinger representations. Eventually, we particularise our general approach given in Chapter 5 to the Heisenberg group. Among other things, we show that using the (Euclidean) Weyl quantization, the analysis of pseudo-differential operators on the Heisenberg group can be reduced to considering scalar-valued symbols parametrised not only by the elements of the Heisenberg group but also by a parameter $\lambda \in \mathbb{R}\backslash\{0\}$; such symbols will be called λ-symbols. The corresponding classes of symbols are of Shubin-type but with an interesting dependence on λ which we explore in detail in this chapter; such classes will be called λ-Shubin classes. Some results of this chapter have been announced in the authors' paper [FR14b], this chapter contains their proofs.

In [BFKG12a], a pseudo-differential calculus on the Heisenberg group was developed with a different approach (but related results) from our work presented here.

There is an important change of notation concerning the Heisenberg group in this chapter. In Example 1.6.4, where the Heisenberg group $\mathbb{H}_{n_o}$ was introduced, we used the index n_o as its subscript because the index n was already used to denote quantities associated with the homogeneous groups. However, throughout Chapter 6, general groups will hardly appear, so we can simplify the notation by denoting the Heisenberg group by $\mathbb{H}_n$ instead of $\mathbb{H}_{n_o}$, so that the notation change is

$$\boxed{\mathbb{H}_{n_o} \longrightarrow \mathbb{H}_n}$$

We emphasise that n is the index here (not the dimension): the topological dimension on $\mathbb{H}_n$ is $2n+1$, and its homogeneous dimension is $2n+2$.

6.1 Preliminaries

In this section, we discuss several aspects of the Heisenberg group, hopefully shedding some light on its importance and general structure.

6.1.1 Descriptions of the Heisenberg group

We remind the reader that the Heisenberg group $\mathbb{H}_n$ was defined in Example 1.6.4 in the following way: the *Heisenberg group* $\mathbb{H}_n$ is the manifold $\mathbb{R}^{2n+1}$ endowed with the law

$$(x,y,t)(x',y',t') := (x+x', y+y', t+t'+\frac{1}{2}(xy'-x'y)), \qquad (6.1)$$

where (x,y,t) and (x',y',t') are in $\mathbb{R}^n \times \mathbb{R}^n \times \mathbb{R} \sim \mathbb{H}_n$.

In the formula above as in the whole chapter, we adopt the following convention: if x and y are two vectors in $\mathbb{R}^n$ for some $n \in \mathbb{N}$, then xy denotes their standard scalar product

$$xy = \sum_{j=1}^{n} x_j y_j \quad \text{if} \quad x = (x_1, \ldots, x_n), \ y = (y_1, \ldots, y_n).$$

First we remark that the factor $\frac{1}{2}$ in the group law given by (6.1) is irrelevant in the following sense. Let $\alpha \in \mathbb{R}^* = \mathbb{R}\backslash\{0\}$. Consider the group $\mathbb{H}_n^{(\alpha)}$ endowed with the law

$$(x,y,t)(x',y',t') := (x+x', y+y', t+t'+\frac{1}{\alpha}(xy'-x'y)).$$

Then the groups $\mathbb{H}_n^{(\alpha)}$ and $\mathbb{H}_n = \mathbb{H}_n^{(2)}$ are isomorphic via

$$\left\{ \begin{array}{ccc} \mathbb{H}_n & \longrightarrow & \mathbb{H}_n^{(\alpha)} \\ (x,y,t) & \longmapsto & (x,y,\frac{2}{\alpha}t) \end{array} \right. .$$

In the same way, consider the *polarised Heisenberg group* $\tilde{\mathbb{H}}_n$ (or $\mathbb{H}_n^{pol}$) endowed with the law

$$(x,y,t)(x',y',t') := (x+x', y+y', t+t'+xy').$$

Then the groups $\tilde{\mathbb{H}}_n$ and $\mathbb{H}_n$ are isomorphic via

$$\left\{ \begin{array}{ccc} \mathbb{H}_n & \longrightarrow & \tilde{\mathbb{H}}_n \\ (x,y,t) & \longmapsto & (x,y,t+\frac{1}{2}xy) \end{array} \right. .$$

Note that the Heisenberg group $\mathbb{H}_n$ can be also viewed as a matrix group. For simplicity, we consider $n = 1$, in which case the group $\tilde{\mathbb{H}}_1$ is isomorphic to T_3, the group of 3-by-3 upper triangular real matrices with 1 on the diagonal:

$$\begin{cases} \tilde{\mathbb{H}}_1 & \longrightarrow & T_3 \\ (x,y,t) & \longmapsto & \begin{bmatrix} 1 & x & t \\ 0 & 1 & y \\ 0 & 0 & 1 \end{bmatrix} \end{cases}.$$

All the statements above can be readily checked by a straightforward computation. Combining two isomorphisms above, we obtain the identification $\mathbb{H}_1 \longrightarrow \tilde{\mathbb{H}}_1 \longrightarrow T_3$ given by

$$\begin{cases} \mathbb{H}_1 & \longrightarrow & T_3 \\ (x,y,t) & \longmapsto & \begin{bmatrix} 1 & x & t+\frac{1}{2}xy \\ 0 & 1 & y \\ 0 & 0 & 1 \end{bmatrix} \end{cases}.$$

Although we will not use it, let us mention a couple of other important appearances of the Heisenberg group. The Heisenberg group can be also realised as a group of transformations; for example, for each

$$h = (x,y,t) \in \mathbb{H}_1,$$

the affine (holomorphic) map given by

$$\phi_h : \mathbb{C} \times \mathbb{C} \ni (z_1, z_2) \longmapsto (z_1 + x + iy, z_2 + t + 2iz_1(x - iy) + i(x^2 + y^2)) \in \mathbb{C} \times \mathbb{C},$$

sends the (Siegel) domain

$$\mathscr{U} := \{(z_1, z_2) \in \mathbb{C} \times \mathbb{C} \ : \ \operatorname{Im} z_2 > |z_1|^2\} \quad (= \mathrm{SU}(2,1)/\mathrm{U}(2))$$

to itself, and the (Shilov) boundary of $\mathscr{U}$,

$$b\mathscr{U} := \{(z_1, z_2) \in \mathbb{C} \times \mathbb{C} \ : \ \operatorname{Im} z_2 = |z_1|^2\},$$

also to itself. One can check that $\mathbb{H}_1 \ni h \mapsto \phi_h$ defines an action of $\mathbb{H}_1$ on $\mathscr{U}$ and on $b\mathscr{U}$. Furthermore, the action of $\mathbb{H}_1$ on $b\mathscr{U}$ is simply transitive. A Cayley type transform

$$(w_1, w_2) \longmapsto (z_1, z_2) \quad \text{with} \quad z_1 = \frac{w_1}{1+w_2}, \quad z_2 = i\frac{1-w_2}{1+w_2},$$

is a biholomorphic bijective mapping which sends $\mathscr{U}$ onto the unit complex ball of $\mathbb{C}^2$. It also send $b\mathscr{U}$ to the unit complex sphere $\mathbb{S}^3$, more precisely onto $\mathbb{S}^3 \backslash \{S\}$ where $S = (0, -1)$ is the south pole (which may be viewed as the image of ∞). Hence the Heisenberg group acts simply transitively on $\mathbb{S}^3 \backslash \{S\}$.

We can also mention here that the group $\mathrm{U}(n)$ acts naturally by automorphisms on $\mathbb{H}_n$ leading to the interpretation of $(\mathrm{U}(n), \mathbb{H}_n)$ as a nilpotent Gelfand pair with strong relation to the theory of commutative convolution algebras. For example, such analysis can be used to characterise Gelfand (spherical) transforms of K-invariant Schwartz functions on $\mathbb{H}_n$ for a group $K \subset \mathrm{U}(n)$ ([BJR98]), or view them as Schwartz functions on the Gelfand spectrum ([ADBR09]).

6.1.2 Heisenberg Lie algebra and the stratified structure

The Lie algebra $\mathfrak{h}_n$ of $\mathbb{H}_n$ is identified with the vector space of left-invariant vector fields. Its canonical basis is given by the left-invariant vector fields

$$X_j = \partial_{x_j} - \frac{y_j}{2}\partial_t, \quad Y_j = \partial_{y_j} + \frac{x_j}{2}\partial_t, \; j = 1, \dots, n, \quad \text{and } T = \partial_t. \tag{6.2}$$

For comparison, the corresponding right-invariant vector fields are

$$\tilde{X}_j = \partial_{x_j} + \frac{y_j}{2}\partial_t, \; \tilde{Y}_j = \partial_{y_j} - \frac{x_j}{2}\partial_t, \; j = 1, \dots, n, \text{ and } \tilde{T} = \partial_t. \tag{6.3}$$

The canonical commutation relations are

$$[X_j, Y_j] = T, \quad j = 1, \dots, n,$$

and T is the centre of $\mathfrak{h}_n$. This shows that the Lie algebra $\mathfrak{h}_n$ and the Lie group $\mathbb{H}_n$ are nilpotent of step 2. Hence the Heisenberg group $\mathbb{H}_n$ described above in Section 6.1.1, that is, $\mathbb{R}^{2n+1}$ endowed with the group law given in (6.1), is the connected simply connected (step-two nilpotent) Lie group whose Lie algebra is $\mathfrak{h}_n$ and which is realised via the exponential mapping together with the canonical basis. This means that the element $(x, y, t) = (x_1, \dots, x_n, y_1, \dots, y_n, t)$ of $\mathbb{H}_n$ can be written as

$$(x, y, t) = \exp_{\mathbb{H}_n}(x_1 X_1 + \ldots + x_n X_n + y_1 Y_1 + \ldots + y_n Y_n + tT).$$

We fix

$$dxdydt = dx_1 \ldots dx_n dy_1 \ldots dy_n dt$$

as the Lebesgue measure on $\mathbb{H}_n$, see Proposition 1.6.6. Therefore, we may be free to write formulae like

$$\int_{\mathbb{H}_n} \cdots \, dxdydt = \int_{\mathbb{R}^{2n+1}} \cdots \, dxdydt.$$

The Heisenberg Lie algebra is stratified via $\mathfrak{h}_n = V_1 \oplus V_2$, where V_1 is linearly spanned by the X_j's and Y_j's, while $V_2 = \mathbb{R}T$. Since the Heisenberg Lie algebra is stratified via $\mathfrak{h}_n = V_1 \oplus V_2$, the natural dilations on the Lie algebra are given by

$$D_r(X_j) = rX_j \quad \text{and} \quad D_r(Y_j) = rY_j, \quad j = 1, \dots, n, \quad \text{and} \quad D_r(T) = r^2 T, \tag{6.4}$$

see Section 3.1.2. We keep the same notation D_r for the dilations on the group $\mathbb{H}_n$. They are therefore given by

$$D_r(x,y,t) = r(x,y,t) = (rx, ry, r^2 t), \quad (x,y,t) \in \mathbb{H}_n, \; r > 0.$$

We also keep the same notation D_r for the dilations on the universal enveloping algebra $\mathfrak{U}(\mathfrak{h}_n)$ induced by Property (6.4).

Note that the homogeneous dimension of $\mathbb{H}_n$ is $Q = 2n + 2$. This is also the homogeneous degree of the Lebesgue measure $dxdydt$.

Example 6.1.1. The sub-Laplacian

$$\begin{aligned} \mathcal{L} &:= \sum_{j=1}^{n} (X_j^2 + Y_j^2) \qquad (6.5) \\ &= \sum_{j=1}^{n} \left(\partial_{x_j} - \frac{y_j}{2}\partial_t\right)^2 + \left(\partial_{y_j} + \frac{x_j}{2}\partial_t\right)^2, \end{aligned}$$

is homogeneous of degree 2 since

$$D_r(\mathcal{L}) = r^2 \mathcal{L}.$$

Remark 6.1.2. The 'canonical' positive Rockland operator in this setting is

$$\mathcal{R} = -\mathcal{L}.$$

We will also use the mapping $\Theta : \mathbb{H}_n \to \mathbb{H}_n$ given by

$$\Theta(x,y,t) := (x, -y, -t).$$

One checks easily that for any $(x,y,t), (x',y',t') \in \mathbb{H}_n$, we have

$$\Theta\big((x,y,t)(x',y',t')\big) = \Theta(x,y,t)\;\Theta(x',y',t') \quad \text{and} \quad \Theta\big(\Theta(x,y,t)\big) = (x,y,t).$$

Therefore, Θ is a group automorphism and an involution. Furthermore, it is clear that it commutes with the dilations:

$$\forall r > 0 \qquad \Theta \circ D_r = D_r \circ \Theta.$$

We keep the same notation for the corresponding Lie algebra morphism and we have

$$\Theta(X_j) = X_j, \; \Theta(Y_j) = -Y_j, \; j = 1, \dots, n, \; \Theta(T) = -T. \qquad (6.6)$$

6.2 Dual of the Heisenberg group

In this section we will analyse the unitary dual of the Heisenberg group $\mathbb{H}_n$. For our purposes, it will be more convenient to work with the Schrödinger representations. This will lead to the group Fourier transform parametrised by λ in (6.19). Such group Fourier transforms yield operators acting on the representation space $L^2(\mathbb{R}^n)$. The latter can be, in turn, analysed using the Weyl quantization on $\mathbb{R}^n$ that appears naturally.

6.2.1 Schrödinger representations π_λ

The Schrödinger representations of the Heisenberg group $\mathbb{H}_n$ are the infinite dimensional unitary representations of $\mathbb{H}_n$, where, as usual, we allow ourselves to identify unitary representations with their unitary equivalence classes. They are parametrised by the co-adjoint orbits (see Section 1.8.1) and more concretely by $\lambda \in \mathbb{R}\backslash\{0\}$. We denote these representations π_λ. Each π_λ acts on the Hilbert space

$$\mathcal{H}_{\pi_\lambda} = L^2(\mathbb{R}^n)$$

in the way we now describe. An element of $L^2(\mathbb{R}^n)$ will very often be denoted as a function h of the variable $u = (u_1, \ldots, u_n) \in \mathbb{R}^n$.

First let us define π_1 corresponding to $\lambda = 1$. It is the representation of the group $\mathbb{H}_n$ acting on $L^2(\mathbb{R}^n)$ via

$$\pi_1(x, y, t)h(u) := e^{i(t+\frac{1}{2}xy)} e^{iyu} h(u + x),$$

for $h \in L^2(\mathbb{R}^n)$ and $(x, y, t) \in \mathbb{H}_n$. Here xy denotes the scalar product in $\mathbb{R}^n$ of x and y, and similarly for yu. Consequently its infinitesimal representation (see Section 1.7) is given by

$$\begin{cases} \pi_1(X_j) &= \partial_{u_j} \quad \text{(differentiate with respect to } u_j), \quad j = 1, \ldots, n, \\ \pi_1(Y_j) &= iu_j, \quad \text{(multiplication by } iu_j), \quad j = 1, \ldots, n, \\ \pi_1(T) &= i\mathrm{I}, \quad \text{(multiplication by } i). \end{cases} \tag{6.7}$$

The Schrödinger representations π_λ on the group are realised in this monograph using

$$\pi_\lambda := \begin{cases} \pi_1 \circ D_{\sqrt{\lambda}} & \text{if } \lambda > 0, \\ \pi_{-\lambda} \circ \Theta & \text{if } \lambda < 0, \end{cases}$$

that is,

$$\pi_\lambda(x, y, t)h(u) = e^{i\lambda(t+\frac{1}{2}xy)} e^{i\sqrt{\lambda}yu} h(u + \sqrt{|\lambda|}x), \tag{6.8}$$

for $h \in L^2(\mathbb{R}^n)$ and $(x, y, t) \in \mathbb{H}_n$ where we use the following convention:

$$\sqrt{\lambda} := \operatorname{sgn}(\lambda)\sqrt{|\lambda|} = \begin{cases} \sqrt{\lambda} & \text{if } \lambda > 0, \\ -\sqrt{|\lambda|} & \text{if } \lambda < 0. \end{cases} \tag{6.9}$$

We observe that for any $\lambda \in \mathbb{R}\backslash\{0\}$ and $r > 0$,

$$\pi_\lambda \circ \Theta = \pi_{-\lambda} \quad \text{and} \quad \pi_\lambda \circ D_r = \pi_{r^2\lambda}, \tag{6.10}$$

and this is true for the group representation π_λ on $\mathbb{H}_n$ and for its corresponding infinitesimal representation on the Lie algebra $\mathfrak{h}_n$ and on the universal enveloping algebra $\mathfrak{U}(\mathfrak{h}_n)$. As usual we keep the same notation, here π_λ for the corresponding infinitesimal representation.

Lemma 6.2.1. *The infinitesimal representation of* π_λ *acts on the canonical basis of* $\mathfrak{h}_n$ *via*

$$\pi_\lambda(X_j) = \sqrt{|\lambda|}\partial_{u_j}, \ \pi_\lambda(Y_j) = i\sqrt{\lambda}u_j, \ j = 1, \ldots, n, \quad \text{and} \quad \pi_\lambda(T) = i\lambda \mathrm{I}, \quad (6.11)$$

using the convention in (6.9).

Proof. Formulae (6.11) can be computed easily from (6.8). Here we show that they also follow from Properties (6.7) and (6.10). Indeed we have for $\lambda > 0$

$$\begin{cases} \pi_\lambda(X_j) &= \pi_1(D_{\sqrt{\lambda}}(X_j)) = \sqrt{\lambda}\pi_1(X_j) = \sqrt{\lambda}\partial_{u_j} \quad j = 1, \ldots, n, \\ \pi_\lambda(Y_j) &= \pi_1(D_{\sqrt{\lambda}}(Y_j)) = \sqrt{\lambda}\pi_1(Y_j) = \sqrt{\lambda}iu_j, \quad j = 1, \ldots, n, \\ \pi_\lambda(T) &= \pi_1(D_{\sqrt{\lambda}}(T)) = \lambda\pi_1(T) = i\lambda, \end{cases}$$

and thus for $\lambda < 0$

$$\begin{cases} \pi_\lambda(X_j) &= \pi_{-\lambda}(\Theta(X_j)) = \pi_{-\lambda}(X_j) = \sqrt{|\lambda|}\partial_{u_j} \quad j = 1, \ldots, n, \\ \pi_\lambda(Y_j) &= \pi_{-\lambda}(\Theta(Y_j)) = -\pi_{-\lambda}(Y_j) = -\sqrt{|\lambda|}iu_j, \quad j = 1, \ldots, n, \\ \pi_\lambda(T) &= \pi_{-\lambda}(\Theta(T)) = -\pi_{-\lambda}(T) = -(-\lambda)i = i\lambda, \end{cases}$$

proving (6.11) in both cases. □

Consequently, the group Fourier transform of the sub-Laplacian

$$\mathcal{L} = \sum_{j=1}^{n}(X_j^2 + Y_j^2)$$

is

$$\pi_\lambda(\mathcal{L}) = |\lambda| \sum_{j=1}^{n}(\partial_{u_j}^2 - u_j^2). \quad (6.12)$$

A direct characterisation implies that the space of smooth vectors of π_λ is

$$\mathcal{H}_{\pi_\lambda}^\infty = \mathcal{S}(\mathbb{R}^n).$$

This is true more generally for any representation of a connected simply connected nilpotent Lie group realised on some $L^2(\mathbb{R}^m)$ via the orbit method, see [CG90, Corollary 4.1.2].

6.2.2 Group Fourier transform on the Heisenberg group

We could have realised the equivalence classes $[\pi_\lambda]$ of Schrödinger representations in various ways. For instance by composing with the unitary operator $U_\lambda : L^2(\mathbb{R}^n) \to L^2(\mathbb{R}^n)$ given by $Uf(x) = |\lambda|^{\frac{n}{2}} f(\sqrt{\lambda}x)$, one would have obtained a slightly different, although equivalent, representation. Another realisation is with the Bargmann representations, see, e.g., [Tay86]. Our choice of representation π_λ

to represent its equivalence class will prove useful in relation with the Weyl-Shubin calculus on $\mathbb{R}^n$ later, see Section 6.5.

The group Fourier transform of a function $\kappa \in L^1(\mathbb{H}_n)$ at π_1 is

$$\mathcal{F}_{\mathbb{H}_n}(\kappa)(\pi_1) = \pi_1(\kappa) = \int_{\mathbb{H}_n} \kappa(x,y,t)\pi_1(x,y,t)^* dxdydt,$$

that is, the operator on $L^2(\mathbb{R}^n)$ given by

$$\pi_1(\kappa)h(u) = \int_{\mathbb{H}_n} \kappa(x,y,t) e^{i(-t+\frac{1}{2}xy)} e^{-iyu} h(u-x) dxdydt.$$

We now fix the notation concerning the Euclidean Fourier transform and recall some facts about the Weyl quantization on $\mathbb{R}^n$.

The Euclidean Fourier transform

In order to give a nicer expression for the operator $\mathcal{F}_{\mathbb{H}_n}(\kappa)(\pi_1)$, we adopt here the following notation for the Euclidean Fourier transform on $\mathbb{R}^N$:

$$\mathcal{F}_{\mathbb{R}^N} f(\xi) = (2\pi)^{-\frac{N}{2}} \int_{\mathbb{R}^N} f(x) e^{-ix\xi} dx, \tag{6.13}$$

where $\xi \in \mathbb{R}^N$ and $f : \mathbb{R}^N \to \mathbb{C}$ is for instance integrable. With our choice of notation and normalisation, the mapping $\mathcal{F}_{\mathbb{R}^N}$ extends unitarily to a mapping on $L^2(\mathbb{R}^N)$ and

$$\mathcal{F}_{\mathbb{R}^N}(f)(x) = \mathcal{F}_{\mathbb{R}^N}^{-1}(f)(-x).$$

Let us also recall the Fourier inversion formula for a (e.g. Schwartz) function $f : \mathbb{R}^n \to \mathbb{C}$:

$$\int_{\mathbb{R}^N} \int_{\mathbb{R}^N} e^{i(u-v)\xi} f(v) dv d\xi = (2\pi)^N f(u). \tag{6.14}$$

In our context N will be equal to $2n+1$.

Unfortunately, due to our choice of notation π for the representations, in the formulae in the sequel π will appear both as a representation and as the constant $\pi = 3.1415926...$ However, as powers of this 2π will appear mostly as constants in front of integrals it should not lead to major confusion.

The (Euclidean) Weyl quantization

Let us also set some notation regarding the *Weyl quantization* on $\mathbb{R}^n$. If a is a symbol, that is, a reasonable function on $\mathbb{R}^n \times \mathbb{R}^n$, then the Weyl quantization associates to a the operator

$$\mathrm{Op}^W(a) \equiv a(D, X)$$

given by

$$\mathrm{Op}^W(a)f(u) = (2\pi)^{-n} \int_{\mathbb{R}^n}\int_{\mathbb{R}^n} e^{i(u-v)\xi} a(\xi, \frac{u+v}{2}) f(v)dvd\xi, \tag{6.15}$$

where $f \in \mathcal{S}(\mathbb{R}^n)$ and $u \in \mathbb{R}^n$.

Example 6.2.2. Particular examples are

$$\mathrm{Op}^W(1) = \mathrm{I}, \quad \mathrm{Op}^W(\xi_j) = \frac{1}{i}\partial_{u_j}, \quad \mathrm{Op}^W(u_j) = u_j,$$

and

$$\mathrm{Op}^W(\xi_k u_j) = \frac{1}{2i}(\partial_{u_k} u_j + u_j \partial_{u_k}).$$

The composition of two Weyl-quantized operators is

$$\mathrm{Op}^W(a) \circ \mathrm{Op}^W(b) = \mathrm{Op}^W(a \star b), \tag{6.16}$$

where (see, e.g., [Ler10])

$$a \star b(\zeta, u) = (2\pi)^{-2n} 4^n \int_{\mathbb{R}^n}\int_{\mathbb{R}^n}\int_{\mathbb{R}^n}\int_{\mathbb{R}^n} e^{-2i\{(\xi-\zeta)(y-u)-(\eta-\zeta)(x-u)\}} a(\xi, x)\ b(\eta, y)\ d\xi d\eta dx dy,$$

and asymptotically

$$a \star b \sim \sum_{m'=0}^{\infty} c_{m',n} \sum_{|\alpha_1|+|\alpha_2|=m'} \frac{(-1)^{|\alpha_2|}}{\alpha_1!\alpha_2!} \left(\left(\frac{1}{i}\partial_\xi\right)^{\alpha_1} \partial_x^{\alpha_2} a\right)\left(\left(\frac{1}{i}\partial_\xi\right)^{\alpha_2} \partial_x^{\alpha_1} b\right), \tag{6.17}$$

with $c_{0,n_0} = 1$ and, in fact,

$$a \star b \sim ab + \frac{1}{2i}\{a, b\} + \dots \quad \text{where} \quad \{a, b\} = \sum_{j=1}^{n}\left(\frac{\partial a}{\partial \xi_j}\frac{\partial b}{\partial u_j} - \frac{\partial a}{\partial u_j}\frac{\partial b}{\partial \xi_j}\right).$$

This formula can already be checked on the basic examples given in Example 6.2.2 and on the following property:

Lemma 6.2.3. *Let a be a symbol. Then we have*

$$\begin{aligned}
(\mathrm{ad} u_j)\left(\mathrm{Op}^W(a)\right) &\equiv u_j \mathrm{Op}^W(a) - \mathrm{Op}^W(a) u_j = \mathrm{Op}^W(i\partial_{\xi_j} a), \\
(\mathrm{ad}\partial_{u_j})\left(\mathrm{Op}^W(a)\right) &\equiv \partial_{u_j}\mathrm{Op}^W(a) - \mathrm{Op}^W(a)\partial_{u_j} = \mathrm{Op}^W(\partial_{u_j} a).
\end{aligned}$$

Proof. Let $f \in \mathcal{S}(\mathbb{R}^n)$ and $u \in \mathbb{R}^n$. Then we have

$$\begin{aligned}
&(\mathrm{ad} u_j)\left(\mathrm{Op}^W(a)\right) f(u) = u_j \mathrm{Op}^W(a) f(u) - \mathrm{Op}^W(a)(u_j f)(u) \\
&= u_j (2\pi)^{-n} \int_{\mathbb{R}^n}\int_{\mathbb{R}^n} e^{i(u-v)\xi} a(\xi, \frac{u+v}{2}) f(v) dv d\xi \\
&\qquad -(2\pi)^{-n} \int_{\mathbb{R}^n}\int_{\mathbb{R}^n} e^{i(u-v)\xi} a(\xi, \frac{u+v}{2}) v_j f(v) dv d\xi \\
&= (2\pi)^{-n} \int_{\mathbb{R}^n}\int_{\mathbb{R}^n} e^{i(u-v)\xi} a(\xi, \frac{u+v}{2})(u_j - v_j) f(v) dv d\xi \\
&= (2\pi)^{-n} \int_{\mathbb{R}^n}\int_{\mathbb{R}^n} \frac{1}{i}\partial_{\xi_j}\left\{e^{i(u-v)\xi}\right\}\ a(\xi, \frac{u+v}{2}) f(v) dv d\xi \\
&= (2\pi)^{-n} \int_{\mathbb{R}^n}\int_{\mathbb{R}^n} e^{i(u-v)\xi} i \partial_{\xi_j}\left\{a(\xi, \frac{u+v}{2})\right\} f(v) dv d\xi,
\end{aligned}$$

after integration by parts. This shows the first equality.

For the second one, we compute

$$\partial_{u_j} \mathrm{Op}^W(a) f(u) = (2\pi)^{-n} \int_{\mathbb{R}^n}\int_{\mathbb{R}^n} \partial_{u_j}\left\{e^{i(u-v)\xi} a(\xi, \frac{u+v}{2})\right\} f(v) dv d\xi.$$

Since

$$\begin{aligned}
\partial_{u_j}\left\{e^{i(u-v)\xi}\ a(\xi, \frac{u+v}{2})\right\} &= -\left\{\partial_{v_j} e^{i(u-v)\xi}\right\} a(\xi, \frac{u+v}{2}) \\
&\quad + \frac{1}{2} e^{i(u-v)\xi}\{\partial_{u_j} a\}(\xi, \frac{u+v}{2}),
\end{aligned}$$

we compute using integration by parts

$$\begin{aligned}
&\int_{\mathbb{R}^n}\int_{\mathbb{R}^n} \partial_{u_j}\left\{e^{i(u-v)\xi} a(\xi, \frac{u+v}{2})\right\} f(v) dv d\xi \\
&= -\int_{\mathbb{R}^n}\int_{\mathbb{R}^n} \left\{\partial_{v_j} e^{i(u-v)\xi}\right\} a(\xi, \frac{u+v}{2}) f(v) dv d\xi \\
&\qquad + \int_{\mathbb{R}^n}\int_{\mathbb{R}^n} e^{i(u-v)\xi} \frac{1}{2}\{\partial_{u_j} a\}(\xi, \frac{u+v}{2}) f(v) dv d\xi \\
&= \int_{\mathbb{R}^n}\int_{\mathbb{R}^n} e^{i(u-v)\xi} \partial_{v_j}\left\{a(\xi, \frac{u+v}{2}) f(v)\right\} dv d\xi \\
&\qquad + \int_{\mathbb{R}^n}\int_{\mathbb{R}^n} e^{i(u-v)\xi} \frac{1}{2}\{\partial_{u_j} a\}(\xi, \frac{u+v}{2}) f(v) dv d\xi.
\end{aligned}$$

Now

$$\partial_{v_j}\left\{a(\xi, \frac{u+v}{2}) f(v)\right\} = \frac{1}{2}\{\partial_{u_j} a\}(\xi, \frac{u+v}{2}) f(v) + a(\xi, \frac{u+v}{2}) \partial_{v_j} f(v),$$

thus

$$\int_{\mathbb{R}^n}\int_{\mathbb{R}^n} \partial_{u_j} \left\{ e^{i(u-v)\xi} a(\xi, \frac{u+v}{2}) \right\} f(v) dv d\xi$$
$$= \int_{\mathbb{R}^n}\int_{\mathbb{R}^n} e^{i(u-v)\xi} \{\partial_{u_j} a\}(\xi, \frac{u+v}{2}) f(v) dv d\xi$$
$$+ \int_{\mathbb{R}^n}\int_{\mathbb{R}^n} e^{i(u-v)\xi} a(\xi, \frac{u+v}{2}) \partial_{v_j} f(v) dv d\xi.$$

We have obtained

$$\partial_{u_j} \mathrm{Op}^W(a) f(u)$$
$$= (2\pi)^{-n} \int_{\mathbb{R}^n}\int_{\mathbb{R}^n} e^{i(u-v)\xi} \{\partial_{u_j} a\}(\xi, \frac{u+v}{2}) f(v) dv d\xi$$
$$+ (2\pi)^{-n} \int_{\mathbb{R}^n}\int_{\mathbb{R}^n} e^{i(u-v)\xi} a(\xi, \frac{u+v}{2}) \partial_{v_j} f(v) dv d\xi.$$

Therefore, we have

$$\left(\mathrm{ad}\partial_{u_j}\right) \left(\mathrm{Op}^W(a)\right) f(u) = \partial_{u_j} \mathrm{Op}^W(a) f(u) - \mathrm{Op}^W(a)(\partial_{u_j} f)(u)$$
$$= (2\pi)^{-n} \int_{\mathbb{R}^n}\int_{\mathbb{R}^n} e^{i(u-v)\xi} \{\partial_{u_j} a\}(\xi, \frac{u+v}{2}) f(v) dv d\xi$$
$$= \mathrm{Op}^W(\partial_{u_j} a) f(u).$$

This shows the second equality. □

The operator $\mathcal{F}_{\mathbb{H}_n}(\kappa)(\pi_1)$

Going back to $\pi_1(\kappa) \equiv \widehat{\kappa}(\pi_1)$ and using the well-known properties of the Euclidean Fourier transform $\mathcal{F}_{\mathbb{R}^{2n+1}}$, for instance see (6.14), it is not difficult to turn into rigorous computations the following calculations:

$$\pi_1(\kappa)h(u) = \int_{\mathbb{R}^{2n+1}} \kappa(x,y,t) e^{i(-t+\frac{1}{2}xy)} e^{-iyu} h(u-x) dx dy dt$$
$$= \int_{\mathbb{R}^{2n+1}}\int_{\mathbb{R}^{2n+1}} (2\pi)^{-\frac{2n+1}{2}} \mathcal{F}_{\mathbb{R}^{2n+1}}(\kappa)(\xi,\eta,\tau) e^{it\tau} e^{iy\eta} e^{ix\xi}$$
$$e^{i(-t+\frac{1}{2}xy)} e^{-iyu} h(u-x) d\xi d\eta d\tau dx dy dt$$
$$= \sqrt{2\pi} \int_{\mathbb{R}^n \times \mathbb{R}^n} \mathcal{F}_{\mathbb{R}^{2n+1}}(\kappa)(\xi, u - \frac{x}{2}, 1) e^{ix\xi} h(u-x) d\xi dx$$
$$= \sqrt{2\pi} \int_{\mathbb{R}^n \times \mathbb{R}^n} \mathcal{F}_{\mathbb{R}^{2n+1}}(\kappa)(\xi, u - \frac{u-v}{2}, 1) e^{i\xi(u-v)} h(v) d\xi dv,$$

after the change of variable $v = u - x$. Comparing this last expression with (6.15), we see that

$$\pi_1(\kappa)h(u) = \sqrt{2\pi} \int_{\mathbb{R}^n}\int_{\mathbb{R}^n} e^{i\xi(u-v)} \mathcal{F}_{\mathbb{R}^{2n+1}}(\kappa)(\xi, \frac{u+v}{2}, 1) h(v) d\xi dv,$$

may be written as

$$\pi_1(\kappa) = (2\pi)^{\frac{2n+1}{2}} \operatorname{Op}^W \left[\mathcal{F}_{\mathbb{R}^{2n+1}}(\kappa)(\cdot,\cdot,1)\right] = (2\pi)^{\frac{2n+1}{2}} \mathcal{F}_{\mathbb{R}^{2n+1}}(\kappa)(D,X,1). \quad (6.18)$$

More generally, we could compute in the same way $\pi_\lambda(\kappa)$ or use the following computational remarks.

Lemma 6.2.4. *Let $\lambda \in \mathbb{R}\backslash\{0\}$. With the convention given in (6.9) we obtain*

$$\begin{aligned} \pi_\lambda(\kappa) &= |\lambda|^{-(n+1)}\pi_{\operatorname{sgn}(\lambda)1}\left(\kappa \circ D_{1/\sqrt{|\lambda|}}\right) && (6.19)\\ &= (2\pi)^{\frac{2n+1}{2}} \operatorname{Op}^W\left[\mathcal{F}_{\mathbb{R}^{2n+1}}(\kappa)(\sqrt{|\lambda|}\,\cdot, \sqrt{\lambda}\,\cdot, \lambda)\right], && (6.20)\end{aligned}$$

or, equivalently,

$$\begin{aligned} &\pi_\lambda(\kappa)h(u)\\ &= \int_{\mathbb{R}^{2n+1}} \kappa(x,y,t)e^{i\lambda(-t+\frac{1}{2}xy)}e^{-i\sqrt{\lambda}yu}h(u-\sqrt{|\lambda|}x)dxdydt && (6.21)\\ &= (2\pi)^{\frac{2n+1}{2}}\int_{\mathbb{R}^n\times\mathbb{R}^n} e^{i(u-v)\xi}\mathcal{F}_{\mathbb{R}^{2n+1}}(\kappa)(\sqrt{|\lambda|}\ \xi, \sqrt{\lambda}\ \frac{u+v}{2}, \lambda)h(v)dvd\xi. && (6.22)\end{aligned}$$

We also have

$$\pi_\lambda(\kappa) = \pi_{-\lambda}(\kappa \circ \Theta), \quad (6.23)$$

and for $r > 0$, $Q = 2n+2$,

$$\pi_\lambda(r^Q\kappa \circ D_r) = \pi_{r^{-2}\lambda}(\kappa). \quad (6.24)$$

For any $X \in \mathfrak{U}(\mathfrak{h}_n)$ and $r > 0$, we have

$$\pi_\lambda(D_{r^{-1}}X) = \pi_{r^{-2}\lambda}(X). \quad (6.25)$$

Here $\mathfrak{U}(\mathfrak{h}_n)$ stands for the universal enveloping algebra of the Lie algebra $\mathfrak{h}_n$, see Section 1.3.

Proof of Lemma 6.2.4. By (6.8), we have for $h \in L^2(\mathbb{R}^n)$ and $(x,y,t) \in \mathbb{H}_n$,

$$\begin{aligned} \pi_\lambda(x,y,t)^*h(u) &= \pi_\lambda\left((x,y,t)^{-1}\right)h(u) = \pi_\lambda(-x,-y,-t)h(u)\\ &= e^{i\lambda(-t+\frac{1}{2}xy)}e^{-i\sqrt{\lambda}yu}h(u-\sqrt{|\lambda|}x). \end{aligned}$$

Thus

$$\begin{aligned} \pi_\lambda(\kappa)h(u) &= \int_{\mathbb{H}_n} \kappa(x,y,t)\ \pi_\lambda(x,y,t)^*h(u)\ dxdydt\\ &= \int_{\mathbb{R}^{2n+1}} \kappa(x,y,t)e^{i\lambda(-t+\frac{1}{2}xy)}e^{-i\sqrt{\lambda}yu}h(u-\sqrt{|\lambda|}x)dxdydt. \end{aligned}$$

This is Formula (6.21).

For Formula (6.23), since by (6.10) we have $\pi_{-\lambda} = \pi_\lambda \circ \Theta$ for any $\lambda \in \mathbb{R}\backslash\{0\}$, we see that

$$\begin{aligned}\pi_\lambda(\kappa) &= \int_{\mathbb{H}_n} \kappa(x,y,t)\pi_\lambda(x,y,t)^* dxdydt \\ &= \int_{\mathbb{H}_n} \kappa(x,y,t)\pi_{-\lambda}\big(\Theta(x,y,t)\big)^* dxdydt \\ &= \int_{\mathbb{H}_n} \kappa\big(\Theta(x,y,t)\big)\pi_{-\lambda}(x,y,t)^* dxdydt = \pi_{-\lambda}(\kappa\circ\Theta),\end{aligned}$$

after the change of variables given by Θ, which has the Jacobian equal to 1. We proceed in the same way for formula (6.24)

$$\begin{aligned}\pi_\lambda(r^Q\kappa\circ D_r) &= \int_{\mathbb{H}_n} \kappa\circ D_r(x,y,t)\pi_\lambda(x,y,t)^* r^Q dxdydt \\ &= \int_{\mathbb{H}_n} \kappa(x,y,t)\pi_\lambda\big(D_r^{-1}(x,y,t)\big)^* dxdydt \\ &= \int_{\mathbb{H}_n} \kappa(x,y,t)\pi_{r^{-2}\lambda}(x,y,t)^* dxdydt = \pi_{r^{-2}\lambda}(\kappa),\end{aligned}$$

after the change of variable given by D_r, using (6.10).

For any $X \in \mathfrak{U}(\mathfrak{h}_n)$ and $\kappa \in \mathcal{S}(G)$, recalling $D_{r^{-1}}X$ from (6.4), then using

$$(X\kappa)\circ D_r = (D_{r^{-1}}X)(\kappa\circ D_r) \tag{6.26}$$

and (6.24), we have

$$\begin{aligned}\pi_{r^{-2}\lambda}(X)\pi_{r^{-2}\lambda}(\kappa) &= \pi_{r^{-2}\lambda}(X\kappa) \\ &= \pi_\lambda(r^Q(X\kappa)\circ D_r) \\ &= \pi_\lambda(r^Q(D_{r^{-1}}X)(\kappa\circ D_r)) \\ &= \pi_\lambda(D_{r^{-1}}X)\pi_\lambda(r^Q\kappa\circ D_r) \\ &= \pi_\lambda(D_{r^{-1}}X)\pi_{r^{-2}\lambda}(\kappa),\end{aligned}$$

and this shows (6.25).

Thus Formulae (6.25), (6.24) and (6.23) hold for any $\lambda \in \mathbb{R}\backslash\{0\}$.

Let us assume $\lambda > 0$. Using $\pi_\lambda = \pi_1 \circ D_{\sqrt{\lambda}}$ we see that

$$\begin{aligned}\pi_\lambda(\kappa) &= \int_{\mathbb{H}_n} \kappa(x,y,t)\pi_1\big(D_{\sqrt{\lambda}}(x,y,t)\big)^* dxdydt \\ &= \int_{\mathbb{H}_n} \kappa\big(D_{1/\sqrt{\lambda}}(x,y,t)\big)\pi_1(x,y,t)^*\lambda^{-(n+1)} dxdydt \\ &= \lambda^{-(n+1)}\pi_1\left(\kappa\circ D_{1/\sqrt{\lambda}}\right),\end{aligned}$$

and this gives Formula (6.19) for $\lambda > 0$. But Formula (6.18) gives here

$$\pi_1\left(\kappa \circ D_{1/\sqrt{\lambda}}\right) = (2\pi)^{n+\frac{1}{2}} \mathrm{Op}^W \left[\mathcal{F}_{\mathbb{R}^{2n+1}}(\kappa \circ D_{1/\sqrt{\lambda}})(\cdot, \cdot, 1)\right].$$

Since a simple change of variable in $\mathbb{R}^{2n+1}$ yields

$$\mathcal{F}_{\mathbb{R}^{2n+1}}\left(\kappa \circ D_{1/\sqrt{\lambda}}\right) = \lambda^{n+1}\left(\mathcal{F}_{\mathbb{R}^{2n+1}}(\kappa)\right) \circ D_{\sqrt{\lambda}}, \tag{6.27}$$

we obtain Formula (6.20) for any $\lambda > 0$.

For $\lambda < 0$, we use Formula (6.23) and the case $\lambda > 0$, that is,

$$\begin{aligned} \pi_\lambda(\kappa) &= \pi_{-\lambda}(\kappa \circ \Theta) \\ &= (-\lambda)^{-(n+1)} \pi_1\left(\kappa \circ \Theta \circ D_{1/\sqrt{-\lambda}}\right) \\ &= (-\lambda)^{-(n+1)} \pi_1\left(\kappa \circ D_{1/\sqrt{-\lambda}} \circ \Theta\right) \\ &= (-\lambda)^{-(n+1)} \pi_{-1}\left(\kappa \circ D_{1/\sqrt{-\lambda}}\right). \end{aligned}$$

Hence Formula (6.19) is proved for any $\lambda < 0$. Here, Formula (6.18) and the relation $\mathcal{F}_{\mathbb{R}^{2n+1}}(\kappa \circ \Theta) = \mathcal{F}_{\mathbb{R}^{2n+1}}(\kappa) \circ \Theta$ with (6.27) give

$$\begin{aligned} \pi_1\left(\kappa \circ \Theta \circ D_{1/\sqrt{-\lambda}}\right) &= (2\pi)^{n+\frac{1}{2}} \mathrm{Op}^W \left[\mathcal{F}_{\mathbb{R}^{2n+1}}(\kappa \circ \Theta \circ D_{1/\sqrt{-\lambda}})(\cdot, \cdot, 1)\right] \\ &= (2\pi)^{n+\frac{1}{2}} (-\lambda)^{n+1} \left(\mathcal{F}_{\mathbb{R}^{2n+1}}(\kappa)\right) \circ \Theta \circ D_{\sqrt{-\lambda}}(\cdot, \cdot, 1), \end{aligned}$$

we obtain Formula (6.20) for any $\lambda < 0$. □

From Lemma 6.2.4 or from (6.11), we see that

$$\pi_\lambda(X_j) = \mathrm{Op}^W(i\sqrt{|\lambda|}\xi_j) \quad \text{and} \quad \pi_\lambda(Y_j) = \mathrm{Op}^W(i\sqrt{\lambda}u_j). \tag{6.28}$$

Remark 6.2.5. This was already noted in [Tay84, BFKG12a]. However in [Tay84], the Fourier transform on $\mathbb{R}^n$ is chosen to be non-unitarily defined by

$$\xi \longmapsto \int_{\mathbb{R}^n} f(x) e^{-ix\xi} dx, \quad f \in \mathcal{S}(\mathbb{R}^n).$$

Remark 6.2.6. The Schwartz space on the Heisenberg group $\mathbb{H}_n$, realised as we have done, is defined as $\mathcal{S}(\mathbb{R}^{2n+1})$, see Section 3.1.9. The characterisation of the Fourier image of the (full) Schwartz space on $\mathbb{H}_n$ is a difficult problem analysed by Geller in [Gel80]. See also the more recent paper [ADBR13].

6.2.3 Plancherel measure

The dual $\widehat{\mathbb{H}}_n$ of the Heisenberg group $\mathbb{H}_n$ may be described together with its Plancherel measure by the orbit method, see Section 1.8.1. Here we obtain a concrete formula for the Plancherel measure μ of the Heisenberg group $\mathbb{H}_n$ using well known properties of Euclidean analysis together with our choice of representatives for the elements of $\widehat{\mathbb{H}}_n$, especially the Schrödinger representations π_λ.

Proposition 6.2.7. *Let $f \in \mathcal{S}(\mathbb{H}_n)$. Then for each $\lambda \in \mathbb{R}\backslash\{0\}$ the operator $\widehat{f}(\pi_\lambda)$ acting on $L^2(\mathbb{R}^n)$ is the Hilbert-Schmidt operator with integral kernel*

$$K_{f,\lambda} : \mathbb{R}^n \times \mathbb{R}^n \longrightarrow \mathbb{C},$$

given by

$$K_{f,\lambda}(u,v) = (2\pi)^{n+\frac{1}{2}} \int_{\mathbb{R}^n} e^{i(u-v)\xi} \mathcal{F}_{\mathbb{R}^{2n+1}}(f)(\sqrt{|\lambda|}\xi, \sqrt{\lambda}\frac{u+v}{2}, \lambda)d\xi,$$

and Hilbert-Schmidt norm

$$\begin{aligned} \|\widehat{f}(\pi_\lambda)\|_{\mathtt{HS}(L^2(\mathbb{R}^n))} &= (2\pi)^{\frac{3n+1}{2}} |\lambda|^{-\frac{n}{2}} \|\mathcal{F}_{\mathbb{R}^{2n+1}}(f)(\cdot,\cdot,\lambda)\|_{L^2(\mathbb{R}^{2n})} \\ &= (2\pi)^{\frac{3n+1}{2}} |\lambda|^{-\frac{n}{2}} \left(\int_{\mathbb{R}^n}\int_{\mathbb{R}^n} |\mathcal{F}_{\mathbb{R}^{2n+1}}(f)(\xi,w,\lambda)|^2 d\xi dw\right)^{\frac{1}{2}}. \end{aligned}$$

Furthermore, we have

$$\int_{\mathbb{H}_n} |f(x,y,t)|^2 dxdydt = c_n \int_{\lambda\in\mathbb{R}\backslash\{0\}} \|\widehat{f}(\pi_\lambda)\|^2_{\mathtt{HS}(L^2(\mathbb{R}^n))} |\lambda|^n d\lambda,$$

where $c_n = (2\pi)^{-(3n+1)}$.

In particular, Proposition 6.2.7 implies that the Plancherel measure μ on the Heisenberg group is supported in $\{[\pi_\lambda],\ \lambda \in \mathbb{R}\backslash\{0\}\}$, see (6.29). Moreover, we have

$$d\mu(\pi_\lambda) \equiv c_n|\lambda|^n d\lambda, \quad \lambda \in \mathbb{R}\backslash\{0\}.$$

The constant c_n depends on our choice of realisation of $\pi_\lambda \in [\pi_\lambda]$.

Proof of Proposition 6.2.7. By (6.22), we have for $h \in L^2(\mathbb{R}^n)$ and $u \in \mathbb{R}^n$,

$$\begin{aligned} \widehat{f}(\pi_\lambda)h(u) &= (2\pi)^{n+\frac{1}{2}} \int_{\mathbb{R}^n}\int_{\mathbb{R}^n} e^{i(u-v)\xi} \mathcal{F}_{\mathbb{R}^{2n+1}}(f)(\sqrt{|\lambda|}\xi, \sqrt{\lambda}\frac{u+v}{2}, \lambda)h(v)dvd\xi \\ &= \int_{\mathbb{R}^n} K_{f,\lambda}(u,v)h(v)dv, \end{aligned}$$

where $K_{f,\lambda}$ is the integral kernel of $\widehat{f}(\pi_\lambda)$ hence given by

$$K_{f,\lambda}(u,v) = (2\pi)^{n+\frac{1}{2}} \int_{\mathbb{R}^n} e^{i(u-v)\xi} \mathcal{F}_{\mathbb{R}^{2n+1}}(f)(\sqrt{|\lambda|}\xi, \sqrt{\lambda}\frac{u+v}{2}, \lambda)d\xi.$$

Using the Euclidean Fourier transform (see (6.13) for our normalisation of $\mathcal{F}_{\mathbb{R}^n}$), we may rewrite this as

$$K_{f,\lambda}(u,v) = (2\pi)^{\frac{3}{2}n+\frac{1}{2}} \mathcal{F}_{\mathbb{R}^n}\left\{\mathcal{F}_{\mathbb{R}^{2n+1}}(f)(\sqrt{|\lambda|}\,\cdot, \sqrt{\lambda}\frac{u+v}{2}, \lambda)\right\}(v-u).$$

The $L^2(\mathbb{R}^n \times \mathbb{R}^n)$-norm of the integral kernel is

$$\begin{aligned}
&\int_{\mathbb{R}^n\times\mathbb{R}^n} |K_{f,\lambda}(u,v)|^2 du dv \\
&\quad = (2\pi)^{3n+1} \int_{\mathbb{R}^n\times\mathbb{R}^n} \left|\mathcal{F}_{\mathbb{R}^n}\left\{\mathcal{F}_{\mathbb{R}^{2n+1}}(f)(\sqrt{|\lambda|}\,\cdot, \sqrt{\lambda}\frac{u+v}{2}, \lambda)\right\}(v-u)\right|^2 du dv \\
&\quad = (2\pi)^{3n+1} \int_{\mathbb{R}^n}\int_{\mathbb{R}^n} \left|\mathcal{F}_{\mathbb{R}^n}\left\{\mathcal{F}_{\mathbb{R}^{2n+1}}(f)(\sqrt{|\lambda|}\,\cdot, w_2, \lambda)\right\}(w_1)\right|^2 |\lambda|^{-\frac{n}{2}} dw_1 dw_2,
\end{aligned}$$

after the change of variable $(w_1, w_2) = (v-u, \sqrt{\lambda}\frac{u+v}{2})$. The (Euclidean) Plancherel formula on $\mathbb{R}^n$ in the variable w_1 (with dual variable ξ_1) then yields

$$\begin{aligned}
&\int_{\mathbb{R}^n\times\mathbb{R}^n} |K_{f,\lambda}(u,v)|^2 du dv \\
&\quad = (2\pi)^{3n+1} \int_{\mathbb{R}^n}\int_{\mathbb{R}^n} |\mathcal{F}_{\mathbb{R}^{2n+1}}(f)(\sqrt{|\lambda|}\xi_1, w_2, \lambda)|^2 |\lambda|^{-\frac{n}{2}} d\xi_1 dw_2 \\
&\quad = (2\pi)^{3n+1} |\lambda|^{-n} \int_{\mathbb{R}^n}\int_{\mathbb{R}^n} |\mathcal{F}_{\mathbb{R}^{2n+1}}(f)(\xi, w_2, \lambda)|^2 d\xi dw_2,
\end{aligned}$$

after the change of variable $\xi = \sqrt{|\lambda|}\xi_1$. Since $f \in \mathcal{S}(\mathbb{H}_n)$, this quantity is finite. Since the integral kernel of $\widehat{f}(\pi_\lambda)$ is square integrable, the operator $\widehat{f}(\pi_\lambda)$ is Hilbert-Schmidt and its Hilbert-Schmidt norm is the L^2-norm of its integral kernel (see, e.g., [RS80, Theorem VI.23]). This shows the first part of the statement.

To finish the proof, we now integrate each side of the last equality against $|\lambda|^n d\lambda$ and then use again the (Euclidean) Plancherel formula on $\mathbb{R}^{2n+1}$ in the variable (ξ, w_2, λ). We obtain

$$\begin{aligned}
&\int_{\mathbb{R}\setminus\{0\}} \int_{\mathbb{R}^n\times\mathbb{R}^n} |K_{f,\lambda}(u,v)|^2 du dv \; |\lambda|^n d\lambda \\
&\quad = (2\pi)^{3n+1} \int_{\mathbb{R}\setminus\{0\}} \int_{\mathbb{R}^n}\int_{\mathbb{R}^n} |\mathcal{F}_{\mathbb{R}^{2n+1}}(f)(\xi, w_2, \lambda)|^2 d\xi dw_2 d\lambda \\
&\quad = (2\pi)^{3n+1} \int_{\mathbb{R}^{2n+1}} |f(x,y,t)|^2 dx dy dt.
\end{aligned}$$

This concludes the proof of Proposition 6.2.7. □

It follows from the Plancherel formula in Proposition 6.2.7 that the Schrödinger representations π_λ, $\lambda \in \mathbb{R}\setminus\{0\}$, are almost all the representations of $\mathbb{H}_n$

modulo unitary equivalence. 'Almost all' here refers to the Plancherel measure $\mu = c_n|\lambda|^n d\lambda$ on $\widehat{\mathbb{H}}_n$. The other representations are finite dimensional and in fact 1-dimensional. They are given by the unitary characters of $\mathbb{H}_n$

$$\chi_w : (x, y, t) \mapsto e^{i(xw_1+yw_2)}, \quad w = (w_1, w_2) \in \mathbb{R}^n \times \mathbb{R}^n \sim \mathbb{R}^{2n}.$$

See also Example 1.8.1 for the link with the orbit method.

We can summarise this paragraph by writing

$$\widehat{\mathbb{H}}_n = \{[\pi_\lambda],\ \lambda \in \mathbb{R}\backslash\{0\}\} \bigcup \{[\chi_w],\ w \in \mathbb{R}^{2n}\} \overset{\mu \text{ a.e.}}{=} \{[\pi_\lambda],\ \lambda \in \mathbb{R}\backslash\{0\}\}. \tag{6.29}$$

6.3 Difference operators

In this section we compute the difference operators Δ_{x_j}, Δ_{y_j}, and Δ_t which are the operators defined via

$$\begin{aligned} \Delta_{x_j}\widehat{\kappa}(\pi_\lambda) &:= \pi_\lambda(x_j\kappa), \\ \Delta_{y_j}\widehat{\kappa}(\pi_\lambda) &:= \pi_\lambda(y_j\kappa), \\ \Delta_t\widehat{\kappa}(\pi_\lambda) &:= \pi_\lambda(t\kappa). \end{aligned}$$

General properties of such difference operators have been analysed in Section 5.2.1. Here we aim at providing explicit expressions for them in the setting of the Heisenberg group $\mathbb{H}_n$.

6.3.1 Difference operators Δ_{x_j} and Δ_{y_j}

We start with the difference operators with respect to x and y.

Lemma 6.3.1. *For any $j = 1, \ldots, n$,*

$$\begin{aligned} \Delta_{x_j}|_{\pi_\lambda} &= \frac{1}{i\lambda}\mathrm{ad}\,(\pi_\lambda(Y_j)) = \frac{1}{\sqrt{|\lambda|}}\mathrm{ad}u_j, \\ \Delta_{y_j}|_{\pi_\lambda} &= -\frac{1}{i\lambda}\mathrm{ad}\,(\pi_\lambda(X_j)) = -\frac{1}{i\sqrt{\lambda}}\mathrm{ad}\partial_{u_j}. \end{aligned}$$

By this we mean that for any κ in some $\mathcal{K}_{a,b}(\mathbb{H}_n)$ such that $x_j\kappa$ is in some $\mathcal{K}_{a',b'}(\mathbb{H}_n)$ or $y_j\kappa$ in some $\mathcal{K}_{a',b'}(\mathbb{H}_n)$ for Δ_{x_j} or Δ_{y_j}, respectively, we have for all $h \in \mathcal{S}(\mathbb{R}^n)$ that

$$\begin{aligned} \left(\Delta_{x_j}\widehat{\kappa}(\pi_\lambda)\right) h\,(u) &= \frac{1}{\sqrt{|\lambda|}}\left(u_j\left(\widehat{\kappa}(\pi_\lambda)h\right)(u) - \left(\widehat{\kappa}(\pi_\lambda)(u_jh)\right)(u)\right), \\ \left(\Delta_{y_j}\widehat{\kappa}(\pi_\lambda)\right) h\,(u) &= \frac{1}{i\sqrt{\lambda}}\left(-\partial_{u_j}\{\widehat{\kappa}(\pi_\lambda)h\}\,(u) + \widehat{\kappa}(\pi_\lambda)\{\partial_{u_j}h\}\,(u)\right). \end{aligned}$$

Proof. Although we could just use direct computations, we prefer to use the following observations. Firstly we have by (6.2) and (6.3) that

$$Y_j - \tilde{Y}_j = x_j \partial_t = \partial_t x_j \quad \text{and} \quad \tilde{X}_j - X_j = y_j \partial_t = \partial_t y_j.$$

Secondly for any κ_1 in some $\mathcal{K}_{a,b}(\mathbb{H}_n)$,

$$\pi_\lambda(\partial_t \kappa_1) = \pi_\lambda(T\kappa_1) = \pi_\lambda(T)\pi_\lambda(\kappa_1) = i\lambda\pi_\lambda(\kappa_1), \tag{6.30}$$

as $T = \partial_t$ and using (6.11). Therefore, these two observations yield

$$\begin{aligned} \pi_\lambda(x_j\kappa) &= \frac{1}{i\lambda}\pi_\lambda(\partial_t x_j \kappa) = \frac{1}{i\lambda}\pi_\lambda\left((Y_j - \tilde{Y}_j)\kappa\right) \\ &= \frac{1}{i\lambda}\left(\pi_\lambda(Y_j\kappa) - \pi_\lambda(\tilde{Y}_j\kappa)\right) \\ &= \frac{1}{i\lambda}\left(\pi_\lambda(Y_j)\pi_\lambda(\kappa) - \pi_\lambda(\kappa)\pi_\lambda(Y_j)\right), \end{aligned}$$

and

$$\begin{aligned} \pi_\lambda(y_j\kappa) &= \frac{1}{i\lambda}\pi_\lambda(\partial_t y_j \kappa) = \frac{1}{i\lambda}\pi_\lambda\left((\tilde{X}_j - X_j)\kappa\right) \\ &= \frac{1}{i\lambda}\left(\pi_\lambda(\kappa)\pi_\lambda(X_j) - \pi_\lambda(X_j)\pi_\lambda(\kappa)\right). \end{aligned}$$

Using Lemma 6.2.1, we have obtained the expressions for Δ_{y_j} and Δ_{x_j} given in the statement. □

Above and also below, we use the formula for the symbols of right derivatives, for example, $\pi_\lambda(\tilde{Y}_j\kappa) = \pi_\lambda(\kappa)\pi_\lambda(Y_j)$, see Proposition 1.7.6, (iv).

Before giving some examples of applications of the difference operators Δ_{x_j} and Δ_{y_j}, let us make a couple of remarks.

Remark 6.3.2. 1. The formulae in Lemma 6.3.1 respect the properties of the automorphism Θ. Indeed, using (6.23) we have

$$\begin{aligned} \left(\Delta_{x_j}\widehat{\kappa}(\pi)\right)|_{\pi=\pi_{-\lambda}} &= (\widehat{x_j\kappa}(\pi))|_{\pi=\pi_{-\lambda}} = \pi_{-\lambda}(x_j\kappa) = \pi_\lambda\left((x_j\kappa)\circ\Theta\right) \\ &= \pi_\lambda\left(x_j \ \kappa\circ\Theta\right) = \Delta_{x_j}\widehat{\kappa\circ\Theta}(\pi_\lambda) = \Delta_{x_j}\left(\widehat{\kappa}(\pi_{-\lambda})\right), \\ \left(\Delta_{y_j}\widehat{\kappa}(\pi)\right)|_{\pi=\pi_{-\lambda}} &= (\widehat{y_j\kappa}(\pi))|_{\pi=\pi_{-\lambda}} = \pi_{-\lambda}(y_j\kappa) = \pi_\lambda\left((y_j\kappa)\circ\Theta\right) \\ &= \pi_\lambda\left(-y_j \ \kappa\circ\Theta\right) = -\Delta_{y_j}\widehat{\kappa\circ\Theta}(\pi_\lambda) = -\Delta_{y_j}\left(\widehat{\kappa}(\pi_{-\lambda})\right). \end{aligned}$$

This can also be viewed directly from the formulae in Lemma 6.3.1:

$$\begin{aligned} \left(\Delta_{x_j}\widehat{\kappa}(\pi)\right)|_{\pi=\pi_{-\lambda}} &= \frac{1}{\sqrt{|-\lambda|}}\mathrm{ad}u_j\left(\widehat{\kappa}(\pi_{-\lambda})\right) = \Delta_{x_j}\left(\widehat{\kappa}(\pi_{-\lambda})\right), \\ \left(\Delta_{y_j}\widehat{\kappa}(\pi)\right)|_{\pi=\pi_{-\lambda}} &= -\frac{1}{i\sqrt{-\lambda}}\mathrm{ad}\partial_{u_j} = -\Delta_{y_j}\left(\widehat{\kappa}(\pi_{-\lambda})\right). \end{aligned}$$

2. The formulae in Lemma 6.3.1 respect the properties of the dilations D_r. This time using (6.24), we have

$$\begin{aligned}\left(\Delta_{x_j}\widehat{\kappa}(\pi)\right)|_{\pi=\pi_{r^{-2}\lambda}} &= (\widehat{x_j\kappa}(\pi))\,|_{\pi=\pi_{r^{-2}\lambda}} = \pi_{r^{-2}\lambda}(x_j\kappa) = \pi_\lambda\left(r^Q(x_j\kappa)\circ D_r\right)\\ &= r\;\pi_\lambda\left(r^Q x_j\;\kappa\circ D_r\right) = r\;\Delta_{x_j}\left(\widehat{\kappa}(\pi_{r^{-2}\lambda})\right).\end{aligned}$$

This can also be viewed directly from the formulae in Lemma 6.3.1:

$$\begin{aligned}\left(\Delta_{x_j}\widehat{\kappa}(\pi)\right)|_{\pi=\pi_{r^{-2}\lambda}} &= \frac{1}{\sqrt{|r^{-2}\lambda|}}\left(\mathrm{ad}u_j\right)\left(\widehat{\kappa}(\pi_{r^{-2}\lambda})\right)\\ &= r\times\left(\frac{1}{\sqrt{|\lambda|}}\left(\mathrm{ad}u_j\right)\left(\widehat{\kappa}(\pi_{r^{-2}\lambda})\right)\right)\\ &= r\;\Delta_{x_j}\left(\widehat{\kappa}(\pi_{r^{-2}\lambda})\right).\end{aligned}$$

In exactly the same two ways we obtain for Δ_{y_j} that

$$\left(\Delta_{y_j}\widehat{\kappa}(\pi)\right)|_{\pi=\pi_{r^{-2}\lambda}} = r\Delta_{y_j}\left(\widehat{\kappa}(\pi_{r^{-2}\lambda})\right).$$

Lemmata 6.3.1 and 6.2.3 imply:

Corollary 6.3.3. *If $\widehat{\kappa}(\pi_\lambda) = \mathrm{Op}^W(a_\lambda)$ and $a_\lambda = \{a_\lambda(\xi,u)\}$, then*

$$\begin{aligned}\Delta_{x_j}\widehat{\kappa}(\pi_\lambda) &= \mathrm{Op}^W\left(\frac{i}{\sqrt{|\lambda|}}\partial_{\xi_j}a_\lambda\right),\\ \Delta_{y_j}\widehat{\kappa}(\pi_\lambda) &= \mathrm{Op}^W\left(\frac{i}{\sqrt{\lambda}}\partial_{u_j}a_\lambda\right).\end{aligned}$$

If $\widehat{\kappa}(\pi_\lambda) = \mathrm{Op}^W(a_\lambda)$ and $a_\lambda = \{a_\lambda(\xi,u)\}$ as in the statement above, we will often say that a_λ is the λ-*symbol.*

Up to now, we analysed the difference operators applied to a 'general' group Fourier transform of a distribution κ (provided that the difference operators made sense, see Definition 5.2.1 and the subsequent discussion). This is equivalent to applying difference operators acting on symbols, see Section 5.1.3. In what follows, we particularise this to some known symbols, mainly to the one in Example 5.1.26, that is, to $\pi(A)$ where A is a left-invariant differential operator such as $A = X_j, Y_j$ or T.

We now give some explicit examples.

Example 6.3.4. We already know that $\Delta_{x_j}\mathrm{I} = 0$, see Example 5.2.8. We can compute

$$\Delta_{x_j}\pi_\lambda(X_k) = -\delta_{jk}\mathrm{I},\quad \Delta_{x_j}\pi_\lambda(Y_k) = 0\quad\text{and}\quad \Delta_{x_j}\pi_\lambda(T) = 0, \tag{6.31}$$

and

$$\Delta_{x_j}\pi_\lambda(\mathcal{L}) = -2\pi_\lambda(X_j). \tag{6.32}$$

Proof. By Lemma 6.3.1,

$$\begin{aligned}\Delta_{x_j}\pi_\lambda(X_k) &= \frac{1}{i\lambda}\mathrm{ad}\,(\pi_\lambda(Y_j))\,\pi_\lambda(X_k) = \frac{1}{i\lambda}[\pi_\lambda(Y_j),\pi_\lambda(X_k)] \\ &= \frac{1}{i\lambda}\pi_\lambda[Y_j,X_k],\end{aligned}$$

since π_λ is a representation of the Lie algebra $\mathfrak{g}$. Similarly,

$$\begin{aligned}\Delta_{x_j}\pi_\lambda(Y_k) &= \frac{1}{i\lambda}\mathrm{ad}\,(\pi_\lambda(Y_j))\,\pi_\lambda(Y_k) = \frac{1}{i\lambda}\pi_\lambda[Y_j,Y_k], \\ \Delta_{x_j}\pi_\lambda(T) &= \frac{1}{i\lambda}\mathrm{ad}\,(\pi_\lambda(Y_j))\,\pi_\lambda(T) = \frac{1}{i\lambda}\pi_\lambda[Y_j,T].\end{aligned}$$

By the canonical commutation relations, we have

$$[Y_j,X_k] = -\delta_{jk}T, \quad [Y_j,Y_k]=0 \quad \text{and} \quad [Y_j,T]=0.$$

Since $\pi_\lambda(T) = i\lambda\mathrm{I}$, we obtain (6.31).

In the same way, we have

$$\Delta_{x_j}\pi_\lambda(X_k)^2 = \frac{1}{i\lambda}\pi_\lambda[Y_j,X_k^2] \quad \text{and} \quad \Delta_{x_j}\pi_\lambda(Y_k^2) = \frac{1}{i\lambda}\pi_\lambda[Y_j,Y_k^2].$$

Using the canonical commutation relations, we see that Y_j and Y_k commute in the Lie algebra $\mathfrak{g}$ thus Y_j and Y_k^2 commute in the enveloping Lie algebra $\mathfrak{U}(\mathfrak{g})$: $[Y_j,Y_k^2]=0$. Again using the canonical commutation relation we compute

$$[Y_j,X_k^2] = -2\delta_{jk}X_kT,$$

since

$$\begin{aligned}Y_jX_k^2 &= Y_jX_kX_k = (-\delta_{jk}T + X_kY_j)X_k \\ &= -\delta_{jk}TX_k + X_k(-\delta_{jk}T + X_kY_j) \\ &= -2\delta_{jk}X_kT + X_k^2Y_j.\end{aligned}$$

Therefore,

$$\begin{aligned}\Delta_{x_j}\pi_\lambda(X_k)^2 &= \frac{1}{i\lambda}\pi_\lambda(-2\delta_{jk}X_kT) = \frac{-2\delta_{jk}}{i\lambda}\pi_\lambda(X_kT) = \frac{-2\delta_{jk}}{i\lambda}\pi_\lambda(X_k)\pi_\lambda(T) \\ &= \frac{-2\delta_{jk}}{i\lambda}\pi_\lambda(X_k)(i\lambda) = -2\delta_{jk}\pi_\lambda(X_k),\end{aligned}$$

and $\Delta_{x_j}\pi_\lambda(Y_k^2) = 0$. This implies (6.32). □

Example 6.3.5. We already know that $\Delta_{y_j}\mathrm{I} = 0$, see Example 5.2.8. We can compute

$$\Delta_{y_j}\pi_\lambda(X_k) = 0, \quad \Delta_{y_j}\pi_\lambda(Y_k) = -\delta_{jk}\mathrm{I} \quad \text{and} \quad \Delta_{y_j}\pi_\lambda(T) = 0, \tag{6.33}$$

and

$$\Delta_{y_j}\pi_\lambda(\mathcal{L}) = -2\pi_\lambda(Y_j). \tag{6.34}$$

Proof. Proceeding as in the proof of Example 6.3.4, we have

$$\begin{aligned}\Delta_{y_j}\pi_\lambda(X_k) &= -\frac{1}{i\lambda}\mathrm{ad}\,(\pi_\lambda(X_j))\,\pi_\lambda(X_k) = -\frac{1}{i\lambda}\pi_\lambda[X_j, X_k],\\ \Delta_{y_j}\pi_\lambda(Y_k) &= -\frac{1}{i\lambda}\mathrm{ad}\,(\pi_\lambda(X_j))\,\pi_\lambda(Y_k) = -\frac{1}{i\lambda}\pi_\lambda[X_j, Y_k],\\ \Delta_{y_j}\pi_\lambda(T) &= -\frac{1}{i\lambda}\mathrm{ad}\,(\pi_\lambda(X_j))\,\pi_\lambda(T) = -\frac{1}{i\lambda}\pi_\lambda[X_j, T],\end{aligned}$$

and this together with the canonical commutation relations and $\pi_\lambda(T) = i\lambda\mathrm{I}$, yield (6.33).

For thc second part of Example 6.3.5, we have

$$\Delta_{y_j}\pi_\lambda(X_k)^2 = -\frac{1}{i\lambda}\pi_\lambda[X_j, X_k^2] \quad \text{and} \quad \Delta_{y_j}\pi_\lambda(Y_k^2) = -\frac{1}{i\lambda}\pi_\lambda[X_j, Y_k^2],$$

and using the canonical commutation relations we compute $[X_j, X_k^2] = 0$ whereas

$$[X_j, Y_k^2] = 2\delta_{jk}Y_kT,$$

since

$$\begin{aligned}X_jY_k^2 &= X_jY_kY_k = (\delta_{jk}T + Y_kX_j)Y_k\\ &= \delta_{jk}TY_k + Y_k(\delta_{jk}T + Y_kX_j)\\ &= 2\delta_{jk}Y_kT + Y_k^2X_j.\end{aligned}$$

Therefore

$$\Delta_{y_j}\pi_\lambda(Y_k)^2 = -\frac{1}{i\lambda}\pi_\lambda(2\delta_{jk}Y_kT) = -2\delta_{jk}\pi_\lambda(Y_k) \quad \text{and} \quad \Delta_{y_j}\pi_\lambda(X_k^2) = 0.$$

This implies (6.34). □

6.3.2 Difference operator Δ_t

Naturally, very important information will be contained in the difference operator corresponding to multiplication by t.

Lemma 6.3.6. *We have*

$$\Delta_t|_{\pi_\lambda} = i\partial_\lambda + \frac{1}{2}\sum_{j=1}^{n}\Delta_{x_j}\Delta_{y_j}|_{\pi_\lambda} + \frac{i}{2\lambda}\sum_{j=1}^{n}\left\{\pi_\lambda(Y_j)\Delta_{y_j}|_{\pi_\lambda} + \Delta_{x_j}|_{\pi_\lambda}\pi_\lambda(X_j)\right\}.$$

By this we mean that for any κ in some $\mathcal{K}_{a,b}(\mathbb{H}_n)$ such that $t\kappa$ is in some $\mathcal{K}_{a',b'}(\mathbb{H}_n)$, we have

$$\begin{aligned}\Delta_t\pi_\lambda(\kappa) &= i\partial_\lambda\pi_\lambda(\kappa) + \frac{1}{2}\sum_{j=1}^{n}\Delta_{x_j}\Delta_{y_j}\pi_\lambda(\kappa)\\ &\quad + \frac{i}{2\lambda}\sum_{j=1}^{n}\left\{\pi_\lambda(Y_j)\Delta_{y_j}\pi_\lambda(\kappa) + \Delta_{x_j}\pi_\lambda(\kappa)\pi_\lambda(X_j)\right\},\end{aligned}$$

or, rewriting this with the equivalent notation $\widehat{\kappa}(\pi_\lambda)$ as before,

$$\begin{aligned}\Delta_t\widehat{\kappa}(\pi_\lambda) &= i\partial_\lambda\widehat{\kappa}(\pi_\lambda)+\frac{1}{2}\sum_{j=1}^n\Delta_{x_j}\Delta_{y_j}\widehat{\kappa}(\pi_\lambda)\\ &\quad+\frac{i}{2\lambda}\sum_{j=1}^n\left\{\pi_\lambda(Y_j)\Delta_{y_j}\widehat{\kappa}(\pi_\lambda)+\Delta_{x_j}\widehat{\kappa}(\pi_\lambda)\pi_\lambda(X_j)\right\}.\end{aligned}$$

Before giving some examples of applications of the difference operator Δ_t, let us make a couple of remarks.

Remark 6.3.7. 1. This lemma shows that the difference operators act on the field of operators $\{\pi_\lambda(\kappa),\ \lambda\in\mathbb{R}\backslash\{0\}\}$, rather than on 'one' $\pi_\lambda(\kappa)$ for an individual λ, see Remark 5.2.2.

2. In a similar way as in Remark 6.3.2, the formula in Lemma 6.3.6 respects the properties of the automorphism Θ and the dilations D_r. Indeed, using (6.23) we have

$$\begin{aligned}(\Delta_t\widehat{\kappa}(\pi))\,|_{\pi=\pi_{-\lambda}} &= \left(\widehat{t\kappa}(\pi)\right)|_{\pi=\pi_{-\lambda}}=\pi_{-\lambda}(t\kappa)=\pi_\lambda\left((t\kappa)\circ\Theta\right)\\ &= \pi_\lambda\left(-t\ \kappa\circ\Theta\right)=-\Delta_t\widehat{\kappa\circ\Theta}(\pi_\lambda)=-\Delta_t\left(\widehat{\kappa}(\pi_{-\lambda})\right),\end{aligned}$$

that is

$$(\Delta_t\widehat{\kappa}(\pi))\,|_{\pi=\pi_{-\lambda}}=-\Delta_t\left(\widehat{\kappa}(\pi_{-\lambda})\right). \tag{6.35}$$

For the dilations, using (6.24), we have

$$\begin{aligned}(\Delta_t\widehat{\kappa}(\pi))\,|_{\pi=\pi_{r^{-2}\lambda}} &= \left(\widehat{t\kappa}(\pi)\right)|_{\pi=\pi_{r^{-2}\lambda}}=\pi_{r^{-2}\lambda}(t\kappa)=\pi_\lambda\left(r^Q(t\kappa)\circ D_r\right)\\ &= r^2\pi_\lambda\left(r^Qt\ \kappa\circ D_r\right)=r^2\Delta_t\left(\widehat{\kappa}(\pi_{r^{-2}\lambda})\right).\end{aligned}$$

that is

$$(\Delta_t\widehat{\kappa}(\pi))\,|_{\pi=\pi_{r^{-2}\lambda}}=r^2\Delta_t\left(\widehat{\kappa}(\pi_{r^{-2}\lambda})\right). \tag{6.36}$$

Formulae (6.35) and (6.36) can also be viewed directly from the formula in Lemma 6.3.6:

$$\begin{aligned}(\Delta_t\widehat{\kappa}(\pi))\,|_{\pi=\pi_{-\lambda}} &= i\partial_{\lambda_1=-\lambda}\{\pi_{\lambda_1}(\kappa)\}+\frac{1}{2}\sum_{j=1}^n\{\Delta_{x_j}\Delta_{y_j}\pi(\kappa)\}_{\pi=\pi_{-\lambda}}\\ &\quad+\frac{i}{-2\lambda}\sum_{j=1}^n\{\pi(Y_j)\Delta_{y_j}\pi(\kappa)+\Delta_{x_j}\pi(\kappa)\pi(X_j)\}_{\pi=\pi_{-\lambda}},\end{aligned} \tag{6.37}$$

$$\begin{aligned}(\Delta_t\widehat{\kappa}(\pi))\,|_{\pi=\pi_{r^{-2}\lambda}} &= i\partial_{\lambda_1=r^{-2}\lambda}\{\pi_{\lambda_1}(\kappa)\}+\frac{1}{2}\sum_{j=1}^n\{\Delta_{x_j}\Delta_{y_j}\pi(\kappa)\}_{\pi=\pi_{r^{-2}\lambda}}\\ &\quad+\frac{i}{2r^{-2}\lambda}\sum_{j=1}^n\{\pi(Y_j)\Delta_{y_j}\pi(\kappa)+\Delta_{x_j}\pi(\kappa)\pi(X_j)\}_{\pi=\pi_{r^{-2}\lambda}}.\end{aligned} \tag{6.38}$$

For the first terms in the right hand side in (6.37) and (6.38) we have easily that

$$\begin{aligned} \partial_{\lambda_1=-\lambda}\pi_{\lambda_1}(\kappa) &= -\partial_\lambda\{\pi_{-\lambda}(\kappa)\}, \\ \partial_{\lambda_1=r^{-2}\lambda}\pi_{\lambda_1}(\kappa) &= r^2\partial_\lambda\{\pi_{r^{-2}\lambda}(\kappa)\}. \end{aligned}$$

From Remark 6.3.2 we know that

$$\left\{ \begin{aligned} \left(\Delta_{x_j}\widehat{\kappa}(\pi)\right)|_{\pi=\pi_{-\lambda}} &= \Delta_{x_j}\left(\widehat{\kappa}(\pi_{-\lambda})\right) \\ \left(\Delta_{y_j}\widehat{\kappa}(\pi)\right)|_{\pi=\pi_{-\lambda}} &= -\Delta_{y_j}\left(\widehat{\kappa}(\pi_{-\lambda})\right) \\ \left(\Delta_{x_j}\widehat{\kappa}(\pi)\right)|_{\pi=\pi_{r^{-2}\lambda}} &= r\Delta_{x_j}\left(\widehat{\kappa}(\pi_{r^{-2}\lambda})\right) \\ \left(\Delta_{y_j}\widehat{\kappa}(\pi)\right)|_{\pi=\pi_{r^{-2}\lambda}} &= r\Delta_{y_j}\left(\widehat{\kappa}(\pi_{r^{-2}\lambda})\right) \end{aligned} \right. \tag{6.39}$$

so we have for the second term of the right hand side in (6.37) and (6.38) respectively:

$$\begin{aligned} \sum_{j=1}^n\{\Delta_{x_j}\Delta_{y_j}\pi(\kappa)\}_{\pi=\pi_{-\lambda}} &= -\sum_{j=1}^n\Delta_{x_j}\Delta_{y_j}\left(\widehat{\kappa}(\pi_{-\lambda})\right), \\ \sum_{j=1}^n\{\Delta_{x_j}\Delta_{y_j}\pi(\kappa)\}_{\pi=\pi_{r^{-2}\lambda}} &= r^2\sum_{j=1}^n\Delta_{x_j}\Delta_{y_j}\left(\widehat{\kappa}(\pi_{r^{-2}\lambda})\right). \end{aligned}$$

Now viewing X_j and Y_j as elements of the Lie algebra and left invariant vector fields, we see using (6.23) and (6.6) that

$$\begin{aligned} \pi_{-\lambda}(X_j) &= \pi_{-\lambda}(\Theta(X_j)) = \pi_{-\lambda}(X_j\circ\Theta) = \pi_\lambda(X_j), \\ \pi_{-\lambda}(Y_j) &= -\pi_{-\lambda}(\Theta(Y_j)) = -\pi_{-\lambda}(Y_j\circ\Theta) = -\pi_\lambda(Y_j), \end{aligned}$$

and, using (6.25) and (6.4), we obtain

$$\begin{aligned} \pi_{r^{-2}\lambda}(X_j) &= \pi_\lambda(D_{r^{-1}}X_j) = r^{-1}\pi_\lambda(X_j), \\ \pi_{r^{-2}\lambda}(Y_j) &= \pi_\lambda(D_{r^{-1}}Y_j) = r^{-1}\pi_\lambda(Y_j). \end{aligned}$$

So from this and (6.39) we obtain for the third terms of the right hand side in (6.35) and in (6.36) that

$$\begin{aligned} &\frac{i}{-2\lambda}\sum_{j=1}^n\{\pi(Y_j)\Delta_{y_j}\pi(\kappa)+\Delta_{x_j}\pi(\kappa)\pi(X_j)\}_{\pi=\pi_{-\lambda}} \\ &\quad= -\frac{i}{2\lambda}\sum_{j=1}^n\pi_{-\lambda}(Y_j)\Delta_{y_j}\pi_{-\lambda}(\kappa)+\Delta_{x_j}\pi_{-\lambda}(\kappa)\pi_{-\lambda}(X_j), \\ &\frac{i}{2r^{-2}\lambda}\sum_{j=1}^n\{\pi(Y_j)\Delta_{y_j}\pi(\kappa)+\Delta_{x_j}\pi(\kappa)\pi(X_j)\}_{\pi=\pi_{r^{-2}\lambda}} \\ &\quad= r^2\frac{i}{2\lambda}\sum_{j=1}^n\pi_{r^{-2}\lambda}(Y_j)\Delta_{y_j}\pi_{r^{-2}\lambda}(\kappa)+\Delta_{x_j}\pi_{r^{-2}\lambda}(\kappa)\pi_{-\lambda}(X_j). \end{aligned}$$

Collecting the new expressions for the three terms of the right hand sides in (6.35) and in (6.36) we obtain a new proof for Equalities (6.35) and (6.36).

Proof of Lemma 6.3.6. Let κ be in some $\mathcal{K}_{a,b}(\mathbb{H}_n)$ and $h \in \mathcal{S}(\mathbb{R}^n)$. We start by differentiating with respect to λ the expression from Lemma 6.2.4:

$$\pi_\lambda(\kappa)h(u) = \int_{\mathbb{H}_n} \kappa(x,y,t)e^{i\lambda(-t+\frac{1}{2}xy)}e^{-i\sqrt{\lambda}yu}h(u-\sqrt{|\lambda|}x)dxdydt,$$

and obtain

$$\partial_\lambda\{\pi_\lambda(\kappa)h(u)\} = \int_{\mathbb{H}_n} \kappa(x,y,t)e^{i\lambda(-t+\frac{1}{2}xy)}e^{-i\sqrt{\lambda}yu}$$
$$\left(\left[i(-t+\frac{1}{2}xy) - i\frac{yu}{2\sqrt{|\lambda|}}\right]h(u-\sqrt{|\lambda|}x) - \frac{1}{2\sqrt{\lambda}}x\nabla h(u-\sqrt{|\lambda|}x)\right)dxdydt;$$

indeed with our convention we have

$$x\nabla h = \sum_{j=1}^n x_j\partial_{u_j}h, \quad \text{and} \quad \partial_\lambda\{\sqrt{\lambda}\} = \frac{1}{2\sqrt{|\lambda|}}, \quad \partial_\lambda\{\sqrt{|\lambda|}\} = \frac{1}{2\sqrt{\lambda}}.$$

We can now interpret the formula above in the light of difference operators as

$$\begin{aligned}
\partial_\lambda\pi_\lambda(\kappa) &= i\pi_\lambda((-t+\frac{1}{2}xy)\kappa) + \sum_{j=1}^n\left\{-\frac{iu_j}{2\sqrt{|\lambda|}}\pi_\lambda(y_j\kappa) - \frac{1}{2\sqrt{\lambda}}\pi_\lambda(x_j\kappa)\partial_{u_j}\right\} \\
&= -i\Delta_t\pi_\lambda(\kappa) + \frac{i}{2}\sum_{j=1}^n\Delta_{x_j}\Delta_{y_j}\pi_\lambda(\kappa) \\
&\quad -\frac{1}{2\lambda}\sum_{j=1}^n\left\{\pi_\lambda(Y_j)\left(\Delta_{y_j}\pi_\lambda(\kappa)\right) + \left(\Delta_{x_j}\pi_\lambda(\kappa)\right)\pi_\lambda(X_j)\right\},
\end{aligned}$$

using (6.11). □

We already know that

$$\Delta_t\mathrm{I} = 0 \quad \text{and} \quad \Delta_t\pi_\lambda(X_k) = \Delta_t\pi_\lambda(Y_k) = 0, \tag{6.40}$$

see Example 5.2.8 and Lemma 5.2.9, but we can also test it with the formula given in Lemma 6.3.6. We also obtain the following (more substantial) examples:

Example 6.3.8. We can compute

$$\Delta_t\pi_\lambda(T) = -\mathrm{I}, \tag{6.41}$$

and

$$\Delta_t\pi_\lambda(\mathcal{L}) = 0. \tag{6.42}$$

Proof. Since

$$\pi_\lambda(T) = i\lambda \mathrm{I}$$

(see Lemma 6.2.1), we compute directly $\partial_\lambda \pi_\lambda(T) = i\mathrm{I}$. By (6.31) and (6.33), we know

$$\Delta_{y_j}\pi_\lambda(T) = \Delta_{x_j}\pi_\lambda(T) = 0,$$

thus we have obtained (6.41) by Lemma 6.3.6. Furthermore, by (6.12), we have

$$\partial_\lambda \pi_\lambda(\mathcal{L}) = \mathrm{sgn}(\lambda)\sum_{j=1}^n \left(\partial_{u_j}^2 - u_j^2\right) = \frac{1}{\lambda}\pi_\lambda(\mathcal{L})$$

and by (6.32) and (6.34)

$$\begin{aligned} &\sum_{j=1}^n \left\{\pi_\lambda(Y_j)\Delta_{y_j}\pi_\lambda(\mathcal{L}) + \Delta_{x_j}\pi_\lambda(\mathcal{L})\pi_\lambda(X_j)\right\} \\ &\qquad = -\sum_{j=1}^n \left\{\pi_\lambda(Y_j)2\pi_\lambda(Y_j) + 2\pi_\lambda(X_j)\pi_\lambda(X_j) = -2\pi_\lambda(\mathcal{L})\right\}, \end{aligned}$$

and also by Example 6.3.4, we get

$$\Delta_{x_j}\Delta_{y_j}\pi_\lambda(\mathcal{L}) = -\Delta_{x_j}2\pi_\lambda(Y_j) = 0.$$

Combining all these equalities together with Lemma 6.3.6 yields (6.42). □

Note that (6.42) can also be obtained from (6.40) and the Leibniz formula (in the sense of (5.28)) for Δ_t.

In terms of λ-symbols, we obtain

Corollary 6.3.9. *If* $\widehat{\kappa}(\pi_\lambda) \equiv \pi_\lambda(\kappa) = \mathrm{Op}^W(a_\lambda)$ *with* $a_\lambda = \{a_\lambda(\xi, u)\}$, *then*

$$\Delta_t \widehat{\kappa}\,(\pi_\lambda) = i\mathrm{Op}^W\left(\tilde{\partial}_{\lambda,\xi,u} a_\lambda\right),$$

where

$$\tilde{\partial}_{\lambda,\xi,u} := \partial_\lambda - \frac{1}{2\lambda}\sum_{j=1}^n \left(u_j \partial_{u_j} + \xi_j \partial_{\xi_j}\right). \tag{6.43}$$

Proof. Using formulae (6.28), Corollary 6.3.3 and the properties of the Weyl calculus (see especially the composition formula in (6.16)), we obtain easily that

$$\begin{aligned} \pi_\lambda(Y_j)\Delta_{y_j}\pi_\lambda(\kappa) &= \mathrm{Op}^W\left(i\sqrt{\lambda}u_j\right)\mathrm{Op}^W\left(\frac{-1}{i\sqrt{\lambda}}\partial_{u_j}a_\lambda\right) \\ &= -\mathrm{Op}^W(u_j)\,\mathrm{Op}^W\left(\partial_{u_j}a_\lambda\right) \\ &= -\mathrm{Op}^W\left(u_j\partial_{u_j}a_\lambda - \frac{1}{2i}\partial_{\xi_j}\partial_{u_j}a_\lambda\right), \end{aligned}$$

and

$$\begin{aligned}
\Delta_{x_j}\pi_\lambda(\kappa)\pi_\lambda(X_j) &= \mathrm{Op}^W\left(\frac{-1}{i\sqrt{|\lambda|}}\partial_{\xi_j}a_\lambda\right)\mathrm{Op}^W\left(i\sqrt{|\lambda|}\xi_j\right) \\
&= -\mathrm{Op}^W\left(\partial_{\xi_j}a_\lambda\right)\mathrm{Op}^W\left(\xi_j\right) \\
&= -\mathrm{Op}^W\left((\partial_{\xi_j}a_\lambda)\xi_j - \frac{1}{2i}\partial_{u_j}\partial_{\xi_j}a_\lambda\right),
\end{aligned}$$

thus

$$\begin{aligned}
&\pi_\lambda(Y_j)\Delta_{y_j}\pi_\lambda(\kappa) + \Delta_{x_j}\pi_\lambda(\kappa)\pi_\lambda(X_j) \\
&= -\mathrm{Op}^W\left(u_j\partial_{u_j}a_\lambda - \frac{1}{2i}\partial_{\xi_j}\partial_{u_j}a_\lambda\right) - \mathrm{Op}^W\left((\partial_{\xi_j}a_\lambda)\xi_j - \frac{1}{2i}\partial_{u_j}\partial_{\xi_j}a_\lambda\right). \\
&= \mathrm{Op}^W\left(-u_j\partial_{u_j}a_\lambda - \xi_j\partial_{\xi_j}a_\lambda + \frac{1}{i}\partial_{\xi_j}\partial_{u_j}a_\lambda\right).
\end{aligned}$$

We also have

$$\begin{aligned}
\Delta_{x_j}\Delta_{y_j}\pi_\lambda(\kappa) &= \mathrm{Op}^W\left(\frac{-1}{i\sqrt{|\lambda|}}\partial_{\xi_j}\frac{-1}{i\sqrt{\lambda}}\partial_{u_j}a_\lambda\right) \\
&= -\frac{1}{\lambda}\mathrm{Op}^W\left(\partial_{\xi_j}\partial_{u_j}a_\lambda\right). \qquad (6.44)
\end{aligned}$$

Bringing these equalities in the formula for Δ_t in Lemma 6.3.6, we obtain

$$\begin{aligned}
\Delta_t\pi_\lambda(\kappa) &= i\partial_\lambda\pi_\lambda(\kappa) + \frac{1}{2}\sum_{j=1}^n \Delta_{x_j}\Delta_{y_j}\pi_\lambda(\kappa) \\
&\quad + \frac{i}{2\lambda}\sum_{j=1}^n \left\{\pi_\lambda(Y_j)\Delta_{y_j}\pi_\lambda(\kappa) + \Delta_{x_j}\pi_\lambda(\kappa)\pi_\lambda(X_j)\right\} \\
&= i\mathrm{Op}^W(\partial_\lambda a_\lambda) + \frac{1}{2}\sum_{j=1}^n -\frac{1}{\lambda}\mathrm{Op}^W\left(\partial_{\xi_j}\partial_{u_j}a_\lambda\right) \\
&\quad + \frac{i}{2\lambda}\sum_{j=1}^n \mathrm{Op}^W\left(-u_j\partial_{u_j}a_\lambda - \xi_j\partial_{\xi_j}a_\lambda + \frac{1}{i}\partial_{\xi_j}\partial_{u_j}a_\lambda\right) \\
&= \mathrm{Op}^W\left(i\partial_\lambda a_\lambda - \frac{i}{2\lambda}\sum_{j=1}^n\left(u_j\partial_{u_j}a_\lambda + \xi_j\partial_{\xi_j}a_\lambda\right)\right).
\end{aligned}$$

This completes the proof. □

6.3.3 Formulae

Here we summarise the formulae obtained so far in Sections 6.3.1 and 6.3.2. Let us recall our convention regarding square roots (6.9) setting

$$\sqrt{\lambda} := \mathrm{sgn}(\lambda)\sqrt{|\lambda|} = \begin{cases} \sqrt{\lambda} & \text{if } \lambda > 0 \\ -\sqrt{|\lambda|} & \text{if } \lambda < 0 \end{cases} .$$

For the Schrödinger infinitesimal representation we have obtained (see (6.11), (6.12) and (6.28)) that

$$\boxed{\begin{array}{rclcl} \pi_\lambda(X_j) & = & \sqrt{|\lambda|}\partial_{u_j} & = & \mathrm{Op}^W\left(i\sqrt{|\lambda|}\xi_j\right) \\ \pi_\lambda(Y_j) & = & i\sqrt{\lambda}u_j & = & \mathrm{Op}^W\left(i\sqrt{\lambda}u_j\right) \\ \pi_\lambda(T) & = & i\lambda \mathrm{I} & = & \mathrm{Op}^W(i\lambda) \\ \pi_\lambda(\mathcal{L}) & = & |\lambda|\sum_j(\partial_{u_j}^2 - u_j^2) & = & \mathrm{Op}^W\left(|\lambda|\sum_j(-\xi_j^2 - u_j^2)\right) \end{array}}$$

while for difference operators (cf. Lemmata 6.3.1 and 6.3.6) we have

$$\boxed{\begin{array}{rclcl} \Delta_{x_j}|_{\pi_\lambda} & = & \frac{1}{i\lambda}\mathrm{ad}\left(\pi_\lambda(Y_j)\right) & = & \frac{1}{\sqrt{|\lambda|}}\mathrm{ad}u_j \\ \Delta_{y_j}|_{\pi_\lambda} & = & -\frac{1}{i\lambda}\mathrm{ad}\left(\pi_\lambda(X_j)\right) & = & -\frac{1}{i\sqrt{\lambda}}\mathrm{ad}\partial_{u_j} \\ \Delta_t|_{\pi_\lambda} & = & i\partial_\lambda + \frac{1}{2}\sum_{j=1}^n \Delta_{x_j}\Delta_{y_j}|_{\pi_\lambda} + \frac{i}{2\lambda}\sum_{j=1}^n\{\pi_\lambda(Y_j)|_{\pi_\lambda}\Delta_{y_j} + \Delta_{x_j}|_{\pi_\lambda}\pi_\lambda(X_j)\} & & \end{array}}$$

and in terms of λ-symbols, that is, with

$$\widehat{\kappa}(\pi_\lambda) \equiv \pi_\lambda(\kappa) = \mathrm{Op}^W(a_\lambda) \text{ and } a_\lambda = \{a_\lambda(\xi, u)\},$$

(cf. Corollaries 6.3.3 and 6.3.9):

$$\boxed{\begin{array}{rcl} \Delta_{x_j}\pi_\lambda(\kappa) & = & i\mathrm{Op}^W\left(\frac{1}{\sqrt{|\lambda|}}\partial_{\xi_j}a_\lambda\right) \\ \Delta_{y_j}\pi_\lambda(\kappa) & = & i\mathrm{Op}^W\left(\frac{1}{\sqrt{\lambda}}\partial_{u_j}a_\lambda\right) \\ \Delta_t\pi_\lambda(\kappa) & = & i\mathrm{Op}^W\left(\tilde{\partial}_{\lambda,\xi,u}a_\lambda\right) \\ & = & i\mathrm{Op}^W\left((\partial_\lambda - \frac{1}{2\lambda}\sum_{j=1}^n\{u_j\partial_{u_j} + \xi_j\partial_{\xi_j}\})a_\lambda\right) \end{array}} \tag{6.45}$$

In Examples 6.3.4, 6.3.5, 6.3.8 together with (6.40), we have also obtained

	$\pi_\lambda(X_k)$	$\pi_\lambda(Y_k)$	$\pi_\lambda(T)$	$\pi_\lambda(\mathcal{L})$
Δ_{x_j}	$-\delta_{j=k}$	0	0	$-2\pi_\lambda(X_j)$
Δ_{y_j}	0	$-\delta_{j=k}$	0	$-2\pi_\lambda(Y_j)$
Δ_t	0	0	$-\mathrm{I}$	0

The equalities given in the following lemma concern another normalisation of the Weyl symbol which is motivated by (6.20) and by the fact that the expressions of the right-hand sides in (6.45), in particular for the operator $\tilde{\partial}_{\lambda,\xi,u}$, become then very simple:

Lemma 6.3.10. *Let $a_\lambda = \{a_\lambda(\xi,u)\}$ be a family of Weyl symbols depending smoothly on $\lambda \neq 0$. If $\tilde{a}_\lambda$ is the renormalisation obtained via*

$$a_\lambda(\xi,u) = \tilde{a}_\lambda(\sqrt{|\lambda|}\xi, \sqrt{\lambda}u), \tag{6.46}$$

then

$$\begin{aligned}
\{\tilde{\partial}_{\lambda,\xi,u} a_\lambda\}(\xi,u) &= \{\partial_\lambda \tilde{a}_\lambda\}(\sqrt{|\lambda|}\xi, \sqrt{\lambda}u), \\
\frac{1}{\sqrt{|\lambda|}}\{\partial_{\xi_j} a_\lambda\}(\xi,u) &= \{\partial_{\xi_j} \tilde{a}_\lambda\}(\sqrt{|\lambda|}\xi, \sqrt{\lambda}u), \\
\frac{1}{\sqrt{\lambda}}\{\partial_{u_j} a_\lambda\}(\xi,u) &= \{\partial_{u_j} \tilde{a}_\lambda\}(\sqrt{|\lambda|}\xi, \sqrt{\lambda}u).
\end{aligned}$$

Proof. We see that

$$\tilde{a}_\lambda(\xi,u) = a_\lambda\left(\frac{1}{\sqrt{|\lambda|}}\xi, \frac{1}{\sqrt{\lambda}}u\right),$$

thus

$$\begin{aligned}
\partial_\lambda \tilde{a}_\lambda(\xi,u) &= (\partial_\lambda a_\lambda)\left(\frac{1}{\sqrt{|\lambda|}}\xi, \frac{1}{\sqrt{\lambda}}u\right) \\
&\quad - \sum_{j=1}^n \frac{\xi_j}{2\lambda\sqrt{|\lambda|}} (\partial_{\xi_j} a_\lambda)\left(\frac{1}{\sqrt{|\lambda|}}\xi, \frac{1}{\sqrt{\lambda}}u\right) \\
&\quad - \sum_{j=1}^n \frac{u_j}{2|\lambda|\sqrt{|\lambda|}} (\partial_{u_j} a_\lambda)\left(\frac{1}{\sqrt{|\lambda|}}\xi, \frac{1}{\sqrt{\lambda}}u\right),
\end{aligned}$$

and

$$\begin{aligned}
\{\partial_\lambda \tilde{a}_\lambda\}\left(\sqrt{|\lambda|}\xi, \sqrt{\lambda}u\right) &= (\partial_\lambda a_\lambda)(\xi,u) \\
&\quad - \sum_{j=1}^n \left(\frac{\sqrt{|\lambda|}\xi_j}{2\lambda\sqrt{|\lambda|}} \partial_{\xi_j} a_\lambda(\xi,u) + \frac{\sqrt{\lambda}u_j}{2|\lambda|\sqrt{|\lambda|}} \partial_{u_j} a_\lambda(\xi,u)\right) \\
&= \partial_\lambda a_\lambda(\xi,u) - \frac{1}{2\lambda}\sum_{j=1}^n \left(\xi_j \partial_{\xi_j} a_\lambda(\xi,u) + u_j \partial_{u_j} a_\lambda(\xi,u)\right) \\
&= \tilde{\partial}_{\lambda,\xi,u} a_\lambda(\xi,u).
\end{aligned}$$

This shows the first stated equality. The other two are easy. □

Lemma 6.3.10 and the formulae already obtained yield

$$\begin{aligned}
\Delta_{x_j} \pi_\lambda(\kappa) &= i\mathrm{Op}^W\left(\partial_{\xi_j} \tilde{a}_\lambda\right), \\
\Delta_{y_j} \pi_\lambda(\kappa) &= i\mathrm{Op}^W\left(\partial_{u_j} \tilde{a}_\lambda\right), \\
\Delta_t \pi_\lambda(\kappa) &= i\mathrm{Op}^W\left(\partial_\lambda \tilde{a}_\lambda\right),
\end{aligned}$$

where the λ-symbol a_λ of $\pi_\lambda(\kappa)$, that is, $\pi_\lambda(\kappa) = \mathrm{Op}^W(a_\lambda)$, has been rescaled via (6.46), i.e.

$$a_\lambda(\xi, u) = \tilde{a}_\lambda(\sqrt{|\lambda|}\xi, \sqrt{\lambda}u).$$

Recall that

$$a_\lambda(\xi, u) = (2\pi)^{\frac{2n+1}{2}} \mathcal{F}_{\mathbb{R}^{2n+1}}(\kappa)(\sqrt{|\lambda|}\xi, \sqrt{\lambda}u, \lambda),$$

see (6.20), so

$$\tilde{a}_\lambda(\xi, u) = (2\pi)^{\frac{2n+1}{2}} \mathcal{F}_{\mathbb{R}^{2n+1}}(\kappa)(\xi, u, \lambda).$$

The above formulae in terms of the rescaled λ-symbols look neat. The drawback of using this rescaling is that one rescales the Weyl quantization:

$$\widehat{\kappa}(\pi_\lambda) = \mathrm{Op}^W(a_\lambda) = \mathrm{Op}^W\left(\tilde{a}_\lambda\left(\sqrt{|\lambda|}\,\cdot, \sqrt{\lambda}\,\cdot\right)\right).$$

Since our aim is to study the group Fourier transform on $\mathbb{H}_n$, it is more natural to study the Weyl-symbol a_λ without any rescaling.

In fact, the following two sections are devoted to understanding $\widehat{\kappa} \equiv \{\pi_\lambda(\kappa)\}$ as a family of Weyl pseudo-differential operators parametrised by $\lambda \in \mathbb{R}\backslash\{0\}$. The Weyl quantization will force us to work on the λ-symbol a_λ directly, and not on its rescaling $\tilde{a}_\lambda$.

This will lead to defining a family of symbol classes parametrised by $\lambda \in \mathbb{R}\backslash\{0\}$ for the λ-symbols a_λ. This will be done via a family of Hörmander metrics parametrised by $\lambda \in \mathbb{R}\backslash\{0\}$. Importantly the structural bounds of these metrics will be uniform with respect to λ. The resulting symbol classes will be called λ-Shubin classes.

6.4 Shubin classes

In this Section, we recall elements of the Weyl-Hörmander pseudo-differential calculus and the associated Sobolev spaces, and we apply this to obtain the Shubin classes of symbols and the associated Sobolev spaces. The dependence in a parameter λ will be of particular importance to us. We will call the resulting symbol classes the λ-Shubin classes.

6.4.1 Weyl-Hörmander calculus

Here we present the main elements of the Weyl-Hörmander calculus that will be relevant for our analysis. For more details on the underlying general theory, we can refer, for instance, to [Ler10].

We consider $\mathbb{R}^n$ and identify its cotangent bundle $T^*\mathbb{R}^n$ with $\mathbb{R}^{2n}$. The canonical symplectic form on $\mathbb{R}^{2n}$ is ω defined by

$$\omega(T, T') = x \cdot \xi' - x' \cdot \xi, \quad T = (\xi, x),\ T' = (\xi', x') \in \mathbb{R}^{2n}.$$

Definition 6.4.1. If q is a positive quadratic form on $\mathbb{R}^{2n}$, then we define its *conjugate* q^ω by

$$\forall T \in \mathbb{R}^{2n} \quad q^\omega(T) := \sup_{T' \in \mathbb{R}^{2n}\setminus\{0\}} \frac{|\omega(T,T')|^2}{q(T')},$$

and its *gain factor* by

$$\Lambda_q := \inf_{T \in \mathbb{R}^{2n}\setminus\{0\}} \frac{q^\omega(T)}{q(T)}.$$

Definition 6.4.2. A *metric* is a family of positive quadratic forms

$$g = \{g_X, X \in \mathbb{R}^{2n}\}$$

depending smoothly on $X \in \mathbb{R}^{2n}$.

- The metric g is *uncertain* when $\forall X \in \mathbb{R}^{2n}$, $\Lambda_{g_X} \geq 1$.
- The metric g is *slowly varying* when there exists a constant $\bar{C} > 0$ such that we have for any $X, X' \in \mathbb{R}^{2n}$:

$$g_X(X - X') \leq \bar{C}^{-1} \Longrightarrow \sup_{T \in \mathbb{R}^{2n}\setminus\{0\}} \left(\frac{g_X(T)}{g_{X'}(T)} + \frac{g_{X'}(T)}{g_X(T)} \right) \leq \bar{C}.$$

- The metric g is *temperate* when there are constants $\bar{C} > 0$ and $\bar{N} > 0$ such that we have for any $X, X' \in \mathbb{R}^{2n}$ and $T \in \mathbb{R}^{2n}\setminus\{0\}$:

$$\frac{g_X(T)}{g_{X'}(T)} \leq \bar{C}(1 + g_X^\omega(X - X'))^{\bar{N}}.$$

A metric g is of *Hörmander type* if it is uncertain, slowly varying and temperate. In this case the constants $\bar{C}$ and $\bar{N}$ appearing above and any constant depending only on them are called *structural.*

Proposition 6.4.3. *A metric* $g = \{g_X, X \in \mathbb{R}^{2n}\}$ *is* slowly varying *if and only if there exist constants* $C, r > 0$ *such that we have for any* $X, Y \in \mathbb{R}^{2n}$ *that*

$$g_X(Y - X) \leq r^2 \Longrightarrow \forall T \quad g_Y(T) \leq C g_X(T). \tag{6.47}$$

Proof. If g is slowly varying then it satisfies (6.47). Conversely, let us assume (6.47). Necessarily $C \geq 1$ since we can take $X = Y$ in (6.47). If $g_X(Y - X) \leq C^{-1}r^2$, then $g_X(Y - X) \leq r^2$ and, applying (6.47) with $T = Y - X$, we obtain

$$g_Y(Y - X) \leq C g_X(Y - X) \leq r^2,$$

thus re-applying (6.47) (but at g_Y), we have $g_X(T) \leq C g_Y(T)$ for all T. This shows that g is slowly varying. □

Remark 6.4.4. If g satisfies (6.47) with constant $C > 1$ and $r > 0$ then g is slowly varying with a constant $\bar{C} = \min(C^{-1}r^2, 2C)$.

Example 6.4.5. Let ϕ be a positive smooth function on $\mathbb{R}^{2n}$ which is Lipschitz on $\mathbb{R}^{2n}$. We denote by $T \mapsto |T|^2$ the canonical (Euclidean) quadratic form on $\mathbb{R}^{2n}$. The metric g given by

$$g_X(T) = \phi(X)^{-2}|T|^2$$

is slowly varying.

Proof. Let us assume $g_X(Y - X) \leq r^2$ for a constant $r > 0$ to be determined. This means $|Y - X| \leq r\phi(X)$. Since ϕ is Lipschitz on $\mathbb{R}^{2n}$, denoting by L its Lipschitz constant, we have

$$\phi(X) \leq \phi(Y) + L|X - Y| \leq \phi(Y) + Lr\phi(X),$$

thus

$$(1 - Lr)\phi(X) \leq \phi(Y).$$

Hence if we choose $r > 0$ so that $1 - Lr > 0$, we have obtained

$$\forall T \qquad g_Y(T) \leq C g_X(T),$$

with $C = (1 - Lr)^{-1}$. This shows that g_X satisfies (6.47) and is therefore slowly varying. □

Remark 6.4.6. If ϕ is L-Lipschitz then g given in Example 6.4.5 satisfies (6.47) with any $r \in (0, L^{-1})$ and a corresponding $C = (1 - Lr)^{-1}$.

Definition 6.4.7. Let g be a metric of Hörmander type. A positive function M defined on $\mathbb{R}^{2n}$ is a *g-weight* when there are structural constants $\bar{C}'$ and $\bar{N}'$ satisfying for any $X, Y \in \mathbb{R}^{2n}$:

$$g_X(X - Y) \leq \bar{C}'^{-1} \Longrightarrow \frac{M(X)}{M(Y)} + \frac{M(Y)}{M(X)} \leq \bar{C}',$$

and

$$\frac{M(X)}{M(Y)} \leq \bar{C}(1 + g_X^\omega(X - Y))^{\bar{N}'}.$$

It is easy to check that the set of g-weights forms a group for the usual multiplication of positive functions.

Definition 6.4.8 (Hörmander symbol class $S(M, g)$). Let g be a metric of Hörmander type and M a g-weight on $\mathbb{R}^{2n}$. The symbol class $S(M, g)$ is the set of functions $a \in C^\infty(\mathbb{R}^{2n})$ such that for each integer $\ell \in \mathbb{N}_0$, the quantity

$$\|a\|_{S(M,g),\ell} := \sup_{\substack{\ell' \leq \ell, X \in \mathbb{R}^{2n} \\ g_X(T_{\ell'}) \leq 1}} \frac{|\partial_{T_1} \ldots \partial_{T_{\ell'}} a(X)|}{M(X)}$$

is finite.

Here $\partial_T a$ denotes the quantity (da, T).

The following properties are well known [Ler10, Chapters 1 and 2]:

Theorem 6.4.9. *Let g be a metric of Hörmander type and let M, M_1, M_2 be g-weights.*

1. *The symbol class $S(M,g)$ is a vector space endowed with a Fréchet topology via the family of seminorms $\|\cdot\|_{S(M,g),\ell}$, $\ell \in \mathbb{N}_0$.*
2. *If $a \in S(M,g)$ then the symbol b defined by*
$$\mathrm{Op}^W b = \left(\mathrm{Op}^W a\right)^*$$
is in $S(M,g)$ as well. Furthermore, for any $\ell \in \mathbb{N}_0$ there exist a constant $C > 0$ and a integer $\ell' \in \mathbb{N}_0$ such that
$$\|b\|_{S(M,g),\ell} \leq C\|a\|_{S(M,g),\ell'}.$$
The constant C and the integer ℓ' may be chosen to depend on ℓ and on the structural constants and to be independent of g, M and a.
3. *If $a_1 \in S(M_1,g)$ and $a_2 \in S(M_2,g)$ then the symbol b defined by*
$$\mathrm{Op}^W b = \left(\mathrm{Op}^W a_1\right)\left(\mathrm{Op}^W a_2\right),$$
is in $S(M_1M_2,g)$. Furthermore, for any $\ell \in \mathbb{N}_0$ there exist a constant $C > 0$ and two integers $\ell_1, \ell_2 \in \mathbb{N}_0$ such that
$$\|b\|_{S(M_1M_2,g),\ell} \leq C\|a_1\|_{S(M_1,g),\ell_1}\|a_2\|_{S(M_2,g),\ell_2}.$$
The constant C and the integers ℓ_1, ℓ_2 may be chosen to depend on ℓ and on the structural constants and to be independent of g, M_1, M_2 and a_1, a_2.

Definition 6.4.10 (Sobolev spaces $H(M,g)$)**.** Let g be a metric of Hörmander type and M a g-weight on $\mathbb{R}^{2n}$. We denote by $H(M,g)$ the set of all tempered distributions f on $\mathbb{R}^n$ such that for any symbol $a \in S(M,g)$ we have $\mathrm{Op}^W(a)f \in L^2(\mathbb{R}^n)$.

Theorem 6.4.11. *Let g be a metric of Hörmander type on $\mathbb{R}^{2n}$.*

1. *The space $H(1,g)$ coincides with $L^2(\mathbb{R}^n)$. Furthermore, there exist a structural constant $C > 0$ and a structural integer $\ell \in \mathbb{N}_0$ such that for any symbol $a \in S(1,g)$, we have*
$$\|\mathrm{Op}^W(a)\|_{\mathscr{L}(L^2(\mathbb{R}^n))} \leq C\|a\|_{S(1,g),\ell}.$$
2. *Let M_1, M_2 be g-weights. For any $a \in S(M_1,g)$, the operator $\mathrm{Op}^W(a)$ maps continuously $H(M_2,g)$ to $H(M_2M_1^{-1},g)$. Furthermore, there exist a constant $C > 0$ and an integer $\ell \in \mathbb{N}_0$ such that*
$$\|\mathrm{Op}^W(a)\|_{\mathscr{L}(H(M_2,g),H(M_2M_1^{-1},g))} \leq C\|a\|_{S(M_1,g),\ell}.$$
The constant C and the integers ℓ may be chosen to depend only on the structural constants of g, M_1, M_2 and to be independent of g, M and a.

6.4.2 Shubin classes $\Sigma_\rho^m(\mathbb{R}^n)$ and the harmonic oscillator

It is well known (and can be readily checked) that the metric

$$\frac{d\xi^2 + du^2}{(1+|u|^2+|\xi|^2)^\rho},$$

is of Hörmander type with corresponding weights $(1+|u|^2+|\xi|^2)^{m/2}$ for $m \in \mathbb{R}$. This will be also shown later in the proof of Proposition 6.4.21. For $m \in \mathbb{R}$ and $\rho \in (0,1]$, we denote by $\Sigma_\rho^m(\mathbb{R}^n)$ the corresponding symbol class, often called the Shubin classes of symbols on $\mathbb{R}^n$:

$$\Sigma_\rho^m(\mathbb{R}^n) := S\left((1+|u|^2+|\xi|^2)^{m/2}, \frac{d\xi^2+du^2}{(1+|u|^2+|\xi|^2)^\rho}\right).$$

This means that a symbol $a \in C^\infty(\mathbb{R}^{2n})$ is in $\Sigma_\rho^m(\mathbb{R}^n)$ if and only if for any $\alpha, \beta \in \mathbb{N}_0^n$ there exists a constant $C = C_{\alpha,\beta} > 0$ such that

$$\forall (\xi,u) \in \mathbb{R}^{2n} \qquad |\partial_\xi^\alpha \partial_u^\beta a(\xi,u)| \leq C\left(1+|\xi|^2+|u|^2\right)^{\frac{m-\rho(|\alpha|+|\beta|)}{2}}.$$

The class $\Sigma_\rho^m(\mathbb{R}^n)$ is a vector subspace of $C^\infty(\mathbb{R}^n \times \mathbb{R}^n)$ which becomes a Fréchet space when endowed with the family of seminorms

$$\|a\|_{\Sigma_\rho^m,N} = \sup_{\substack{(\xi,u)\in\mathbb{R}^n\times\mathbb{R}^n \\ |\alpha|,|\beta|\leq N}} \left(1+|\xi|^2+|u|^2\right)^{-\frac{m-\rho(|\alpha|+|\beta|)}{2}} |\partial_\xi^\alpha \partial_u^\beta a(\xi,u)|,$$

where $N \in \mathbb{N}_0$. We denote by

$$\Psi\Sigma_\rho^m(\mathbb{R}^n) := \mathrm{Op}^W(\Sigma_\rho^m(\mathbb{R}^n))$$

the corresponding class of operators and by $\|\cdot\|_{\Psi\Sigma_\rho^m,N}$ the corresponding seminorms.

We have the inclusions

$$\rho_1 \geq \rho_2 \quad \text{and} \quad m_1 \leq m_2 \Longrightarrow \Psi\Sigma_{\rho_1}^{m_1}(\mathbb{R}^n) \subset \Psi\Sigma_{\rho_2}^{m_2}(\mathbb{R}^n).$$

Example 6.4.12. The operators $\partial_{u_j} = \mathrm{Op}^W(i\xi_j)$, $j = 1, \ldots, n$, or multiplication by $u_k = \mathrm{Op}^W(u_k)$, $k = 1, \ldots, n$, are two operators in $\Psi\Sigma_1^1(\mathbb{R}^n)$.

Standard computations also show:

Example 6.4.13. For each $m \in \mathbb{R}$, the symbol b^m, where

$$b(\xi,u) = \sqrt{1+|u|^2+|\xi|^2},$$

is in $\Sigma_1^m(\mathbb{R}^n)$.

The following is well known and can be viewed more generally as a consequence of the Weyl-Hörmander calculus (see Theorem 6.4.9)

Theorem 6.4.14. • *The class of operators $\cup_{m\in\mathbb{R}}\Psi\Sigma_\rho^m(\mathbb{R}^n)$ forms an algebra of operators stable by taking the adjoint. Furthermore, the operations*

$$\begin{array}{rcl}\Psi\Sigma_\rho^m(\mathbb{R}^n) & \longrightarrow & \Psi\Sigma_\rho^m(\mathbb{R}^n)\\ A & \longmapsto & A^*\end{array}$$

and

$$\begin{array}{rcl}\Psi\Sigma_\rho^{m_1}(\mathbb{R}^n)\times\Psi\Sigma_\rho^{m_2}(\mathbb{R}^n) & \longrightarrow & \Psi\Sigma_\rho^{m_1+m_2}(\mathbb{R}^n)\\ (A,B) & \longmapsto & AB\end{array}$$

are continuous.

- *The operators in $\Psi\Sigma_\rho^0(\mathbb{R}^n)$ extend boundedly to $L^2(\mathbb{R}^n)$. Furthermore, there exist $C>0$ and $N\in\mathbb{N}$ such that if $A\in\Psi\Sigma_\rho^0(\mathbb{R}^n)$ then*

$$\|A\|_{\mathscr{L}(L^2(\mathbb{R}^n))}\leq C\|A\|_{\Psi\Sigma_\rho^m,N}.$$

From Example 6.4.12, it follows that the (positive) *harmonic oscillator*

$$Q:=\sum_{j=1}^n(-\partial_{u_j}^2+u_j^2), \tag{6.48}$$

is in $\Psi\Sigma_1^2(\mathbb{R}^n)$.

Note that from now on Q denotes the harmonic oscillator and not the homogeneous dimension as in all previous chapters.

We keep the same notation for Q and for its self-adjoint extension as an unbounded operator on $L^2(\mathbb{R}^n)$. The harmonic oscillator Q is a positive (unbounded) operator on $L^2(\mathbb{R}^n)$. Its spectrum is

$$\{2|\ell|+n,\ell\in\mathbb{N}_0^n\},$$

where $|\ell|=\ell_1+\ldots+\ell_n$. The eigenfunctions associated with the eigenvalues $2|\ell|+n$ are

$$h_\ell:x=(x_1,\ldots,x_n)\longmapsto h_{\ell_1}(x_1)\ldots h_{\ell_n}(x_n),$$

where each h_j, $j=0,1,2\ldots$, is a Hermite function, that is,

$$h_j(\tau)=(-1)^j\frac{e^{\frac{\tau^2}{2}}}{\sqrt{2^j j!\sqrt{\pi}}}\frac{d^j}{d\tau^j}e^{-\tau^2},\qquad \tau\in\mathbb{R}.$$

The Hermite functions are Schwartz, i.e. $h_j\in\mathcal{S}(\mathbb{R})$. With our choice of normalisation, the functions h_j, $j=0,1,\ldots$, form an orthonormal basis of $L^2(\mathbb{R})$. Therefore, the functions h_ℓ form an orthonormal basis of $L^2(\mathbb{R}^n)$. For each $s\in\mathbb{R}$, we define the operator $(\mathrm{I}+Q)^{s/2}$ using the functional calculus, that is, in this case, the domain of $(\mathrm{I}+Q)^{s/2}$ is the space of functions

$$\mathrm{Dom}(\mathrm{I}+Q)^{s/2} = \{h \in L^2(\mathbb{R}^n) : \sum_{\ell\in\mathbb{N}_0^n} (2|\ell|+n)^s |(h_\ell, h)_{L^2(\mathbb{R}^n)}|^2 < \infty\},$$

and if $h \in \mathrm{Dom}(\mathrm{I}+Q)^{s/2}$ then

$$(\mathrm{I}+Q)^{s/2}h = \sum_{\ell\in\mathbb{N}_0^n} (2|\ell|+n)^{s/2} (h_\ell, h)_{L^2(\mathbb{R}^n)} h_\ell.$$

6.4.3 Shubin Sobolev spaces

In this section, we study Shubin Sobolev spaces. Many of their properties, especially their equivalent characterisations, are well known. Their proofs are quite easy but often omitted in the literature. Thus we have chosen to sketch their demonstrations.

The Shubin Sobolev spaces below are a special case of Sobolev spaces for measurable fields on representation spaces, see Definition 5.1.6.

Our starting point will be the following definition for the Shubin Sobolev spaces:

Definition 6.4.15. Let $s \in \mathbb{R}$. The *Shubin Sobolev space* $\mathcal{Q}_s(\mathbb{R}^n)$ is the subspace of $\mathcal{S}'(\mathbb{R}^n)$ which is the completion of $\mathrm{Dom}(\mathrm{I}+Q)^{s/2}$ for the norm

$$\|h\|_{\mathcal{Q}_s} := \|(\mathrm{I}+Q)^{s/2}h\|_{L^2(\mathbb{R}^n)}.$$

They satisfy the following properties:

Theorem 6.4.16. 1. *The space $\mathcal{Q}_s(\mathbb{R}^n)$ is a Hilbert space endowed with the sesquilinear form*

$$(g,h)_{\mathcal{Q}_s} = \left((\mathrm{I}+Q)^{s/2}g, (\mathrm{I}+Q)^{s/2}h\right)_{L^2(\mathbb{R}^n)}.$$

We have the inclusions

$$\mathcal{S}(\mathbb{R}^n) \subset \mathcal{Q}_{s_1}(\mathbb{R}^n) \subset \mathcal{Q}_{s_2}(\mathbb{R}^n) \subset \mathcal{S}'(\mathbb{R}^n), \quad s_1 > s_2.$$

We also have

$$L^2(\mathbb{R}^n) = \mathcal{Q}_0(\mathbb{R}^n) \quad \textit{and} \quad \mathcal{S}(\mathbb{R}^n) = \bigcap_{s\in\mathbb{R}} \mathcal{Q}_s(\mathbb{R}^n).$$

2. *The dual of $\mathcal{Q}_s(\mathbb{R}^n)$ may be identified with $\mathcal{Q}_{-s}(\mathbb{R}^n)$ via the distributional duality form $\langle g,h\rangle = \int_{\mathbb{R}^n} gh$.*

3. *If $s \in \mathbb{N}_0$, $\mathcal{Q}_s(\mathbb{R}^n)$ coincides with*

$$\mathcal{Q}_s(\mathbb{R}^n) = \{h \in L^2(\mathbb{R}^n) : u^\alpha \partial_u^\beta h \in L^2(\mathbb{R}^n) \quad \forall \alpha, \beta \in \mathbb{N}_0^n, \ |\alpha| + |\beta| \le s\}.$$

Furthermore, the norm given by

$$\|h\|_{\mathcal{Q}_s}^{(int)} = \sum_{|\alpha|+|\beta|\le s} \|u^\alpha \partial_u^\beta h\|_{L^2(\mathbb{R}^n)},$$

is equivalent to $\|\cdot\|_{\mathcal{Q}_s}$.

4. *For any $s \in \mathbb{R}$, $\mathcal{Q}_s(\mathbb{R}^n)$ coincides with the completion (in $\mathcal{S}'(\mathbb{R}^n)$) of the Schwartz space $\mathcal{S}(\mathbb{R}^n)$ for the norm*

$$\|h\|_{\mathcal{Q}_s}^{(b)} = \|\mathrm{Op}^W(b^s)\, h\|_{L^2(\mathbb{R}^n)},$$

where b was given in Example 6.4.13. The norm $\|\cdot\|_{\mathcal{Q}_s}^{(b)}$ extended to $\mathcal{Q}_s(\mathbb{R}^n)$ is equivalent to $\|\cdot\|_{\mathcal{Q}_s}$.

5. *For any $s \in \mathbb{R}$, the Shubin Sobolev space $\mathcal{Q}_s(\mathbb{R}^n)$ coincides with the Sobolev space associated with the following metric weight (see Definition 6.4.10)*

$$\mathcal{Q}_s(\mathbb{R}^n) = H\left((1+|u|^2+|\xi|^2)^{s/2}, \frac{d\xi^2 + du^2}{1+|u|^2+|\xi|^2}\right).$$

6. *For any $s \in \mathbb{R}$, the operators $\mathrm{Op}^W(b^{-s})(\mathrm{I}+Q)^{s/2}$ and $(\mathrm{I}+Q)^{s/2}\mathrm{Op}^W(b^{-s})$ are bounded and invertible on $L^2(\mathbb{R}^n)$.*

7. *The complex interpolation between the spaces $\mathcal{Q}_{s_0}(\mathbb{R}^n)$ and $\mathcal{Q}_{s_1}(\mathbb{R}^n)$ is*

$$(\mathcal{Q}_{s_0}(\mathbb{R}^n), \mathcal{Q}_{s_1}(\mathbb{R}^n))_\theta = \mathcal{Q}_{s_\theta}(\mathbb{R}^n), \qquad s_\theta = (1-\theta)s_0 + \theta s_1, \ \theta \in (0,1).$$

Before giving the proof of Theorem 6.4.16, let us recall the definition of complex interpolation:

Definition 6.4.17 (Complex interpolation). Let X_0 and X_1 be two subspaces of a vector space Z. We assume that X_0 and X_1 are Banach spaces with norms denoted by $|\cdot|_j$, $j = 0, 1$.

Let $\mathscr{X}$ be the space of the functions f defined on the strip $\bar{S} = \{0 \le \operatorname{Re} z \le 1\}$ and valued in $X_0 + X_1$ such that f is continuous on $\bar{S}$ and holomorphic in $S = \{0 < \operatorname{Re} z < 1\}$. For $f \in \mathscr{X}$ we define the quantity (possibly infinite)

$$\|f\|_{\mathscr{X}} := \sup_{y \in \mathbb{R}} \{|f(iy)|_0, |f(1+iy)|_1\}.$$

The complex interpolation space of exponent $\theta \in (0,1)$ is the space $(X_0, X_1)_\theta$ of vectors $v \in X_0 + X_1$ such that there exists $f \in \mathscr{X}$ satisfying $f(\theta) = v$ and $\|f\|_{\mathscr{X}} < \infty$.

The space $(X_0, X_1)_\theta$ is a subspace of Z; it is a Banach space when endowed with the norm given by

$$|v|_\theta := \inf\{\|f\|_{\mathscr{X}} : f \in \mathscr{X} \quad \text{and} \quad f(\theta) = v\}.$$

We also refer to Appendix A.6 for the notion of analytic interpolation.

Proof of Theorem 6.4.16. From Definition 6.4.15, it is easy to prove that the space $\mathcal{Q}_s(\mathbb{R}^n)$ is a Hilbert space, that it is included in $\mathcal{S}'(\mathbb{R}^n)$ and that $\mathcal{Q}_0(\mathbb{R}^n) = L^2(\mathbb{R}^n)$. It is a routine exercise left to the reader that the dual of $\mathcal{Q}_s(\mathbb{R}^n)$ is $\mathcal{Q}_{-s}(\mathbb{R}^n)$ via the distributional duality (Part (2)) and that the spaces $\mathcal{Q}_s(\mathbb{R}^n)$ decrease with $s \in \mathbb{R}$.

Let us prove the complex interpolation property of Part (7). We may assume $s_1 > s_0$. For $h \in \mathcal{Q}_{s_\theta}$, we consider the function

$$f(z) := (\mathrm{I} + Q)^{\frac{-(zs_1+(1-z)s_0)+s_\theta}{2}} h,$$

and we check easily that

$$f(\theta) = h, \quad \|f(iy)\|_{\mathcal{Q}_{s_0}} = \|f(1+iy)\|_{\mathcal{Q}_{s_1}} = \|h\|_{\mathcal{Q}_{s_\theta}} \quad \forall y \in \mathbb{R}.$$

This shows that $\mathcal{Q}_{s_\theta}$ is continuously included in $(\mathcal{Q}_{s_0}(\mathbb{R}^n), \mathcal{Q}_{s_1}(\mathbb{R}^n))_\theta$. By duality of the complex interpolation and of the $\mathcal{Q}_s(\mathbb{R}^n)$, we obtain the reverse inclusion and Part (7) is proved.

Let us prove Part (4). For any $s \in \mathbb{R}$, the operator $\mathrm{Op}^W(b^s)$ maps $\mathcal{S}(\mathbb{R}^n)$ to itself and the mapping $\|\cdot\|^{(b)}_{\mathcal{Q}_s}$ as defined in Part (4) is a norm on $\mathcal{S}(\mathbb{R}^n)$. We denote its completion in $\mathcal{S}'(\mathbb{R}^n)$ by $\mathcal{Q}^{(b)}_s(\mathbb{R}^n)$. From the properties of the calculus it is again a routine exercise left to the reader that the dual of $\mathcal{Q}^{(b)}_s(\mathbb{R}^n)$ is $\mathcal{Q}^{(b)}_{-s}(\mathbb{R}^n)$ via the distributional duality and that the spaces $\mathcal{Q}^{(b)}_s(\mathbb{R}^n)$ decrease with $s \in \mathbb{R}$.

We can prove the following property about interpolation between the $\mathcal{Q}^{(b)}(\mathbb{R}^n)$ spaces which is analogous to Part (7):

$$(\mathcal{Q}^{(b)}_{s_0}(\mathbb{R}^n), \mathcal{Q}^{(b)}_{s_1}(\mathbb{R}^n))_\theta = \mathcal{Q}^{(b)}_{s_\theta}(\mathbb{R}^n), \qquad s_\theta = (1-\theta)s_0 + \theta s_1,\ \theta \in (0,1). \quad (6.49)$$

Indeed we may assume $s_1 > s_0$. For $h \in \mathcal{Q}^{(b)}_{s_\theta}$, we consider the function

$$f(z) = e^{z(s_z - s_\theta)} \mathrm{Op}^W\left(b^{-s_z+s_\theta}\right) h \quad \text{where} \quad s_z = (1-z)s_0 + zs_1.$$

Clearly $f(\theta) = h$. Furthermore,

$$\begin{aligned} \|f(iy)\|^{(b)}_{\mathcal{Q}_{s_1}} &= |e^{iy(s_{iy}-s_\theta)}| \|\mathrm{Op}^W(b^{s_1})\mathrm{Op}^W\left(b^{-s_{iy}+s_\theta}\right) h\|_{L^2(\mathbb{R}^n)} \\ &\leq e^{-y^2(s_1-s_0)} \|\mathrm{Op}^W(b^{s_1})\mathrm{Op}^W\left(b^{-s_{iy}+s_\theta}\right) \mathrm{Op}^W(b^{-s_\theta})\|_{\mathscr{L}(L^2(\mathbb{R}^n))} \\ &\qquad \|h\|^{(b)}_{\mathcal{Q}_{s_\theta}}, \end{aligned} \quad (6.50)$$

and

$$\begin{aligned} \|f(1+iy)\|^{(b)}_{\mathcal{Q}_{s_0}} &= |e^{(1+iy)(s_{1+iy}-s_\theta)}| \|\mathrm{Op}^W(b^{s_0})\mathrm{Op}^W\left(b^{-s_{1+iy}+s_\theta}\right) h\|_{L^2(\mathbb{R}^n)} \\ &\leq e^{s_1-s_\theta-y^2(s_1-s_0)} \|\mathrm{Op}^W(b^{s_0})\mathrm{Op}^W\left(b^{-s_{1+iy}+s_\theta}\right) \mathrm{Op}^W(b^{-s_\theta})\|_{\mathscr{L}(L^2(\mathbb{R}^n))} \\ &\qquad \|h\|^{(b)}_{\mathcal{Q}_{s_\theta}}. \end{aligned} \quad (6.51)$$

From the calculus we obtain that the two operator norms on $L^2(\mathbb{R}^n)$ in (6.50) and (6.51) are bounded by a constant of the form $C(1+|y|)^N$ where $C>0$ and $N\in\mathbb{N}_0$ are independent of y. This shows that $\mathcal{Q}^{(b)}_{s_\theta}$ is continuously included in $(\mathcal{Q}^{(b)}_{s_0}(\mathbb{R}^n),\mathcal{Q}^{(b)}_{s_1}(\mathbb{R}^n))_\theta$. By duality of the complex interpolation and of the spaces $\mathcal{Q}_s(\mathbb{R}^n)$, we obtain the reverse inclusion and (6.49) is proved.

Let us show that the spaces $\mathcal{Q}^{(b)}_s(\mathbb{R}^n)$ and $\mathcal{Q}_s(\mathbb{R}^n)$ coincide. First let us assume $s\in 2\mathbb{N}_0$. We have for any $h\in\mathcal{Q}^{(b)}_s(\mathbb{R}^n)$:

$$\|h\|_{\mathcal{Q}_s}\leq\|(\mathrm{I}+\mathcal{Q})^{s/2}\mathrm{Op}^W(b^{-s})\|_{\mathcal{L}(L^2(\mathbb{R}^n))}\|h\|^{(b)}_{\mathcal{Q}_s}.$$

As $\mathcal{Q}\in\Psi\Sigma^2_1(\mathbb{R}^n)$, by Theorem 6.4.14, the operator $(\mathrm{I}+\mathcal{Q})^{s/2}\mathrm{Op}^W(b^{-s})$ is in $\Psi\Sigma^0_1$ and thus is bounded on $L^2(\mathbb{R}^n)$. We have obtained a continuous inclusion of $\mathcal{Q}^{(b)}_s(\mathbb{R}^n)$ into $\mathcal{Q}_s(\mathbb{R}^n)$. Conversely, we have for any $h\in\mathcal{Q}_s(\mathbb{R}^n)$ that

$$\|h\|^{(b)}_{\mathcal{Q}_s}\leq\|\mathrm{Op}^W(b^s)(\mathrm{I}+\mathcal{Q})^{-s/2}\|_{\mathcal{L}(L^2(\mathbb{R}^n))}\|h\|_{\mathcal{Q}_s}.$$

The inverse of $\mathrm{Op}^W(b^s)(\mathrm{I}+\mathcal{Q})^{-s/2}$ is $(\mathrm{I}+\mathcal{Q})^{s/2}(\mathrm{Op}^W(b^s))^{-1}$ since the operators $\mathrm{I}+\mathcal{Q}$ and $\mathrm{Op}^W(b^s)$ are invertible. Moreover, for the same reason as above, $(\mathrm{I}+\mathcal{Q})^{s/2}(\mathrm{Op}^W(b^s))^{-1}$ is bounded on $L^2(\mathbb{R}^n)$. By the inverse mapping theorem, $\mathrm{Op}^W(b^s)(\mathrm{I}+\mathcal{Q})^{-s/2}$ is bounded on $L^2(\mathbb{R}^n)$. This shows the reverse continuous inclusion. We have proved

$$\mathcal{Q}^{(b)}_s(\mathbb{R}^n)=\mathcal{Q}_s(\mathbb{R}^n)$$

with equivalence of norms for $s\in 2\mathbb{N}_0$ and this implies that this is true for any $s\in\mathbb{R}$ by the properties of duality and interpolation for $\mathcal{Q}^{(b)}_s(\mathbb{R}^n)$ and $\mathcal{Q}_s(\mathbb{R}^n)$. This shows Part (4) and implies Parts (5) and (6).

Let us show that, for each $s\in\mathbb{N}_0$, the space $\mathcal{Q}_s(\mathbb{R}^n)$ coincides with the space $\mathcal{Q}^{(int)}_s(\mathbb{R}^n)$ of functions $h\in L^2(\mathbb{R}^n)$ such that the tempered distributions $u^\alpha\partial_u^\beta h$ are in $L^2(\mathbb{R}^n)$ for every $\alpha,\beta\in\mathbb{N}_0^n$ such that $|\alpha|+|\beta|\leq s$. Endowed with the norm $\|\cdot\|^{(int)}_{\mathcal{Q}_s}$ defined in Part (3), $\mathcal{Q}^{(int)}_s(\mathbb{R}^n)$ is a Banach space. We have for any $h\in\mathcal{Q}_s(\mathbb{R}^n)=\mathcal{Q}^{(b)}_s(\mathbb{R}^n)$

$$\|h\|^{(int)}_{\mathcal{Q}_s}\leq\sum_{|\alpha|+|\beta|\leq s}\|u^\alpha\partial_u^\beta\mathrm{Op}^W(b^{-s})\|_{\mathcal{L}(L^2(\mathbb{R}^n))}\|h\|^{(b)}_{\mathcal{Q}_s}.$$

Since the operators $u^\alpha\partial_u^\beta\mathrm{Op}^W(b^{-s})$ are in $\Psi\Sigma_1^{|\alpha|+|\beta|-s}(\mathbb{R}^n)$ thus continuous on $L^2(\mathbb{R}^n)$ when $|\alpha|+|\beta|\leq s$, we see that $\mathcal{Q}_s(\mathbb{R}^n)$ is continuously included in $\mathcal{Q}^{(int)}_s(\mathbb{R}^n)$. For the converse, we separate the cases s even and odd. If $s\in 2\mathbb{N}_0$ then we have easily that

$$\begin{aligned}\|h\|_{\mathcal{Q}_s}&=\|(\mathrm{I}+\sum_j(-\partial_{u_j}^2+u_j^2))^{s/2}h\|_{L^2(\mathbb{R}^n)}\\&\leq C_s\sum_{|\alpha|+|\beta|\leq s}\|u^\alpha\partial_u^\beta h\|_{L^2(\mathbb{R}^n)}=C_s\|h\|^{(int)}_{\mathcal{Q}_s}.\end{aligned}$$

Now if $s \in 2\mathbb{N}_0 + 1$, we have, since $\mathrm{Op}^W(b^{-1})(\mathrm{I}+Q)^{1/2}$ is bounded and invertible (see Part (6) already proven),

$$\begin{aligned}\|h\|_{\mathcal{Q}_s} &= \|(\mathrm{I}+Q)^{s/2}h\|_{L^2(\mathbb{R}^n)} \leq C\|\mathrm{Op}^W(b^{-1})(\mathrm{I}+Q)^{1/2}(\mathrm{I}+Q)^{s/2}h\|_{L^2(\mathbb{R}^n)} \\ &\leq C\|\mathrm{Op}^W(b^{-1})(\mathrm{I}+\sum_j -\partial_{u_j}^2 + u_j^2)^{(s+1)/2}h\|_{L^2(\mathbb{R}^n)} \\ &\leq C_s \sum_{|\alpha|+|\beta|\leq s+1} \|\mathrm{Op}^W(b^{-1})x^\alpha \partial_x^\beta h\|_{L^2(\mathbb{R}^n)} \\ &\leq C_s \sum_{|\alpha'|+|\beta'|\leq s} \|u^{\alpha'}\partial_u^{\beta'}h\|_{L^2(\mathbb{R}^n)} = C_s\|h\|_{\mathcal{Q}_s}^{(int)},\end{aligned}$$

by the property of the calculus. Therefore, for s even and odd, $\mathcal{Q}_s^{(int)}(\mathbb{R}^n)$ is continuously included in $\mathcal{Q}_s(\mathbb{R}^n)$. As we have already proven the reverse inclusion, the equality holds and Part (3) is proved. This implies

$$\bigcap_{s\in\mathbb{R}} \mathcal{Q}_s(\mathbb{R}^n) = \mathcal{S}(\mathbb{R}^n)$$

and Part (1) is now completely proved. □

These Sobolev spaces enable us to characterise the operators in the calculus. We allow ourselves to use the shorthand notation

$$(\mathrm{ad}u)^{\alpha_1} := (\mathrm{ad}u_1)^{\alpha_{11}} \ldots (\mathrm{ad}u_n)^{\alpha_{1n}},$$

and

$$(\mathrm{ad}\partial_u)^{\alpha_2} := (\mathrm{ad}\partial_{u_1})^{\alpha_{21}} \ldots (\mathrm{ad}\partial_{u_n})^{\alpha_{2n}}.$$

Theorem 6.4.18. *We assume that $\rho \in (0,1]$. Let $A : \mathcal{S}(\mathbb{R}^n) \to \mathcal{S}'(\mathbb{R}^n)$ be a linear continuous operator such that all the operators*

$$(\mathrm{ad}u)^{\alpha_1}(\mathrm{ad}\partial_u)^{\alpha_2}A, \quad \alpha_1, \alpha_2 \in \mathbb{N}_0^n,$$

are in $\mathscr{L}(L^2(\mathbb{R}^n), \mathcal{Q}_{-m+\rho(|\alpha_1|+|\alpha_2|)})$ in the sense that they extend to continuous operators from $L^2(\mathbb{R}^n)$ to $\mathcal{Q}_{-m+\rho(|\alpha_1|+|\alpha_2|)}$. Then $A \in \Psi\Sigma_\rho^m(\mathbb{R}^n)$. Moreover, for any $\ell \in \mathbb{N}$, there exist a constant C and an integer ℓ', both independent of A, such that

$$\|A\|_{\Psi\Sigma_\rho^m,\ell} \leq C \sum_{|\alpha_1|+|\alpha_2|\leq \ell'} \|(\mathrm{ad}u)^{\alpha_1}(\mathrm{ad}\partial_u)^{\alpha_2}A\|_{\mathscr{L}(L^2(\mathbb{R}^n),\mathcal{Q}_{-m+\rho(|\alpha_1|+|\alpha_2|)})}.$$

Note that the converse is true, that is, given $A \in \Psi\Sigma_\rho^m$ then

$$\forall \alpha_1, \alpha_2 \in \mathbb{N}_0^n \qquad (\mathrm{ad}u)^{\alpha_1}(\mathrm{ad}\partial_u)^{\alpha_2}A \in \mathscr{L}(L^2(\mathbb{R}^n), \mathcal{Q}_{-m+\rho(|\alpha_1|+|\beta|)},).$$

This is just a consequence of the properties of the calculus.

The proof of Theorem 6.4.18 relies on the following characterisation of the class of symbols

$$\Sigma_0^0(\mathbb{R}^n) := S(1, d\xi^2 + du^2).$$

Theorem 6.4.19 (Beals' characterisation of $\Sigma_0^0(\mathbb{R}^n)$). *Let $A : \mathcal{S}(\mathbb{R}^n) \to \mathcal{S}'(\mathbb{R}^n)$ be a linear continuous operator such that all the operators*

$$(\mathrm{ad}u)^{\alpha_1}(\mathrm{ad}\partial_u)^{\alpha_2}A, \quad \alpha_1, \alpha_2 \in \mathbb{N}_0^n,$$

are in $\mathscr{L}(L^2(\mathbb{R}^n))$ in the sense that they extend to continuous operators on $L^2(\mathbb{R}^n)$. Then there exits a unique function $a = \{a(\xi, x)\} \in \Sigma_0^0(\mathbb{R}^n)$ such that $A = \mathrm{Op}^W(a)$. Moreover, for any $\ell \in \mathbb{N}$, there exist a constant C and an integer ℓ', both independent of A, such that

$$\|a\|_{\Sigma_0^0,\ell} \leq C \sum_{|\alpha_1|+|\alpha_2|\leq \ell'} \|(\mathrm{ad}u)^{\alpha_1}(\mathrm{ad}\partial_u)^{\alpha_2}A\|_{\mathscr{L}(L^2(\mathbb{R}^n))}.$$

The converse is true, that is, given $a \in \Sigma_0^0(\mathbb{R}^n)$ then $A = \mathrm{Op}^W(a)$ satisfies

$$\forall \alpha_1, \alpha_2 \in \mathbb{N}_0^n \qquad (\mathrm{ad}u)^{\alpha_1}(\mathrm{ad}\partial_u)^{\alpha_2}A \in \mathscr{L}(L^2(\mathbb{R}^n)).$$

We admit Beals' theorem stated in Theorem 6.4.19, see the original article [Bea77a] for the proof.

For the sake of completeness we prove Theorem 6.4.18. This proof can also be found in [Hel84a, Théorème 1.21.1].

Sketch of the proof of Theorem 6.4.18. Let A be as in the statement and b as in Example 6.4.13. We write

$$B_s := \mathrm{Op}^W(b^s)$$

and

$$A_{\alpha_1,\alpha_2} := (\mathrm{ad}u)^{\alpha_1}(\mathrm{ad}\partial_u)^{\alpha_2}A, \quad \alpha_1, \alpha_2 \in \mathbb{N}_0^n.$$

We set $s := m - \rho(|\alpha_1| + |\alpha_2|)$. Then $B_s^{-1}A_{\alpha_1,\alpha_2} \in \mathscr{L}(L^2(\mathbb{R}^n))$. Moreover, we have

$$\mathrm{ad}\partial_{u_1}\left(B_s^{-1}A_{\alpha_1,\alpha_2}\right) = \left(\mathrm{ad}\partial_{u_1}\left(B_s^{-1}\right)\right)A_{\alpha_1,\alpha_2} + B_s^{-1}\mathrm{ad}\partial_{u_1}\left(A_{\alpha_1,\alpha_2}\right);$$

the first operator of the right-hand side is in $\mathscr{L}(L^2(\mathbb{R}^n), \mathcal{Q}_1(\mathbb{R}^n))$ whereas the second is in $\mathscr{L}(L^2(\mathbb{R}^n), \mathcal{Q}_\rho(\mathbb{R}^n))$. Proceeding recursively, we obtain that the operator $B^{-1}_{m-\rho(|\alpha_1|+|\alpha_2|)}A_{\alpha_1,\alpha_2}$ satisfies the hypothesis of Beals' Theorem (Theorem 6.4.19). Therefore, there exists $c_{\alpha_1,\alpha_2} \in \Sigma_0^0(\mathbb{R}^n)$ such that

$$B^{-1}_{m-\rho(|\alpha_1|+|\alpha_2|)}A_{\alpha_1,\alpha_2} = \mathrm{Op}^W(c_{\alpha_1,\alpha_2})$$

or, equivalently,

$$A_{\alpha_1,\alpha_2} = \mathrm{Op}^W(a_{\alpha_1,\alpha_2}) \quad \text{with} \quad a_{\alpha_1,\alpha_2} = b_{m-\rho(|\alpha_1|+|\alpha_2|)} \star c_{\alpha_1,\alpha_2}.$$

We have $A = \mathrm{Op}^W(a_{0,0})$ and

$$\begin{aligned} \mathrm{Op}^W(a_{\alpha_1,\alpha_2}) &= A_{\alpha_1,\alpha_2} = (\mathrm{ad}u)^{\alpha_1}(\mathrm{ad}\partial_u)^{\alpha_2} A \\ &= (\mathrm{ad}u)^{\alpha_1}(\mathrm{ad}\partial_u)^{\alpha_2}\mathrm{Op}^W(a_{0,0}) \\ &= \mathrm{Op}^W\left(i^{|\alpha_1|}\partial_\xi^{\alpha_1}\partial_u^{\alpha_2} a_{0,0}\right), \end{aligned}$$

by Lemma 6.2.3, thus

$$a_{\alpha_1,\alpha_2} = i^{|\alpha_1|}\partial_\xi^{\alpha_1}\partial_u^{\alpha_2} a_{0,0}.$$

Consequently $a \in \Sigma_\rho^m$. □

Looking back at the proof, we see that it can be slightly improved in the following way:

Corollary 6.4.20. *We assume that $\rho \in (0,1]$. Let $A : \mathcal{S}(\mathbb{R}^n) \to \mathcal{S}'(\mathbb{R}^n)$ be a linear continuous operator.*

The operator A is in $\Psi\Sigma_\rho^m(\mathbb{R}^n)$ if and only if there exists $\gamma_o \in \mathbb{R}$ such that for each $\alpha_1, \alpha_2 \in \mathbb{N}_0^n$ we have

$$(\mathrm{ad}u)^{\alpha_1}(\mathrm{ad}\partial_u)^{\alpha_2} A \in \mathscr{L}(\mathcal{Q}_{\gamma_o}(\mathbb{R}^n), \mathcal{Q}_{-m+\rho(|\alpha_1|+|\alpha_2|)+\gamma_o}).$$

In this case this property is true for every $\gamma \in \mathbb{R}$, that is, for each $\gamma \in \mathbb{R}$ and $\alpha_1, \alpha_2 \in \mathbb{N}_0^n$, we have

$$(\mathrm{ad}u)^{\alpha_1}(\mathrm{ad}\partial_u)^{\alpha_2} A \in \mathscr{L}(\mathcal{Q}_{\gamma}(\mathbb{R}^n), \mathcal{Q}_{-m+\rho(|\alpha_1|+|\alpha_2|)+\gamma}).$$

Moreover, for any $\ell \in \mathbb{N}$, there exist a constant C and an integer ℓ', both independent of A, such that

$$\|A\|_{\Psi\Sigma_\rho^m,\ell} \le C \sum_{|\alpha_1|+|\alpha_2|\le\ell'} \|(\mathrm{ad}u)^{\alpha_1}(\mathrm{ad}\partial_u)^{\alpha_2} A\|_{\mathscr{L}(\mathcal{Q}_\gamma(\mathbb{R}^n),\mathcal{Q}_{-m+\rho(|\alpha_1|+|\alpha_2|)+\gamma})}.$$

Sketch of the proof of Corollary 6.4.20. We keep the notation of the proof of Theorem 6.4.18. Let A be as in the statement and let $s := m - \rho(|\alpha_1| + |\alpha_2|)$. Then $B_{s+\gamma_o}^{-1} A_{\alpha_1,\alpha_2} B_{\gamma_o} \in \mathscr{L}(L^2(\mathbb{R}^n))$. Moreover, we have

$$\begin{aligned} \mathrm{ad}\partial_{u_1}\left(B_{s+\gamma_o}^{-1} A_{\alpha_1,\alpha_2} B_{\gamma_o}\right) &= \left(\mathrm{ad}\partial_{u_1}\left(B_{s+\gamma_o}^{-1}\right)\right) A_{\alpha_1,\alpha_2} B_{\gamma_o} \\ &\quad + B_{s+\gamma_o}^{-1}\,\mathrm{ad}\partial_{u_1}\left(A_{\alpha_1,\alpha_2}\right) B_{\gamma_o} \\ &\quad + B_{s+\gamma_o}^{-1} A_{\alpha_1,\alpha_2} B_{\gamma_o}\ B_{\gamma_o}^{-1}\left(\mathrm{ad}\partial_{u_1} B_{\gamma_o}\right); \end{aligned}$$

the first operator of the right-hand side is in $\mathscr{L}(L^2(\mathbb{R}^n), \mathcal{Q}_1(\mathbb{R}^n))$, the second is in $\mathscr{L}(L^2(\mathbb{R}^n), \mathcal{Q}_\rho(\mathbb{R}^n))$ and the third is in $\mathscr{L}(L^2(\mathbb{R}^n))$. Proceeding recursively, we obtain that $B_{s+\gamma_o}^{-1} A_{\alpha_1,\alpha_2} B_{\gamma_o}$ satisfies the hypothesis of Theorem 6.4.19. We then conclude as in the proof of Theorem 6.4.18. □

6.4.4 The λ-Shubin classes $\Sigma^m_{\rho,\lambda}(\mathbb{R}^n)$

The Shubin metric depending on a parameter $\lambda \in \mathbb{R}\backslash\{0\}$ is the metric $g^{(\lambda)}$ on $\mathbb{R}^{2n}$ defined via

$$g^{(\rho,\lambda)}_{\xi,u}(d\xi, du) := \left(\frac{|\lambda|}{1+|\lambda|(1+|\xi|^2+|u|^2)}\right)^\rho (d\xi^2+du^2).$$

The associated positive function $M^{(\lambda)}$ on $\mathbb{R}^{2n}$ is defined via

$$M^{(\lambda)}(\xi,u) := \left(1+|\lambda|(1+|\xi|^2+|u|^2)\right)^{\frac{1}{2}}.$$

These λ-families of metrics and weights were first introduced in [BFKG12a] in the case $\rho = 1$. The authors of [BFKG12a] realised that, placing λ as above, the structural constants may be chosen independently of λ:

Proposition 6.4.21. *For each $\lambda \in \mathbb{R}\backslash\{0\}$, the metric $g^{(\rho,\lambda)}$ is of Hörmander type (see Definition 6.4.2) and the function $M^{(\lambda)}$ is a $g^{(\rho,\lambda)}$-weight (see Definition 6.4.7). Furthermore, if $\rho \in (0,1]$ is fixed, then the structural constants for $g^{(\rho,\lambda)}$ and for $M^{(\lambda)}$ can be chosen independent of λ.*

The proof of Proposition 6.4.21 follows the proof of the case $\rho = 1$ given in [BFKG12a, Proposition 1.20].

Proof of Proposition 6.4.21. The conjugate of $g^{(\rho,\lambda)}_{\xi,u}$ is $(g^{(\rho,\lambda)}_{\xi,u})^\omega$ given by

$$(g^{(\rho,\lambda)}_{\xi,u})^\omega(d\xi, du) = \left(\frac{1+|\lambda|(1+|\xi|^2+|u|^2)}{|\lambda|}\right)^\rho (d\xi^2+du^2).$$

The gain is then

$$\Lambda_{g^{(\rho,\lambda)}_{\xi,u}} = \left(\frac{1+|\lambda|(1+|\xi|^2+|u|^2)}{|\lambda|}\right)^{2\rho}.$$

We have for any ρ, λ, ξ, u:

$$\Lambda_{g^{(\rho,\lambda)}_{\xi,u}} \geq \left(\frac{1+|\lambda|}{|\lambda|}\right)^{2\rho} \geq 1.$$

This proves the uniform uncertain property in Definition 6.4.2.

To show that the metric $g^{\rho,\lambda}$ is slowly varying, we notice that it is of the form $\phi(X)^{-2}|T|^2$ as in Example 6.4.5 with

$$\phi(X) = \left(\frac{1+|\lambda|(1+|X|^2)}{|\lambda|}\right)^{\rho/2}.$$

We compute the gradient of ϕ and obtain

$$\begin{aligned} |\nabla_X \phi| &= \rho|\lambda|^{1-\frac{\rho}{2}}|X|(1+|\lambda|(1+|X|^2))^{\frac{\rho}{2}-1} \\ &\leq \begin{cases} \rho\left(\frac{|\lambda|}{1+|\lambda|}\right)^{1-\frac{\rho}{2}} \leq \rho & \text{if } |X| \leq 1, \\ \rho\left(\frac{|\lambda||X|^2}{1+|\lambda||X|^2}\right)^{1-\frac{\rho}{2}} |X|^{1-2(1-\frac{\rho}{2})} \leq \rho & \text{if } |X| > 1. \end{cases} \end{aligned}$$

So ϕ is ρ-Lipschitz on $\mathbb{R}^{2n}$. Therefore, $g^{\rho,\lambda}$ is slowly varying with a constant $\bar{C}$ independent of λ (see Example 6.4.5 as well as Remarks 6.4.4 and 6.4.6).

Let us prove that $g^{\rho,\lambda}$ is temperate. For any $X, Y \in \mathbb{R}^{2n}$ we have

$$|Y|^2 \leq 2|X|^2 + 2|X-Y|^2;$$

thus

$$\frac{1+|\lambda|(1+|Y|^2)}{1+|\lambda|(1+|X|^2)} \leq 2 + 2\frac{|\lambda|}{1+|\lambda|(1+|X|^2)}|X-Y|^2. \tag{6.52}$$

Now

$$|\lambda| \leq 1+|\lambda|(1+|X|^2) \quad \text{thus} \quad \left(\frac{|\lambda|}{1+|\lambda|(1+|X|^2)}\right)^{1+\rho} \leq 1,$$

and

$$\frac{|\lambda|}{1+|\lambda|(1+|X|^2)} \leq \left(\frac{1+|\lambda|(1+|X|^2)}{|\lambda|}\right)^{\rho}.$$

Plugging this into (6.52), we obtain

$$\frac{1+|\lambda|(1+|Y|^2)}{1+|\lambda|(1+|X|^2)} \leq 2 + 2\left(\frac{1+|\lambda|(1+|X|^2)}{|\lambda|}\right)^{\rho} |X-Y|^2.$$

Taking the ρth power yields

$$\begin{aligned} \frac{g_X^{(\rho,\lambda)}(T)}{g_Y^{(\rho,\lambda)}(T)} &= \left(\frac{1+|\lambda|(1+|Y|^2)}{1+|\lambda|(1+|X|^2)}\right)^{\rho} \\ &\leq 2^{\rho}\left(1+\left(\frac{1+|\lambda|(1+|X|^2)}{|\lambda|}\right)^{\rho}|X-Y|^2\right)^{\rho} \\ &= 2^{\rho}\left(1+(g_X^{(\rho,\lambda)})^{\omega}(X-Y)\right)^{\rho}. \end{aligned}$$

This shows that $g^{(\rho,\lambda)}$ is temperate with constant independent of λ.

So far we have shown that $g^{(\rho,\lambda)}$ is a metric of Hörmander type. Following the same computations, it is not difficult to show that $M^{(\lambda)}$ are g-weights with constants independent of λ. This concludes the proof of Proposition 6.4.21. □

Let $\rho \in (0,1]$ be a fixed parameter.

For each parameter $\lambda \in \mathbb{R}\backslash\{0\}$, we define the λ-*Shubin classes* by

$$\Sigma^m_{\rho,\lambda}(\mathbb{R}^n) := S\left(\left(M^{(\lambda)}\right)^m, g^{(\rho,\lambda)}\right),$$

where we have used the Hörmander notation to define a class of symbols in terms of a metric and a weight, see Definition 6.4.8.

Here this means that $\Sigma^m_{\rho,\lambda}(\mathbb{R}^n)$ is the class of functions $a \in C^\infty(\mathbb{R}^n \times \mathbb{R}^n)$ such that for each $N \in \mathbb{N}_0$, the quantity

$$\|a\|_{\Sigma^m_{\rho,\lambda},N} := \sup_{\substack{(\xi,u)\in\mathbb{R}^n\times\mathbb{R}^n \\ |\alpha|,|\beta|\le N}} |\lambda|^{-\rho\frac{|\alpha|+|\beta|}{2}} \left(1+|\lambda|(1+|\xi|^2+|u|^2)\right)^{-\frac{m-\rho(|\alpha|+|\beta|)}{2}} |\partial^\alpha_\xi \partial^\beta_u a(\xi,u)|,$$

is finite. This also means that a symbol $a = \{a(\xi,u)\}$ is in $\Sigma^m_{\rho,\lambda}(\mathbb{R}^n)$ if and only if it satisfies

$$\begin{gathered}\forall \alpha,\beta\in\mathbb{N}^n_0 \qquad \exists C = C_{\alpha,\beta} > 0 \qquad \forall (\xi,u)\in\mathbb{R}^n\times\mathbb{R}^n \\ |\partial^\alpha_\xi \partial^\beta_u a(\xi,u)| \le C|\lambda|^{\rho\frac{|\alpha|+|\beta|}{2}} \left(1+|\lambda|(1+|\xi|^2+|u|^2)\right)^{\frac{m-\rho(|\alpha|+|\beta|)}{2}}. \end{gathered} \tag{6.53}$$

The class of symbols $\Sigma^m_{\rho,\lambda}(\mathbb{R}^n)$ is a vector subspace of $C^\infty(\mathbb{R}^n \times \mathbb{R}^n)$ which becomes a Fréchet space when endowed with the family of seminorms $\|\cdot\|_{\Sigma^m_{\rho,\lambda},N}$, $N \in \mathbb{N}_0$. We denote by

$$\Psi\Sigma^m_{\rho,\lambda}(\mathbb{R}^n) := \mathrm{Op}^W(\Sigma^m_{\rho,\lambda}(\mathbb{R}^n))$$

the corresponding class of operators, and by $\|\cdot\|_{\Psi\Sigma^m_{\rho,\lambda},N}$ the corresponding seminorms on the Fréchet space $\Psi\Sigma^m_{\rho,\lambda}(\mathbb{R}^n)$.

It is clear that all the spaces of the same order m and parameter ρ coincide in the sense that

$$\forall \lambda \ne 0 \qquad \Sigma^m_{\rho,\lambda}(\mathbb{R}^n) = \Sigma^m_{\rho,1}(\mathbb{R}^n) = \Sigma^m_\rho(\mathbb{R}^n), \tag{6.54}$$

and the same is true for $\Psi\Sigma^m_{\rho,\lambda}(\mathbb{R}^n) = \Psi\Sigma^m_\rho(\mathbb{R}^n)$. However, the seminorms

$$\|\cdot\|_{\Sigma^m_{\rho,\lambda},N} \quad \text{and} \quad \|\cdot\|_{\Psi\Sigma^m_{\rho,\lambda},N}$$

carry the dependence on λ. This dependence on λ will be crucial for our purposes. From the general properties of metrics of Hörmander type (see Theorem 6.4.9 and Proposition 6.4.21), we readily obtain the following 'λ-uniform' calculus.

Proposition 6.4.22. *1. If, for each $\lambda \in \mathbb{R}\backslash\{0\}$, we are given a symbol $a_\lambda = \{a_\lambda(\xi,u)\}$ in $\Sigma^m_{\rho,\lambda}(\mathbb{R}^n)$ such that*

$$\forall N \in \mathbb{N}_0 \qquad \sup_{\lambda\ne 0} \|a_\lambda\|_{\Sigma^m_{\rho,\lambda},N} < \infty, \tag{6.55}$$

then each symbol b_λ *defined by*

$$\mathrm{Op}^W b_\lambda = \left(\mathrm{Op}^W a_\lambda\right)^*$$

is in $\Sigma^m_{\rho,\lambda}(\mathbb{R}^n)$ *as well. Furthermore, for any* $\ell \in \mathbb{N}_0$ *there exist a constant* $C > 0$ *and a integer* $\ell' \in \mathbb{N}_0$ *such that for any* $\lambda \neq 0$

$$\|b_\lambda\|_{\Sigma^m_{\rho,\lambda},\ell} \leq C\|a_\lambda\|_{\Sigma^m_{\rho,\lambda},\ell'}.$$

The constant C *and the integer* ℓ' *may be chosen to depend on* ℓ, m, n *and to be independent of* λ *and* a.

2. *If, for each* $\lambda \in \mathbb{R}\backslash\{0\}$, *we are given two symbols* $a_{1,\lambda} = \{a_{1,\lambda}(\xi, u)\}$ *in* $\Sigma^{m_1}_{\rho,\lambda}(\mathbb{R}^n)$ *and* $a_{2,\lambda} = \{a_{2,\lambda}(\xi, u)\}$ *in* $\Sigma^{m_2}_{\rho,\lambda}(\mathbb{R}^n)$ *such that*

$$\forall N \in \mathbb{N}_0 \qquad \sup_{\lambda\neq 0} \|a_{1,\lambda}\|_{\Sigma^{m_1}_{\rho,\lambda},N} < \infty \quad \textit{and} \quad \sup_{\lambda\neq 0} \|a_{2,\lambda}\|_{\Sigma^{m_2}_{\rho,\lambda},N} < \infty,$$

then each symbol b_λ *defined by*

$$\mathrm{Op}^W b_\lambda = \left(\mathrm{Op}^W a_{1,\lambda}\right)\left(\mathrm{Op}^W a_{2,\lambda}\right),$$

is in $\Sigma^{m_1+m_2}_{\rho,\lambda}(\mathbb{R}^n)$. *Furthermore, for any* $\ell \in \mathbb{N}_0$ *there exist a constant* $C > 0$ *and two integers* $\ell_1, \ell_2 \in \mathbb{N}_0$ *such that*

$$\|b_\lambda\|_{\Sigma^{m_1+m_2}_{\lambda},\ell} \leq C\|a_{1,\lambda}\|_{\Sigma^{m_1}_{\rho,\lambda},\ell_1}\|a_{2,\lambda}\|_{\Sigma^{m_2}_{\rho,\lambda},\ell_2}.$$

The constant C *and the integers* ℓ_1, ℓ_2 *may be chosen to depend on* ℓ, m_1, m_2, n *and to be independent of* λ *and* $a_{1,\lambda}, a_{2,\lambda}$.

We will say that a family of symbols $a_\lambda = \{a_\lambda(\xi, u)\}$, $\lambda \in \mathbb{R}\backslash\{0\}$, which satisfies Property (6.55) is λ-*uniform* in $\Sigma^m_{\rho,\lambda}(\mathbb{R}^n)$. The corresponding family of operators via the Weyl quantization is said to be λ-uniform in $\Psi\Sigma^m_{\rho,\lambda}(\mathbb{R}^n)$.

Let us give some useful examples of such families of operators.

Example 6.4.23. The families of symbols given by

$$\pi_\lambda(X_j) = i\sqrt{|\lambda|}\xi_j, \quad \pi_\lambda(Y_j) = i\sqrt{\lambda}u_j \quad \text{and} \quad \pi_\lambda(T) = i\lambda$$

are λ-uniform in $\Sigma^1_{1,\lambda}(\mathbb{R}^n)$, $\Sigma^1_{1,\lambda}(\mathbb{R}^n)$, and $\Sigma^2_{1,\lambda}(\mathbb{R}^n)$, respectively.

In particular, the constant operator $\pi_\lambda(T) = i\lambda$ has to be considered as being of order 2 because of the dependence on λ.

Proof. We want to estimate the supremum over $\lambda \neq 0$ of each of the seminorms

$$\|\pi_\lambda(X_j)\|_{\Psi\Sigma^1_{1,\lambda},N} = \|i\sqrt{|\lambda|}\xi_j\|_{\Sigma^1_{1,\lambda},N} \quad \text{and} \quad \|\pi_\lambda(Y_j)\|_{\Psi\Sigma^1_{1,\lambda},N} = \|i\sqrt{\lambda}u_j\|_{\Sigma^1_{1,\lambda},N}.$$

We compute directly for $N = 0$:

$$\sup_{\lambda\neq 0} \|i\sqrt{|\lambda|}\xi_j\|_{\Sigma^1_{1,\lambda},0} = \sup_{\lambda\neq 0,(\xi,u)\in\mathbb{R}^n\times\mathbb{R}^n} \frac{\sqrt{|\lambda|}|\xi_j|}{\sqrt{1+|\lambda|(1+|\xi|^2+|u|^2)}} < \infty,$$

$$\sup_{\lambda\neq 0} \|i\sqrt{\lambda}u_j\|_{\Sigma^1_{1,\lambda},0} = \sup_{\lambda\neq 0,(\xi,u)\in\mathbb{R}^n\times\mathbb{R}^n} \frac{\sqrt{|\lambda|}|u_j|}{\sqrt{1+|\lambda|(1+|\xi|^2+|u|^2)}} < \infty,$$

and

$$\sup_{\substack{|\alpha|+|\beta|=1\\(\xi,u)\in\mathbb{R}^n\times\mathbb{R}^n}} |\partial_\xi^\alpha\partial_u^\beta\{\sqrt{|\lambda|}\xi_j\}| = \sup_{\substack{|\alpha|+|\beta|=1\\(\xi,u)\in\mathbb{R}^n\times\mathbb{R}^n}} |\partial_\xi^\alpha\partial_u^\beta\{\sqrt{\lambda}u_j\}| = \sqrt{|\lambda|},$$

therefore

$$\sup_{\lambda\neq 0} \|i\sqrt{|\lambda|}\xi_j\|_{\Sigma^1_{1,\lambda},1} < \infty \quad \text{and} \quad \sup_{\lambda\neq 0} \|i\sqrt{\lambda}u_j\|_{\Sigma^1_{1,\lambda},1} < \infty.$$

Since all the higher derivatives $\partial_\xi^\alpha\partial_u^\beta$ with $|\alpha|+|\beta|>1$ of the symbols $i\sqrt{|\lambda|}\xi_j$ and $i\sqrt{\lambda}u_j$ are zero, we obtain that the families of symbols given by $\pi_\lambda(X_j)$, $\pi_\lambda(Y_j)$, are λ-uniform in $\Sigma^1_{1,\lambda}(\mathbb{R}^n)$.

For $\pi_\lambda(T) = \mathrm{Op}^W(i\lambda)$, we see that

$$\|i\lambda\|_{\Sigma^2_{1,\lambda},0} = \sup_{(\xi,u)\in\mathbb{R}^n\times\mathbb{R}^n} \frac{|i\lambda|}{1+|\lambda|(1+|\xi|^2+|u|^2)} < \infty,$$

and since $i\lambda$ is a constant, its derivatives are zero and the family of symbols given by $\pi_\lambda(T)$, is λ-uniform in $\Sigma^2_{1,\lambda}(\mathbb{R}^n)$. □

As a consequence of Example 6.4.23 and Proposition 6.4.22, we also have

Example 6.4.24. The family of operators

$$\pi_\lambda(\mathcal{L}) = \sum_{j=1}^n \{\pi_\lambda(X_j)^2 + \pi_\lambda(Y_j)^2\} = -|\lambda|Q$$

is λ-uniform in $\Psi\Sigma^2_{1,\lambda}(\mathbb{R}^n)$.

Standard computations also show:

Example 6.4.25. For each $m \in \mathbb{R}$, the family of symbols b_λ^m, $\lambda \in \mathbb{R}\backslash\{0\}$, where

$$b_\lambda(\xi, u) = \sqrt{1+|\lambda|(1+|u|^2+|\xi|^2)},$$

is λ-uniform in $\Psi\Sigma^m_{1,\lambda}(\mathbb{R}^n)$.

6.4.5 Commutator characterisation of λ-Shubin classes

In this section, we characterise the λ-Shubin classes in terms of commutators and continuity on the Shubin Sobolev spaces.

First we need to understand some properties of the Sobolev spaces associated with the λ-dependent metric used to define the λ-Shubin symbols.

Proposition 6.4.26. *1. For each $\lambda \in \mathbb{R}\backslash\{0\}$ and $s \in \mathbb{R}$, the Sobolev space corresponding to $g^{(1,\lambda)}$ and $(M^{(\lambda)})^s$ coincides with the Shubin Sobolev space:*

$$H\left((M^{(\lambda)})^s, g^{(1,\lambda)}\right) = \mathcal{Q}_s(\mathbb{R}^n).$$

2. The following define norms on $\mathcal{Q}_s(\mathbb{R}^n)$ equivalent to $\|\cdot\|_{\mathcal{Q}_s}$:

$$\begin{aligned}\|h\|_{\mathcal{Q}_{s,\lambda}} &:= \|(\mathrm{I}+|\lambda|Q)^{s/2}h\|_{L^2(\mathbb{R}^n)},\\ \|h\|^{(b_\lambda)}_{\mathcal{Q}_{s,\lambda}} &:= \|\mathrm{Op}^W(b_\lambda^s)\,h\|_{L^2(\mathbb{R}^n)},\end{aligned}$$

where b_λ was defined in Example 6.4.25. Moreover, in the case $s \in \mathbb{N}_0$, we also have an equivalent norm

$$\|h\|^{(int)}_{\mathcal{Q}_{s,\lambda}} := \sum_{|\alpha|+|\beta|\le s} |\lambda|^{\frac{|\alpha|+|\beta|}{2}} \|u^\alpha \partial_u^\beta h\|_{L^2(\mathbb{R}^n)}.$$

3. Furthermore, for each $s \in \mathbb{R}$ there exists a constant $C_1 = C_{1,s} > 0$ such that

$$\forall \lambda \in \mathbb{R}\backslash\{0\},\ h \in \mathcal{Q}_s(\mathbb{R}^n) \qquad C_1^{-1}\|h\|_{\mathcal{Q}_{s,\lambda}} \le \|h\|^{(b_\lambda)}_{\mathcal{Q}_{s,\lambda}} \le C_1\|h\|_{\mathcal{Q}_{s,\lambda}},$$

and for each $s \in \mathbb{N}_0$ there exists a constant $C_2 = C_{2,s} > 0$ such that

$$\forall \lambda \in \mathbb{R}\backslash\{0\},\ h \in \mathcal{Q}_s(\mathbb{R}^n) \qquad C_2^{-1}\|h\|_{\mathcal{Q}_{s,\lambda}} \le \|h\|^{(int)}_{\mathcal{Q}_{s,\lambda}} \le C_2\|h\|_{\mathcal{Q}_{s,\lambda}}.$$

Naturally, in Part (2), the constants in the equivalences between each of the norms $\|\cdot\|_{\mathcal{Q}_{s,\lambda}}$, $\|\cdot\|^{(int)}_{\mathcal{Q}_{s,\lambda}}$, $\|\cdot\|^{(b_\lambda)}_{\mathcal{Q}_{s,\lambda}}$, and the norm $\|\cdot\|_{\mathcal{Q}_s}$, depend on λ.

Proof of Proposition 6.4.26. Part (1) follows easily from (6.54), Definition 6.4.10, Theorem 6.4.16 especially Part (5).

Using the Shubin calculus $\cup_m \Psi\Sigma_1^m$, it is not difficult to see that the norms $\|\cdot\|^{(b)}_{\mathcal{Q}_s}$ and $\|\cdot\|^{(b_\lambda)}_{\mathcal{Q}_{s,\lambda}}$ are equivalent.

The fact that the norms $\|\cdot\|_{\mathcal{Q}_{s,\lambda}}$, $\|\cdot\|^{(b_\lambda)}_{\mathcal{Q}_{s,\lambda}}$ and, if $s \in \mathbb{N}_0$, $\|\cdot\|^{(int)}_{\mathcal{Q}_{s,\lambda}}$, are equivalent with λ-uniform constants comes from following the same proof as Theorem 6.4.16 but using the seminorms of $\cup_m \Sigma^m_{1,\lambda}$. This is left to the reader and concludes the proof of Proposition 6.4.26. $\square$

Theorem 6.4.27. *We assume that $\rho \in (0,1]$. Let $A_\lambda : \mathcal{S}(\mathbb{R}^n) \to \mathcal{S}'(\mathbb{R}^n)$, $\lambda \in \mathbb{R}\backslash\{0\}$, be a family of linear continuous operators.*

We assume that for every $\alpha_1, \alpha_2 \in \mathbb{N}_0^n$ all the operators

$$|\lambda|^{-\frac{|\alpha_1|+|\alpha_2}{2}} (\mathrm{ad} u)^{\alpha_1} (\mathrm{ad}\partial_u)^{\alpha_2} A_\lambda, \quad \lambda \in \mathbb{R}\backslash\{0\},$$

are λ-uniformly in $\mathscr{L}(L^2(\mathbb{R}^n), \mathcal{Q}_{-m+\rho(|\alpha_1|+|\alpha_2|)})$. This means that

$$\sup_{\lambda \in \mathbb{R}\backslash\{0\}} |\lambda|^{-\frac{|\alpha_1|+|\alpha_2}{2}} \|(\mathrm{ad} u)^{\alpha_1} (\mathrm{ad}\partial_u)^{\alpha_2} A_\lambda\|_{\mathscr{L}(L^2(\mathbb{R}^n), \mathcal{Q}_{-m+\rho(|\alpha_1|+|\alpha_2|)})} < \infty. \quad (6.56)$$

Then $A_\lambda \in \Psi\Sigma^m_{\rho,\lambda}(\mathbb{R}^n)$. Moreover, for any $\ell \in \mathbb{N}$, there exist a constant C and an integer ℓ', both independent of $\{A_{\lambda'}\}$ and λ, such that

$$\|A_\lambda\|_{\Psi\Sigma^m_{\rho,\lambda},\ell} \le C \sum_{|\alpha_1|+|\alpha_2| \le \ell'} |\lambda|^{-\frac{|\alpha_1|+|\alpha_2}{2}} \|(\mathrm{ad} u)^{\alpha_1} (\mathrm{ad}\partial_u)^{\alpha_2} A_\lambda\|_{\mathscr{L}(L^2(\mathbb{R}^n), \mathcal{Q}_{-m+\rho(|\alpha_1|+|\alpha_2|)})}.$$

Proof. The proof follows exactly the same steps as the proof of Theorem 6.4.18 using the calculi $\cup_m \Sigma^m_{\rho,\lambda}(\mathbb{R}^n)$ to give the uniformity in λ. This is left to the reader. □

The converse is true from the λ-Shubin calculus: if $A_\lambda : \mathcal{S}(\mathbb{R}^n) \to \mathcal{S}'(\mathbb{R}^n)$, $\lambda \in \mathbb{R}\backslash\{0\}$, is uniformly in $\Psi\Sigma^m_{\rho,\lambda}(\mathbb{R}^n)$ in the sense that

$$\forall N \in \mathbb{N}_0 \qquad \sup_{\lambda \in \mathbb{R}\backslash\{0\}} \|A_\lambda\|_{\Psi\Sigma^m_{\rho,\lambda},N} < \infty, \quad (6.57)$$

then (6.56) holds for every $\alpha_1, \alpha_2 \in \mathbb{N}_0^n$.

Proceeding as for Corollary 6.4.20, we obtain

Corollary 6.4.28. *We assume that $\rho \in (0,1]$. Let $A_\lambda : \mathcal{S}(\mathbb{R}^n) \to \mathcal{S}'(\mathbb{R}^n)$, $\lambda \in \mathbb{R}\backslash\{0\}$, be a family of linear continuous operators.*

The family of operators $\{A_\lambda, \lambda \in \mathbb{R}\backslash\{0\}\}$ is uniformly in $\Psi\Sigma^m_{\rho,\lambda}(\mathbb{R}^n)$ in the sense of (6.57) if and only if there exists $\gamma_o \in \mathbb{R}$ such that for each $\alpha_1, \alpha_2 \in \mathbb{N}_0^n$,

$$\sup_{\lambda \in \mathbb{R}\backslash\{0\}} |\lambda|^{-\frac{|\alpha_1|+|\alpha_2}{2}} \|(\mathrm{ad} u)^{\alpha_1} (\mathrm{ad}\partial_u)^{\alpha_2} A_\lambda\|_{\mathscr{L}(\mathcal{Q}_{\gamma_o}(\mathbb{R}^n), \mathcal{Q}_{-m+\rho(|\alpha_1|+|\alpha_2|)+\gamma_o})} < \infty.$$

In this case this property is also true for every $\gamma \in \mathbb{R}$. Moreover, for any $\gamma \in \mathbb{R}$ and $\ell \in \mathbb{N}$, there exist a constant C and an integer ℓ', both independent of $\{A_{\lambda'}\}$ and λ, such that

$$\|A_\lambda\|_{\Psi\Sigma^m_{\rho,\lambda},\ell} \le C \sum_{|\alpha|+|\alpha_2| \le \ell'} |\lambda|^{-\frac{|\alpha_1|+|\alpha_2}{2}} \|(\mathrm{ad} u)^{\alpha_1} (\mathrm{ad}\partial_u)^{\alpha_2} A_\lambda\|_{\mathscr{L}(\mathcal{Q}_\gamma(\mathbb{R}^n), \mathcal{Q}_{-m+\rho(|\alpha_1|+|\alpha_2|)+\gamma})}.$$

6.5 Quantization and symbol classes $S^m_{\rho,\delta}$ on the Heisenberg group

We recall that in Section 5.2.2 we have introduced symbol classes $S^m_{\rho,\delta}(G)$ for general graded Lie groups G. In particular, this yields symbol classes $S^m_{\rho,\delta}(\mathbb{H}_n)$ for the particular case of $G = \mathbb{H}_n$. In this section, working with Schrödinger representations π_λ, we obtain a characterisation of these symbol classes $S^m_{\rho,\delta}(\mathbb{H}_n)$ in terms of scalar-valued symbols which will depend on the parameter $\lambda \in \mathbb{R}\backslash\{0\}$; these symbols will be called λ-symbols. The dependence on λ will be of crucial importance here.

We start by adapting the notation of the general construction described in Chapter 5 to the case of the Heisenberg group $\mathbb{H}_n$. It will be convenient to change slightly the notation with respect to the general case. Firstly we want to keep the letter x for denoting part of the coordinates of the Heisenberg group and we choose to denote the general element of the Heisenberg group by, e.g.,

$$g = (x, y, t) \in \mathbb{H}_n.$$

Secondly we may define a symbol as parametrised by

$$\sigma(g, \lambda) := \sigma(g, \pi_\lambda), \quad (g, \lambda) \in \mathbb{H}_n \times \mathbb{R}\backslash\{0\}.$$

Thirdly we modify the indices $\alpha \in \mathbb{N}_0^{2n+1}$ in order to write them as

$$\alpha = (\alpha_1, \alpha_2, \alpha_3),$$

with

$$\alpha_1 = (\alpha_{1,1}, \ldots, \alpha_{1,n}) \in \mathbb{N}_0^n, \quad \alpha_2 = (\alpha_{2,1}, \ldots, \alpha_{2,n}) \in \mathbb{N}_0^n, \quad \alpha_3 \in \mathbb{N}_0.$$

The homogeneous degree of α is then

$$[\alpha] = |\alpha_1| + |\alpha_2| + 2\alpha_3.$$

6.5.1 Quantization on the Heisenberg group

Here we summarise the quantization formula of Section 5.1.3 and its consequences in the particular setting of the Heisenberg group $\mathbb{H}_n$.

As introduced in Definition 5.1.33, a symbol is given by a field of operators

$$\sigma = \{\sigma(g, \lambda) : \mathcal{S}(\mathbb{R}^n) \to L^2(\mathbb{R}^n), (g, \lambda) \in \mathbb{H}_n \times (\mathbb{R}\backslash\{0\})\},$$

satisfying (quite weak) properties so that the quantization makes sense. More rigorously, we require that, for each $\beta \in \mathbb{N}_0^{2n+1}$, the map $g \longmapsto \partial_g^\beta \sigma(g, \lambda)$ is continuous from $\mathbb{H}_n$ to some $L^\infty_{a,b}(\widehat{\mathbb{H}}_n)$.

Recall now, that on the Heisenberg group $\mathbb{H}_n$, the Plancherel measure is given by $c_n|\lambda|^n d\lambda$ (see Proposition 6.2.7). By Theorem 5.1.39, the quantization of a symbol σ as above is the operator

$$A = \mathrm{Op}(\sigma)$$

given by

$$A\phi(g) = c_n \int_{\mathbb{R}\setminus\{0\}} \mathrm{Tr}\left(\pi_\lambda(g)\ \sigma(g,\lambda)\ \widehat{\phi}(\pi_\lambda)\right) |\lambda|^n d\lambda, \tag{6.58}$$

for any $\phi \in \mathcal{S}(\mathbb{H}_n)$ and $g = (x, y, t) \in \mathbb{H}_n$.

Note that, by (1.5), we have

$$\widehat{\varphi}(\pi_\lambda)\, \pi_\lambda(g) = \mathcal{F}_{\mathbb{H}_n}(\varphi(g\,\cdot))(\pi_\lambda),$$

thus the properties of the trace imply that

$$\mathrm{Tr}\left(\pi_\lambda(g)\sigma(g,\lambda)\widehat{\phi}(\pi_\lambda)\right) = \mathrm{Tr}\left(\sigma(g,\lambda)\ \mathcal{F}_{\mathbb{H}_n}(\varphi(g\,\cdot))(\pi_\lambda)\right). \tag{6.59}$$

Furthermore, by (6.20), we have

$$\mathcal{F}_{\mathbb{H}_n}(\varphi(g\,\cdot))(\pi_\lambda) = (2\pi)^{\frac{2n+1}{2}} \mathrm{Op}^W \left[\mathcal{F}_{\mathbb{R}^{2n+1}}(\varphi(g\,\cdot))(\sqrt{|\lambda|}\,\cdot, \sqrt{\lambda}\,\cdot, \lambda)\right]. \tag{6.60}$$

This formula shows that the Weyl quantization is playing an important role in the quantization (6.58) due to its close relation to the group Fourier transform on the Heiseneberg group.

Now, for each $(g, \lambda) \in \mathbb{H}_n \times (\mathbb{R}\setminus\{0\})$, each operator $\sigma(g,\lambda) : \mathcal{S}(\mathbb{R}^n) \to L^2(\mathbb{R}^n)$ in the symbol σ can also be written as the Weyl quantization of some symbol on the Euclidean space $\mathbb{R}^n$, depending on (g, λ). In other words, we can think of the symbol σ as

$$\sigma(g,\lambda) = \mathrm{Op}^W(a_{g,\lambda}), \tag{6.61}$$

where $a = \{a(g,\lambda,\xi,u) = a_{g,\lambda}(\xi,u)\}$ is a function on $\mathbb{H}_n \times \mathbb{R}\setminus\{0\} \times \mathbb{R}^n \times \mathbb{R}^n$. This scalar-valued symbol a will be called the λ-*symbol* of the operator A in (6.58).

In other words, the symbol of the operator A acting on the Heisenberg group is σ, related to A by the quantization formula (6.58). For each (g, λ), the symbol $\sigma_{g,\lambda}$ is itself an operator mapping the Schwartz space $\mathcal{S}(\mathbb{R}^n)$ to $L^2(\mathbb{R}^n)$. So, the λ-symbol a of the operator A is given by the collection of the Weyl symbols $a_{g,\lambda}$ of $\sigma(g,\lambda)$.

Note that if $A \in \Psi^m_{\rho,\delta}$, then its symbol acts on smooth vectors so $\sigma_{g,\lambda}$ is itself an operator mapping the Schwartz space $\mathcal{S}(\mathbb{R}^n)$ to itself, for each (g, λ).

Consequently, using (6.59), we can rewrite our quantization given in (6.58), now using only Euclidean objects, as

$$\begin{aligned} A\varphi(g) & \qquad\qquad (6.62) \\ &= c_n' \int_{\mathbb{R}\backslash\{0\}} \mathrm{Tr}\left(\pi_\lambda(g)\,\mathrm{Op}^W(a_{g,\lambda})\,\mathrm{Op}^W\left[\mathcal{F}_{\mathbb{R}^{2n+1}}(\varphi)(\sqrt{|\lambda|}\,\cdot,\sqrt{\lambda}\,\cdot,\lambda)\right]\right)|\lambda|^n d\lambda \\ &= c_n' \int_{\mathbb{R}\backslash\{0\}} \mathrm{Tr}\left(\mathrm{Op}^W(a_{g,\lambda})\,\mathrm{Op}^W\left[\mathcal{F}_{\mathbb{R}^{2n+1}}(\varphi(g\,\cdot))(\sqrt{|\lambda|}\,\cdot,\sqrt{\lambda}\,\cdot,\lambda)\right]\right)|\lambda|^n d\lambda, \end{aligned}$$

with $c_n' = c_n(2\pi)^{n+\frac{1}{2}} = (2\pi)^{-2n-\frac{1}{2}}$.

In Definition 5.2.11 we have introduced the symbol classes $S^m_{\rho,\delta}(G)$ for general graded Lie groups G. Now, in the particular case $G = \mathbb{H}_n$ of the Heisenberg group, using the relation (6.61) between symbols σ and a, we can ask the following question:

what does the condition $\sigma \in S^m_{\rho,\delta}(\mathbb{H}_n)$ mean in terms of the λ-symbol $a_{g,\lambda}$?

This question will be answered in the following sections.

6.5.2 An equivalent family of seminorms on $S^m_{\rho,\delta} = S^m_{\rho,\delta}(\mathbb{H}_n)$

We now follow Definition 5.2.11 to define the symbol class

$$S^m_{\rho,\delta} = S^m_{\rho,\delta}(\mathbb{H}_n).$$

As positive Rockland operator, we will use $\mathcal{R} = -\mathcal{L}$ where $\mathcal{L}$ is the (canonical) sub-Laplacian given in (6.5). We realise almost all the elements of $\widehat{\mathbb{H}}_n$ via their representatives given by the Schrödinger representations π_λ, $\lambda \in \mathbb{R}\backslash\{0\}$, which all act on

$$\mathcal{H}_{\pi_\lambda} = L^2(\mathbb{R}^n),$$

see Section 6.2. Therefore, our symbol class on $\mathbb{H}_n$ is defined by the following family of seminorms

$$\|\sigma\|_{S^m_{\rho,\delta},a,b,c} := \sup_{\lambda\in\mathbb{R}\backslash\{0\},\, g\in\mathbb{H}_n} \|\sigma(g,\lambda)\|_{S^m_{\rho,\delta},a,b,c}, \quad a,b,c \in \mathbb{N}_0,$$

where

$$\begin{aligned} &\|\sigma(g,\lambda)\|_{S^m_{\rho,\delta},a,b,c} \\ &\quad := \sup_{\substack{[\alpha]\le a \\ [\beta]\le b,\, |\gamma|\le c}} \|\pi_\lambda(\mathrm{I}-\mathcal{L})^{\frac{\rho[\alpha]-m-\delta[\beta]+\gamma}{2}} X_g^\beta \Delta^\alpha \sigma(g,\lambda)\pi_\lambda(\mathrm{I}-\mathcal{L})^{-\frac{\gamma}{2}}\|_{\mathscr{L}(L^2(\mathbb{R}^n))}. \end{aligned}$$

Here the difference operators Δ^α correspond to the family of operators $\Delta_{\tilde{q}_\alpha}$ where the q_α's are the polynomials appearing in the Taylor expansion. See Example 5.2.4 for some explicit formulae.

By Remark 5.2.13 (4), we can also use the canonical basis

$$x^{\alpha_1} y^{\alpha_2} t^{\alpha_3}, \quad \alpha = (\alpha_1, \alpha_2, \alpha_3) \in \mathbb{N}^{2n+1} = \mathbb{N}_0^n \times \mathbb{N}_0^n \times \mathbb{N}_0,$$

where

$$x^{\alpha_1} = x_1^{\alpha_{11}} \dots x_n^{\alpha_{1n}}, \quad y^{\alpha_2} = y_1^{\alpha_{21}} \dots y_n^{\alpha_{2n}}.$$

We define

$$\Delta'^{\alpha} := \Delta_{x^{\alpha_1} y^{\alpha_2} t^{\alpha_3}}, \quad \alpha \in \mathbb{N}^{2n+1}.$$

In this case, for any $\alpha, \beta \in \mathbb{N}_0^{2n+1}$, we have

$$\Delta'^{\alpha+\beta} = \Delta'^{\alpha} \Delta'^{\beta}.$$

An equivalent family of seminorms on $S^m_{\rho,\delta}$ using the difference operators Δ'^{α} is given by

$$\|\sigma\|'_{S^m_{\rho,\delta},a,b,c} := \sup_{\lambda \in \mathbb{R}\setminus\{0\},\, g \in \mathbb{H}_n} \|\sigma(g,\lambda)\|'_{S^m_{\rho,\delta},a,b,c}, \quad a,b,c \in \mathbb{N}_0,$$

where

$$\|\sigma(g,\lambda)\|'_{S^m_{\rho,\delta},a,b,c}$$
$$:= \sup_{\substack{[\alpha]\leq a \\ [\beta]\leq b,\ |\gamma|\leq c}} \|\pi_\lambda(\mathrm{I}-\mathcal{L})^{\frac{\rho[\alpha]-m-\delta[\beta]+\gamma}{2}} X_g^\beta \Delta'^{\alpha} \sigma(g,\lambda) \pi_\lambda(\mathrm{I}-\mathcal{L})^{-\frac{\gamma}{2}}\|_{\mathscr{L}(L^2(\mathbb{R}^n))}.$$

Although the difference operators which intervene in the asymptotic expansions of the composition and the adjoint properties are the difference operators Δ^α, the operators Δ'_α are more handy for the computations to follow.

6.5.3 Characterisation of $S^m_{\rho,\delta}(\mathbb{H}_n)$

In this section we describe the symbol classes $S^m_{\rho,\delta}(\mathbb{H}_n)$ from Section 5.2.2 (more specifically, from Definition 5.2.11) in terms of scalar-valued λ-symbols. More precisely, we show that the symbols $\sigma = \{\sigma(g,\lambda)\}$ in $S^m_{\rho,\delta}$ are all of the form

$$\sigma(g,\lambda) = \mathrm{Op}^W(a_{g,\lambda}(\xi,u)), \tag{6.63}$$

with the λ-symbol $a_{g,\lambda}$ satisfying some properties described below in terms of the family of λ-Shubin classes described in Section 6.4.4 and of the operator $\tilde{\partial}_{\lambda,\xi,u}$ defined in (6.43).

Theorem 6.5.1. *Let $m, \rho, \delta \in \mathbb{R}$ with $1 \geq \rho \geq \delta \geq 0$, $\rho \neq 0$, $\delta \neq 1$. If $\sigma = \{\sigma(g, \lambda)\}$ is in $S^m_{\rho,\delta}$ then there exists a unique smooth function $a = \{a(g, \lambda, \xi, u) = a_{g,\lambda}(\xi, u)\}$ on $\mathbb{H}_n \times \mathbb{R}\backslash\{0\} \times \mathbb{R}^n \times \mathbb{R}^n$ such that*

$$\sigma(g, \lambda) = \mathrm{Op}^W (a_{g,\lambda}), \tag{6.64}$$

with $\tilde{\partial}^{\alpha_3}_{\lambda,\xi,u} X^\beta_g a_{g,\lambda} \in \Sigma^{m-2\rho\alpha_3+\delta[\beta]}_{\rho,\lambda}(\mathbb{R}^n)$ for each $(g, \lambda) \in \mathbb{H}_n \times \mathbb{R}\backslash\{0\}$ satisfying

$$\sup_{(g,\lambda)\in\mathbb{H}_n\times\mathbb{R}\backslash\{0\}} \|\tilde{\partial}^{\alpha_3}_{\lambda,\xi,u} X^\beta_g a_{g,\lambda}\|_{\Sigma^{m-2\rho\alpha_3+\delta[\beta]}_{\rho,\lambda}(\mathbb{R}^n),N} < \infty, \tag{6.65}$$

for every $N \in \mathbb{N}_0$. More precisely, for every $N \in \mathbb{N}_0$ there exist $C > 0$ and a, b, c such that

$$\sup_{(g,\lambda)\in\mathbb{H}_n\times\mathbb{R}\backslash\{0\}} \|\tilde{\partial}^{\alpha_3}_{\lambda,\xi,u} X^\beta_g a_{g,\lambda}\|_{\Sigma^{m-2\rho\alpha_3+\delta[\beta]}_{\rho,\lambda}(\mathbb{R}^n),N} \leq C\|\sigma\|_{S^m_{\rho,\lambda}(\mathbb{H}_n),a,b,c}.$$

Conversely, if $a = \{a(g, \lambda, \xi, u) = a_{g,\lambda}(\xi, u)\}$ is a smooth function on $\mathbb{H}_n \times \mathbb{R}\backslash\{0\} \times \mathbb{R}^n \times \mathbb{R}^n$ satisfying (6.65) for every $N \in \mathbb{N}_0$, then there exists a unique symbol $\sigma \in S^m_{\rho,\delta}$ such that (6.64) holds. Furthermore, for every a, b, c there exists $C > 0$ and $N \in \mathbb{N}_0$ such that

$$\|\sigma\|_{S^m_{\rho,\lambda}(\mathbb{H}_n),a,b,c} \leq C \sup_{(g,\lambda)\in\mathbb{H}_n\times\mathbb{R}\backslash\{0\}} \|\tilde{\partial}^{\alpha_3}_{\lambda,\xi,u} X^\beta_g a_{g,\lambda}\|_{\Sigma^{m-2\rho\alpha_3+\delta[\beta]}_{\rho,\lambda}(\mathbb{R}^n),N}.$$

In other words, Theorem 6.5.1 shows that

$$\sigma \in S^m_{\rho,\delta}(\mathbb{H}_n)$$

is equivalent to

$$\sigma(g, \lambda) = \mathrm{Op}^W(a_{g,\lambda}),$$

for each (g, λ) with $a_{g,\lambda} \in C^\infty(\mathbb{R}^{2n})$ satisfying

$$\forall\alpha \in \mathbb{N}^{2n+1}_0 \quad \exists C > 0 \quad \forall(g, \lambda) \in \mathbb{H}_n{\times}(\mathbb{R}\backslash\{0\}) \quad \forall(\xi, u) \in \mathbb{R}^{2n}$$
$$|\partial^{\alpha_1}_\xi \partial^{\alpha_2}_u \tilde{\partial}^{\alpha_3}_{\lambda,\xi,u} X^\beta_g a_{g,\lambda}(\xi, u)| \leq C|\lambda|^{\rho\frac{|\alpha_1|+|\alpha_2|}{2}} \left(1 + |\lambda|(1 + |\xi|^2 + |u|^2)\right)^{\frac{m-\rho[\alpha]+\delta[\beta]}{2}}.$$

Choosing a rescaled Weyl symbol as in Lemma 6.3.10, we see that

$$\sigma \in S^m_{\rho,\delta}(\mathbb{H}_n)$$

is equivalent to

$$\sigma(g, \lambda) = \mathrm{Op}^W \left(\tilde{a}_{g,\lambda}(\sqrt{|\lambda|}\xi, \sqrt{\lambda}u)\right),$$

for each (g, λ) with $\tilde{a}_{g,\lambda} \in C^\infty(\mathbb{R}^{2n})$ satisfying

$$\forall\alpha \in \mathbb{N}^{2n+1}_0 \quad \exists C > 0 \quad \forall(g, \lambda) \in \mathbb{H}_n{\times}(\mathbb{R}\backslash\{0\}) \quad \forall(\xi, u) \in \mathbb{R}^{2n}$$
$$|\partial^{\alpha_1}_\xi \partial^{\alpha_2}_u \partial^{\alpha_3}_\lambda X^\beta_g \tilde{a}_{g,\lambda}(\xi, u)| \leq C \left(1 + |\lambda| + |\xi|^2 + |u|^2\right)^{\frac{m-\rho[\alpha]+\delta[\beta]}{2}}.$$

Note that, by (6.20),

$$\tilde{a}_{g,\lambda}(\xi,u)=(2\pi)^{\frac{2n+1}{2}}\mathcal{F}_{\mathbb{R}^{2n+1}}(\kappa_g)(\xi,u,\lambda)$$

where $\{\kappa_g(x,y,t)\}$ is the kernel of the symbol $\{\sigma(g,\lambda)\}$, i.e.

$$\sigma(g,\lambda)=\pi_\lambda(\kappa_g),$$

(see Definition 5.1.36).

Proof of Theorem 6.5.1. Let $\sigma\in S^m_{\rho,\delta}$. This means that for each $\alpha,\beta\in\mathbb{N}_0^{2n+1}$ and $\gamma\in\mathbb{R}$ we have

$$\pi_\lambda(\mathrm{I}-\mathcal{L})^{\frac{\rho[\alpha]-m-\delta[\beta]+\gamma}{2}}X_g^\beta\Delta'^\alpha\sigma(g,\lambda)\pi_\lambda(\mathrm{I}-\mathcal{L})^{-\frac{\gamma}{2}}\in\mathscr{L}(L^2(\mathbb{R}^n)),$$

with operator norm uniformly bounded with respect to λ, or equivalently, (see the formulae in Section 6.3.3),

$$|\lambda|^{-\frac{|\alpha_1|+|\alpha_2|}{2}}\|(\mathrm{ad}u)^{\alpha_1}(\mathrm{ad}\partial_u)^{\alpha_2}X_g^\beta\Delta_3'^{\alpha_3}\sigma(g,\lambda)h\|_{\mathcal{Q}_{\rho[\alpha]-m-\delta[\beta]+\gamma,\lambda}}\leq C\|h\|_{\mathcal{Q}_{\gamma,\lambda}}$$

with $C=C_{\alpha,\beta,\gamma}$ independent of λ. Taking $\gamma=0$, we see that the λ-family of $X_g^\beta\Delta_3'^{\alpha_3}\sigma(g,\lambda)$ satisfies the hypotheses of Theorem 6.4.27. For $\beta=\alpha_3=0$, this shows that $\sigma(g,\lambda)=\mathrm{Op}^W(a_{g,\lambda})$ with $a_{g,\lambda}\in\Sigma^m_{\rho,\lambda}$ uniformly in λ. For any β and α_3, this shows that the λ-family of

$$X_g^\beta\Delta_3'^{\alpha_3}\sigma(g,\lambda)=i^{\alpha_3}\mathrm{Op}^W(X_g^\beta\tilde{\partial}^{\alpha_3}_{\lambda,\xi,u}a_{g,\lambda})$$

(see the formulae in Section 6.3.3, or equivalently Corollary 6.3.9) also satisfies the hypotheses of Theorem 6.4.27. Therefore, $X_g^\beta\tilde{\partial}^{\alpha_3}_{\lambda,\xi,u}a_{g,\lambda}$ is in $\Sigma^{m-2\rho[\alpha_3]+\delta[\beta]}_{\rho,\lambda}$ uniformly in λ. This proves the first part of the statement.

The converse follows from the Shubin calculi depending on λ. □

The proof above shows that we can always assume $\gamma=0$ in the definition of a class of symbols. But we could have fixed any γ and use Corollary 6.4.28 instead of Theorem 6.4.27 in the proof above. This shows:

Corollary 6.5.2. *A symbol $\sigma=\{\sigma(g,\lambda)\}$ is in $S^m_{\rho,\delta}$ if and only if there exists* one *$\gamma\in\mathbb{R}$ such that for every $\alpha,\beta\in\mathbb{N}_0^{2n+1}$ the quantity*

$$\sup_{\lambda\in\mathbb{R}\backslash\{0\}}\|\pi_\lambda(\mathrm{I}-\mathcal{L})^{\frac{\rho[\alpha]-m-\delta[\beta]+\gamma}{2}}X_g^\beta\Delta'^\alpha\sigma(g,\lambda)\pi_\lambda(\mathrm{I}-\mathcal{L})^{-\frac{\gamma}{2}}\|_{\mathscr{L}(L^2(\mathbb{R}^n))}\qquad(6.66)$$

is finite.

In this case the quantity (6.66) is finite for every $\gamma\in\mathbb{R}$ and $\alpha,\beta\in\mathbb{N}_0^n$.

Furthermore, for any $\gamma_o\in\mathbb{R}$ fixed, an equivalent family of seminorms for $S^m_{\rho,\delta}$ is given by

$$\sigma\longmapsto\sup_{\lambda\in\mathbb{R}\backslash\{0\},[\alpha]\leq a,\,[\beta]\leq b}\|\pi_\lambda(\mathrm{I}-\mathcal{L})^{\frac{\rho[\alpha]-m-\delta[\beta]+\gamma_o}{2}}X_g^\beta\Delta'^\alpha\sigma(g,\lambda)\pi_\lambda(\mathrm{I}-\mathcal{L})^{-\frac{\gamma_o}{2}}\|_{\mathscr{L}(L^2(\mathbb{R}^n))}$$

with $a,b\in\mathbb{N}_0$.

6.6 Parametrices

In this section, we present conditions for the ellipticity and hypoellipticity in the setting of the Heisenberg group as a special case of those presented in Sections 5.8.1 and 5.8.3. In particular, we can also derive conditions in terms of the λ-symbols discussed in Section 6.5.3.

6.6.1 Condition for ellipticity

We start by providing conditions on the λ-symbol ensuring that the assumptions for the ellipticity in Definition 5.8.1 and in Theorem 5.8.7 are satisfied.

Theorem 6.6.1. *Let $m \in \mathbb{R}$ and $1 \geq \rho > \delta \geq 0$. Let $\sigma = \{\sigma(g,\lambda)\}$ be in $S^m_{\rho,\delta}(\mathbb{H}_n)$ with*

$$\sigma(g,\lambda) = \mathrm{Op}^W(a_{g,\lambda})$$

as in Theorem 6.5.1. Assume that there are $R \in \mathbb{R}$ and $C > 0$ such that for any $(\xi, u) \in \mathbb{R}^{2n}$ and $\lambda \neq 0$ satisfying $|\lambda|(|\xi|^2 + |u|^2) \geq R$ we have

$$|a_{g,\lambda}(\xi,u)| \geq C\left(1 + |\lambda|(1 + |\xi|^2 + |u|^2)\right)^{\frac{m}{2}}. \tag{6.67}$$

Then there exists Λ such that σ is $(-\mathcal{L}, \Lambda, m)$-elliptic in the sense of Definition 5.8.1. Thus it satisfies the hypotheses of Theorem 5.8.7 and we can construct a left parametrix $B \in \Psi^{-m}_{\rho,\delta}$ for the operator $A = \mathrm{Op}(\sigma)$, that is, there exists $B \in \Psi^{-m}_{\rho,\delta}$ such that

$$BA - \mathrm{I} \in \Psi^{-\infty}.$$

Proof. Let $\chi \in C^\infty(\mathbb{R})$ be such that $0 \leq \chi \leq 1$ with $\chi = 0$ on $(-\infty, R)$ and $\chi = 1$ on $[2R, +\infty)$. We set for any $(\xi, u) \in \mathbb{R}^{2n}$ and $\lambda \neq 0$

$$b_{\lambda,g}(\xi,u) := \frac{\chi(|\lambda|(|\xi|^2 + |u|^2))}{a_{g,\lambda}(\xi,u)}.$$

Using the properties of a, one check easily that this defines a symbol $b_{\lambda,g}$ with $b_{\lambda,g} \in \Sigma^{-m}_{\rho,\lambda}$, and more precisely for every $N \in \mathbb{N}_0$ there exist $C > 0$ and a, b, c all independent on λ or g such that

$$\sup_{(g,\lambda)\in\mathbb{H}_n\times\mathbb{R}\backslash\{0\}} \|\tilde{\partial}^{\alpha_3}_{\lambda,\xi,u} X^\beta_g b_{g,\lambda}\|_{\Sigma^{-m-2\rho\alpha_3+\delta[\beta]}_{\rho,\lambda}(\mathbb{R}^n),N} \leq C\|\sigma\|_{S^m_{\rho,\lambda},a,b,c} := C'.$$

By the properties of uniform families of Weyl-Hörmander metrics (see Proposition 6.4.22), we have

$$\mathrm{Op}^W(b_{\lambda,g})\mathrm{Op}^W(a_{g,\lambda}) = \mathrm{Op}^W(\chi(|\lambda|(|\xi|^2 + |u|^2))) + E_{\lambda,g} = \mathrm{I} + \tilde{E}_{\lambda,g} \tag{6.68}$$

with

$$\|\tilde{\partial}^{\alpha_3}_{\lambda,\xi,u} X^\beta_g E_{\lambda,g}\|_{\Psi\Sigma^{-\rho-2\rho\alpha_3+\delta[\beta]}_{\rho,\lambda}(\mathbb{R}^n),N} \leq C_1\|\sigma\|_{S^m,a_1,b_1,c_1} := C_1', \tag{6.69}$$

and similarly for the 'error' term $\tilde{E}_{\lambda,g}$. Also we have

$$\|\pi_\lambda(\mathrm{I}-\mathcal{L})^{\frac{m}{2}}\mathrm{Op}^W(b_{\lambda,g})\|_{\mathscr{L}(L^2(\mathbb{R}^n))} \leq C_2\|\sigma\|_{S^m,a_2,b_2,c_2} := C_2'. \tag{6.70}$$

In Estimates (6.69) and (6.70), the constants C_1 and C_2, and the parameters $a_1, b_1, c_1, a_2, b_2, c_2$ do not depend on λ, g or σ. By (6.70), we have

$$C_2'\|\mathrm{Op}^W(a_{g,\lambda})u\|_{L^2(\mathbb{R}^n)} \geq \|\pi_\lambda(\mathrm{I}-\mathcal{L})^{\frac{m}{2}}\mathrm{Op}^W(b_{\lambda,g})\mathrm{Op}^W(a_{g,\lambda})u\|_{L^2(\mathbb{R}^n)}.$$

We now use (6.68) on the right hand-side and the reverse triangle inequality to obtain

$$\begin{aligned}
&\|\pi_\lambda(\mathrm{I}-\mathcal{L})^{\frac{m}{2}}\mathrm{Op}^W(b_{\lambda,g})\mathrm{Op}^W(a_{g,\lambda})u\|_{L^2(\mathbb{R}^n)} \\
&\quad= \|\pi_\lambda(\mathrm{I}-\mathcal{L})^{\frac{m}{2}}\left(\mathrm{I}+\tilde{E}_{\lambda,g}\right)u\|_{L^2(\mathbb{R}^n)} \\
&\quad\geq \|\pi_\lambda(\mathrm{I}-\mathcal{L})^{\frac{m}{2}}u\|_{L^2(\mathbb{R}^n)} - \|\pi_\lambda(\mathrm{I}-\mathcal{L})^{\frac{m}{2}}\tilde{E}_{\lambda,g}u\|_{L^2(\mathbb{R}^n)}.
\end{aligned}$$

We can write the last term as

$$\|\pi_\lambda(\mathrm{I}-\mathcal{L})^{\frac{m}{2}}\tilde{E}_{\lambda,g}u\|_{L^2(\mathbb{R}^n)} = \|U_{\lambda,g}\pi_\lambda(\mathrm{I}-\mathcal{L})^{\frac{m-\rho}{2}}u\|_{L^2(\mathbb{R}^n)}.$$

with $U_{\lambda,g} := \pi_\lambda(\mathrm{I}-\mathcal{L})^{\frac{m}{2}}\tilde{E}_{\lambda,g}\pi_\lambda(\mathrm{I}-\mathcal{L})^{\frac{-m+\rho}{2}}$ of order 0 and, therefore, bounded on $L^2(\mathbb{R}^n)$ satisfying

$$\|U_{\lambda,g}\|_{\mathscr{L}(L^2(\mathbb{R}^n))} \leq C_3\|\sigma\|_{S^m_{\rho,\delta},a_3,b_3,c_3} := C_3'.$$

Let us consider $\Lambda \in \mathbb{R}$ and $u \in \mathcal{S}(\mathbb{R}^n)$ with $u \in E_{\pi_\lambda}(\Lambda,\infty)L^2(\mathbb{R}^n)$, then

$$\|\pi_\lambda(\mathrm{I}-\mathcal{L})^{\frac{m-\rho}{2}}u\|_{L^2(\mathbb{R}^n)} \leq (1+\max(\Lambda,0))^{-\frac{\rho}{2}}\|\pi_\lambda(\mathrm{I}-\mathcal{L})^{\frac{m}{2}}u\|_{L^2(\mathbb{R}^n)},$$

thus

$$\begin{aligned}
&\|U_{\lambda,g}\pi_\lambda(\mathrm{I}-\mathcal{L})^{\frac{m-\rho}{2}}u\|_{L^2(\mathbb{R}^n)} \\
&\quad\leq C_3'\|\pi_\lambda(\mathrm{I}-\mathcal{L})^{\frac{m-\rho}{2}}u\|_{L^2(\mathbb{R}^n)} \\
&\quad\leq C_3'(1+\max(\Lambda,0))^{-\frac{\rho}{2}}\|\pi_\lambda(\mathrm{I}-\mathcal{L})^{\frac{m}{2}}u\|_{L^2(\mathbb{R}^n)}.
\end{aligned}$$

We choose $\Lambda \in \mathbb{R}$ such that

$$C_3'(1+\max(\Lambda,0))^{-\frac{\rho}{2}} \leq \frac{1}{2},$$

for example for $\Lambda > 0$, the smallest Λ satisfying the equality. We have obtained

$$\|\pi_\lambda(\mathrm{I}-\mathcal{L})^{\frac{m}{2}}\tilde{E}_{\lambda,g}u\|_2 = \|U_{\lambda,g}\pi_\lambda(\mathrm{I}-\mathcal{L})^{\frac{m-\rho}{2}}u\|_2 \leq \frac{1}{2}\|\pi_\lambda(\mathrm{I}-\mathcal{L})^{\frac{m}{2}}u\|_2.$$

Collecting the estimates, we obtain

$$\begin{aligned}
C_2'\|\mathrm{Op}^W(a_{g,\lambda})u\|_2 &\geq \|\pi_\lambda(\mathrm{I}-\mathcal{L})^{\frac{m}{2}}\mathrm{Op}^W(b_{\lambda,g})\mathrm{Op}^W(a_{g,\lambda})u\|_2 \\
&\geq \|\pi_\lambda(\mathrm{I}-\mathcal{L})^{\frac{m}{2}}u\|_2 - \|\pi_\lambda(\mathrm{I}-\mathcal{L})^{\frac{m}{2}}\tilde{E}_{\lambda,g}u\|_2 \geq \frac{1}{2}\|\pi_\lambda(\mathrm{I}-\mathcal{L})^{\frac{m}{2}}u\|_2.
\end{aligned}$$

This shows that σ satisfies (5.79) for $-\mathcal{L}$, Λ and m. □

From the proof, it follows that the choice of Λ depends on ρ, δ, and a bound for a (computable) seminorm of σ in $S^m_{\rho,\delta}$.

We have already proved that, for instance, $\mathrm{I}-\mathcal{L}$ is elliptic for $-\mathcal{L}$, see Proposition 5.8.2.

Here is another example.

Example 6.6.2. On $\mathbb{H}_1$, if $m \in 2\mathbb{N}$ is an even integer, then the operator $X^m + iY^m + T^{m/2} \in \Psi^m$ is elliptic with respect to $-\mathcal{L}$ and of elliptic order m.

Proof. The symbol of $X^m + iY^m + T^{m/2}$ is

$$\begin{aligned}\sigma(\lambda) &= \pi_\lambda(X)^m + i\pi_\lambda(Y)^m + \pi_\lambda(T)^{\frac{m}{2}} \\ &= \left(\mathrm{Op}^W(i\sqrt{|\lambda|}\xi)\right)^m + i\left(\mathrm{Op}^W(i\sqrt{\lambda}u)\right)^m + (i\lambda)^{\frac{m}{2}},\end{aligned}$$

by (6.28) and (6.11). Hence its λ-symbol is

$$\begin{aligned}a_\lambda(\xi,u) &= \left(i\sqrt{|\lambda|}\xi\right)^m + i\left(i\sqrt{\lambda}u\right)^m + (i\lambda)^{\frac{m}{2}} \\ &= (-1)^{\frac{m}{2}}|\lambda|^{\frac{m}{2}}\left(\xi^m + iu^m + (-(\mathrm{sgn}\lambda)i)^{\frac{m}{2}}\right).\end{aligned}$$

Clearly a_λ satisfies the condition of Theorem 6.6.1. □

6.6.2 Condition for hypoellipticity

We have also proved a general result regarding hypoellipticity in Theorem 5.8.9 (in the sense of the existence of a left parametrix). In the case of the Heisenberg group, we obtain the following sufficient condition on the scalar-valued symbol:

Theorem 6.6.3. *Let $m \in \mathbb{R}$ and $1 \geq \rho > \delta \geq 0$. Let $\sigma = \{\sigma(g,\lambda)\}$ be in $S^m_{\rho,\delta}(\mathbb{H}_n)$ with*

$$\sigma(g,\lambda) = \mathrm{Op}^W(a_{g,\lambda})$$

as in Theorem 6.5.1.

We assume that there is $m_o < m$ such that σ satisfies for a given R, for any $(\xi,u) \in \mathbb{R}^{2n}$ such that $|\lambda|(|\xi|^2 + |u|^2) \geq R$, the inequalities

$$|a_{g,\lambda}(\xi,u)| \geq C\left(1 + |\lambda|(1 + |\xi|^2 + |u|^2)\right)^{\frac{m_o}{2}} \tag{6.71}$$

and

$$\begin{aligned}&|\partial_\xi^{\alpha_1}\partial_u^{\alpha_2}\tilde{\partial}_{\lambda,\xi,u}^{\alpha_3}X_g^\beta a_{g,\lambda}(\xi,u)| \\ &\quad \leq C_{\alpha,\beta}|\lambda|^{\rho\frac{|\alpha_1|+|\alpha_2|}{2}}\left(1 + |\lambda|(1 + |\xi|^2 + |u|^2)\right)^{\frac{-\rho[\alpha]+\delta[\beta]}{2}}|a_{g,\lambda}(\xi,u)|. \quad (6.72)\end{aligned}$$

Then $\sigma(g,\lambda)$ satisfies the hypotheses of Theorem 5.8.9 for $-\mathcal{L}$ and m_o. Therefore, we can construct a left parametrix $B \in \Psi^{-m_o}_{\rho,\delta}$ for the operator $A = \mathrm{Op}(\sigma)$, that is, there exists $B \in \Psi^{-m_o}_{\rho,\delta}$ such that

$$BA - \mathrm{I} \in \Psi^{-\infty}.$$

In (6.71) and (6.72), the constants C and $C_{\alpha,\beta}$ are assumed to be independent of λ, ξ, u or g.

For each fixed $\lambda \in \mathbb{R}\backslash\{0\}$, the conditions (6.71) and (6.72) are very close to Shubin's in [Shu87, §25.1]. However Theorem 6.6.3 asks for these conditions to be satisfied uniformly in $\lambda \in \mathbb{R}\backslash\{0\}$.

The proof is in essence an adaptation of the proof of Theorem 6.6.1.

Proof. We choose χ and define $b_{\lambda,g}$ as in the proof of Theorem 6.6.1. This time, $b_{\lambda,g}$ is in $\Sigma_{\rho,\delta}^{-m_o}$, with

$$\sup_{(g,\lambda)\in\mathbb{H}_n\times\mathbb{R}\backslash\{0\}} \|\tilde{\partial}_{\lambda,\xi,u}^{\alpha_3} X_g^{\beta} b_{g,\lambda}\|_{\Sigma_{\rho,\lambda}^{-m_o-2\rho\alpha_3+\delta[\beta]}(\mathbb{R}^n),N} \leq C\|\sigma\|_{S^m_{\rho,\lambda},a,b,c},$$

and

$$\|\pi_\lambda(\mathrm{I}-\mathcal{L})^{\frac{m_o}{2}}\mathrm{Op}^W(b_{\lambda,g})\|_{\mathscr{L}(L^2(\mathbb{R}^n))} \leq C_2\|\sigma\|_{S^m_{\rho,\lambda},a_2,b_2,c_2} := C_2'. \tag{6.73}$$

In the proof of Theorem 6.6.1, we developed the product $\mathrm{Op}^W(b_{\lambda,g})\mathrm{Op}^W(a_{g,\lambda})$ at order 0, but here we now develop it up to order M such that the error term is of strictly negative order:

$$\mathrm{Op}^W(b_{\lambda,g})\mathrm{Op}^W(a_{g,\lambda}) = \sum_{m'=0}^{M} \mathrm{Op}^W(d_{m',\lambda,g}) + E_{\lambda,g}, \tag{6.74}$$

where (see (6.17))

$$d_{m',\lambda,g} := c_{m',n} \sum_{|\alpha_1|+|\alpha_2|=m'} \frac{(-1)^{|\alpha_2|}}{\alpha_1!\alpha_2!} \left(\left(\frac{1}{i}\partial_\xi\right)^{\alpha_1} \partial_x^{\alpha_2} b_{\lambda,g}\right)\left(\left(\frac{1}{i}\partial_\xi\right)^{\alpha_2} \partial_x^{\alpha_1} a_{g,\lambda}\right).$$

To fix the idea, we choose $M \in \mathbb{N}_0$ the smallest integer such that

$$m - m_o - 2(M+1)\rho \leq -\rho.$$

Using (6.17) and the properties of uniform families of Weyl-Hörmander metrics (see Proposition 6.4.22), the error term satisfies

$$\|\tilde{\partial}_{\lambda,\xi,u}^{\alpha_3} X_g^{\beta} E_{g,\lambda}\|_{\Psi\Sigma_{\rho,\lambda}^{-\rho-2\rho\alpha_3+\delta[\beta]}(\mathbb{R}^n),N} \leq C_1\|\sigma\|_{S^m,a_1,b_1,c_1} := C_1'. \tag{6.75}$$

For the term of order 0, we see that

$$d_{0,\lambda,g} = \chi(|\lambda|(|\xi|^2+|u|^2) = 1 + (\chi-1)(|\lambda|(|\xi|^2+|u|^2),$$

and clearly the symbol $(\chi-1)(|\lambda|(|\xi|^2+|u|^2)$ is smoothing. For the term of positive order $m'>0$, we can write

$$d_{m',\lambda,g} = c_{m',n}\tilde{d}_{m',\lambda,g} + r_{m',\lambda,g},$$

where

$$\tilde{d}_{m',\lambda,g} := \chi(|\lambda|(|\xi|^2+|u|^2) \sum_{|\alpha_1|+|\alpha_2|=m'} \frac{(-1)^{|\alpha_2|}}{\alpha_1!\alpha_2!} \left(\left(\frac{1}{i}\partial_\xi\right)^{\alpha_1} \partial_x^{\alpha_2} \left\{ \frac{1}{a_{g,\lambda}} \right\} \right) \left(\left(\frac{1}{i}\partial_\xi\right)^{\alpha_2} \partial_x^{\alpha_1} a_{g,\lambda} \right),$$

and the small reminder contains all the χ-derivatives, that is, is of the form

$$r_{m',\lambda,g} = \sum_{\substack{\alpha_1'',\alpha_2'' \\ 0<\alpha_1''+\alpha_2''\le 2M}} \left(\left(\partial_\xi^{\alpha_1''}\partial_x^{\alpha_2''}\right) \chi(|\lambda|(|\xi|^2+|u|^2)) \left(\dots\right) \right).$$

Clearly the derivatives of the χ's are smoothing. One can check that the conditions on the symbol a imply that $\tilde{d}_{m',\lambda,g}$ is of order $-2m'\rho$. For example,

$$\left| \partial_{\xi_1} a_{g,\lambda} \partial_{\xi_1} \frac{1}{a_{g,\lambda}} \right| = \left| \frac{\partial_{\xi_1} a_{g,\lambda}}{a_{g,\lambda}} \right|^2 \le C_{1,0}|\lambda|^\rho \left(1+|\lambda|(1+|\xi|^2+|u|^2)\right)^{-\rho}.$$

We also write

$$\chi(|\lambda|(|\xi|^2+|u|^2) = 1 + (\chi-1)(|\lambda|(|\xi|^2+|u|^2),$$

and the symbol $(\chi-1)(|\lambda|(|\xi|^2+|u|^2)$ is smoothing.

We now incorporate all the terms of order $\le -\rho$ in a new error term. Indeed, the considerations above show that we can now write

$$\mathrm{Op}^W(b_{\lambda,g})\mathrm{Op}^W(a_{g,\lambda}) = \mathrm{I} + \tilde{E}_{\lambda,g},$$

with $\tilde{E}_{\lambda,g}$ satisfying similar estimates to (6.69).

The end of the proof is now identical to the one of Theorem 6.6.1 with m replaced by m_o. □

Modifying Example 6.6.2, we have the following example of hypoelliptic operators in the sense that they satisfy the hypotheses of Theorem 5.8.9, and therefore admit a left parametrix.

Example 6.6.4. On $\mathbb{H}_1$, if $m, m_o \in 2\mathbb{N}$ are two even integers such that $m \ge m_0$, then the operators

$$X^m + iY^{m_o} + T^{m_o/2} \in \Psi^m \quad \text{and} \quad X^{m_o} + iY^m + T^{m_o/2} \in \Psi^m$$

satisfy the hypotheses of Theorem 5.8.9 for $-\mathcal{L}$ and m_o.

Proof. The symbols of

$$A_1 := X^m + iY^{m_o} + T^{m_o/2} \quad \text{and} \quad A_2 := X^{m_o} + iY^m + T^{m_o/2},$$

are

$$\begin{aligned}\sigma_{A_1}(\lambda) &= \pi_\lambda(X)^m + i\pi_\lambda(Y)^{m_o} + \pi_\lambda(T)^{\frac{m_o}{2}} \\ &= \left(\mathrm{Op}^W(i\sqrt{|\lambda|}\xi)\right)^m + i\left(\mathrm{Op}^W(i\sqrt{\lambda}u)\right)^{m_o} + (i\lambda)^{\frac{m_o}{2}}, \\ \sigma_{A_2}(\lambda) &= \pi_\lambda(X)^{m_o} + i\pi_\lambda(Y)^m + \pi_\lambda(T)^{\frac{m_o}{2}} \\ &= \left(\mathrm{Op}^W(i\sqrt{|\lambda|}\xi)\right)^{m_o} + i\left(\mathrm{Op}^W(i\sqrt{\lambda}u)\right)^m + (i\lambda)^{\frac{m_o}{2}},\end{aligned}$$

by (6.28) and (6.11). Hence their λ-symbols are

$$\begin{aligned}a_{A_1,\lambda}(\xi,x) &= \left(i\sqrt{|\lambda|}\xi\right)^m + i\left(i\sqrt{\lambda}u\right)^{m_o} + (i\lambda)^{\frac{m_o}{2}}, \\ a_{A_2,\lambda}(\xi,x) &= \left(i\sqrt{|\lambda|}\xi\right)^{m_o} + i\left(i\sqrt{\lambda}u\right)^m + (i\lambda)^{\frac{m_o}{2}}.\end{aligned}$$

From this, it is not difficult to see that $a_{A_j,\lambda}$, $j = 1, 2$ satisfy

$$|\lambda| \max(|\xi|, |u|) \geq 1 \Longrightarrow |a_{A_j,\lambda}(\xi, u)| \geq C|\lambda|^{m_o} \left(\max(|\xi|, |u|)^{m_o} + 1\right),$$

thus they also satisfy (6.71). The other condition in (6.72) of Theorem 6.6.3 is easy to check. □

6.6.3 Subelliptic estimates and hypoellipticity

The sufficient conditions for ellipticity in Theorem 6.6.1, or at least the existence of left parametrix (see Theorem 6.6.3) yield sufficient conditions for subelliptic estimates and hypoellipticity. More precisely, Corollary 5.8.12 and Propositions 5.8.13 and 5.8.15 imply:

Corollary 6.6.5. *Let $m \in \mathbb{R}$ and $1 \geq \rho > \delta \geq 0$. Let $\sigma = \{\sigma(g,\lambda)\}$ be in $S^m_{\rho,\delta}(\mathbb{H}_n)$ with $\sigma(g,\lambda) = \mathrm{Op}^W(a_{g,\lambda})$ as in Theorem 6.5.1.*

(i) Assume that there are $R \in \mathbb{R}$ and $C > 0$ such that for any $(\xi, u) \in \mathbb{R}^{2n}$ and $\lambda \neq 0$ satisfying $|\lambda|(|\xi|^2 + |u|^2) \geq R$, we have (6.67), that is,

$$|a_{g,\lambda}(\xi,u)| \geq C\left(1 + |\lambda|(1 + |\xi|^2 + |u|^2)\right)^{\frac{m}{2}}.$$

Then $A = \mathrm{Op}(\sigma) = \mathrm{Op}(\mathrm{Op}^W(a_{g,\lambda}))$ is (locally) hypoelliptic. It is also globally hypoelliptic in the sense of Proposition 5.8.15. The operator A also satisfies the following subelliptic estimates

$$\forall s \in \mathbb{R} \quad \forall N \in \mathbb{R} \quad \exists C > 0 \quad \forall f \in \mathcal{S}(\mathbb{H}_n)$$
$$\|f\|_{L^2_{s+m}} \leq C\Big(\|Af\|_{L^2_s} + \|f\|_{L^2_{-N}}\Big).$$

(ii) *We assume that there is $m_o < m$ such that σ satisfies for a given R, for any $(\xi, u) \in \mathbb{R}^{2n}$ such that $|\lambda|(|\xi|^2 + |u|^2) \geq R$, the inequalities (6.71) and (6.72), that is,*

$$|a_{g,\lambda}(\xi, u)| \geq C \left(1 + |\lambda|(1 + |\xi|^2 + |u|^2)\right)^{\frac{m_o}{2}},$$

and

$$|\partial_\xi^{\alpha_1} \partial_u^{\alpha_2} \tilde{\partial}_{\lambda,\xi,u}^{\alpha_3} X_g^\beta a_{g,\lambda}(\xi, u)| \\ \leq C_{\alpha,\beta} |\lambda|^{\rho \frac{|\alpha_1|+|\alpha_2|}{2}} \left(1 + |\lambda|(1 + |\xi|^2 + |u|^2)\right)^{\frac{-\rho[\alpha]+\delta[\beta]}{2}} |a_{g,\lambda}(\xi, u)|.$$

Then $A = \mathrm{Op}(\sigma) = \mathrm{Op}(\mathrm{Op}^W(a_{g,\lambda}))$ is (locally) hypoelliptic. It is also globally hypoelliptic in the sense of Proposition 5.8.15. The operator A also satisfies the following subelliptic estimates

$$\forall s \in \mathbb{R} \quad \forall N \in \mathbb{R} \quad \exists C > 0 \quad \forall f \in \mathcal{S}(\mathbb{H}_n) \\ \|f\|_{L^2_{s+m_o}} \leq C\Big(\|Af\|_{L^2_s} + \|f\|_{L^2_{-N}}\Big).$$

(iii) *In the case $(\rho, \delta) = (1, 0)$, assume that $A \in \Psi^m$ is either elliptic of order $m_0 = m$ or is elliptic of some order m_0 and satisfies the hypotheses of Parts (i) or (ii), respectively. Then A satisfies the subelliptic estimates*

$$\forall s \in \mathbb{R} \quad \forall N \in \mathbb{R} \quad \forall p \in (1, \infty) \quad \exists C > 0 \quad \forall f \in \mathcal{S}(\mathbb{H}_n) \\ \|f\|_{L^p_{s+m_o}} \leq C\Big(\|Af\|_{L^p_s} + \|f\|_{L^p_{-N}}\Big).$$

In the estimates above, $\|\cdot\|_{L^p_s}$ denotes any (fixed) Sobolev norm, for example obtained from a (fixed) positive Rockland operator $\mathcal{R}$, such as $\mathcal{R} = -\mathcal{L}$.

Examples

We proceed by giving examples, applying Corollary 5.8.12 to obtain subelliptic estimates for some of the examples of operators encountered in previous sections. First, naturally, we can apply Corollary 6.6.5 to Examples 6.6.2 and 6.6.4, which we now continue.

Example 6.6.2, continued: On $\mathbb{H}_1$, if $m \in 2\mathbb{N}$ is an even integer, then the operator $X^m + iY^m + T^{m/2}$ is hypoelliptic and satisfies the following estimate

$$\forall p \in (1, \infty) \quad \forall s \in \mathbb{R} \quad \forall N \in \mathbb{R} \quad \exists C > 0 \quad \forall f \in \mathcal{S}(\mathbb{H}_1) \\ \|f\|_{L^p_{s+m}} \leq C \left(\|(X^m + iY^m + T^{m/2})f\|_{L^p_s} + \|f\|_{L^p_{-N}}\right).$$

Example 6.6.4, continued: Let $m, m_o \in 2\mathbb{N}$ be two even integers such that $m \geq m_0$. Then the differential operators $X^m + iY^{m_o} + T^{m_o/2}$ and $X^{m_o} + iY^m + T^{m_o/2}$ on

$\mathbb{H}_1$ are hypoelliptic and satisfy the following subelliptic estimates

$$\forall p \in (1,\infty) \quad \forall s \in \mathbb{R} \quad \forall N \in \mathbb{R} \quad \exists C > 0 \quad \forall f \in \mathcal{S}(\mathbb{H}_1)$$
$$\|f\|_{L^p_{s+m}} \leq C \left(\|(X^m + iY^{m_o} + T^{m_o/2})f\|_{L^p_s} + \|f\|_{L^p_{-N}} \right).$$

and

$$\forall p \in (1,\infty) \quad \forall s \in \mathbb{R} \quad \forall N \in \mathbb{R} \quad \exists C > 0 \quad \forall f \in \mathcal{S}(\mathbb{H}_1)$$
$$\|f\|_{L^p_{s+m}} \leq C \left(\|(X^{m_o} + iY^m + T^{m_o/2})f\|_{L^p_s} + \|f\|_{L^p_{-N}} \right).$$

We can also obtain the hypoellipticity and subelliptic estimates for the elliptic operators in Corollary 5.8.16 choosing first the Rockland operator $\mathcal{R} = -\mathcal{L}$:

Corollary 6.6.6. *As usual, $\mathcal{L}$ denotes the canonical sub-Laplacian on the Heisenberg group $\mathbb{H}_n$ (see (6.5)).*

1. *If f_1 and f_2 are complex-valued smooth functions on $\mathbb{H}_n$ such that*

$$\inf_{x\in\mathbb{H}_n, \lambda\geq\Lambda} \frac{|f_1(x) + f_2(x)\lambda|}{1+\lambda} > 0 \quad \text{for some } \Lambda \geq 0,$$

and such that $X^{\alpha_1} f_1$, $X^{\alpha_2} f_2$ are bounded on $\mathbb{H}_n$ for each $\alpha_1, \alpha_2 \in \mathbb{N}_0^n$, then the differential operator $f_1(x) - f_2(x)\mathcal{L}$ is (locally) hypoelliptic. It is also globally hypoelliptic in the sense of Proposition 5.8.15. This operator also satisfies the following subelliptic estimates

$$\forall p \in (1,\infty) \quad \forall s \in \mathbb{R} \quad \forall N \in \mathbb{R} \quad \exists C > 0 \quad \forall \varphi \in \mathcal{S}(\mathbb{H}_n)$$
$$\|\varphi\|_{L^p_{s+2}} \leq C\Big(\|f_1\varphi - f_2\mathcal{L}\varphi\|_{L^p_s} + \|\varphi\|_{L^p_{-N}} \Big).$$

2. *Let $\psi \in C^\infty(\mathbb{R})$ be such that*

$$\psi|_{(-\infty,\Lambda_1]} = 0 \quad \text{and} \quad \psi|_{[\Lambda_2,\infty)} = 1,$$

for some real numbers Λ_1, Λ_2 satisfying $0 < \Lambda_1 < \Lambda_2$. Let also f_1 be a continuous complex-valued function on $\mathbb{H}_n$ such that $\inf_{\mathbb{H}_n} |f_1| > 0$ and that $X^\alpha f_1$ is bounded on $\mathbb{H}_n$ for each $\alpha \in \mathbb{N}_0^n$. Then the operator $f_1(x)\psi(-\mathcal{L})\mathcal{L}$ is (locally) hypoelliptic. It is also globally hypoelliptic in the sense of Proposition 5.8.15. This operator also satisfies the following subelliptic estimates

$$\forall p \in (1,\infty) \quad \forall s \in \mathbb{R} \quad \forall N \in \mathbb{R} \quad \exists C > 0 \quad \forall \varphi \in \mathcal{S}(\mathbb{H}_n)$$
$$\|\varphi\|_{L^p_{s+2}} \leq C\Big(\|f_1\psi(-\mathcal{L})\mathcal{L}\varphi\|_{L^p_s} + \|\varphi\|_{L^p_{-N}} \Big).$$

We could also use Corollary 5.8.16 with other Rockland operators, such as $\mathcal{R} = \mathcal{L}^2$ or $\mathcal{R} = \mathcal{L}^2 + T^2$. In this case, it would yield:

Corollary 6.6.7. *Let $\mathcal{R} = \mathcal{L}^2$ or $\mathcal{R} = \mathcal{L}^2 + T^2$ where $\mathcal{L}$ denotes the canonical sub-Laplacian on the Heisenberg group $\mathbb{H}_n$ and T is the central derivative.*

1. *If f_1 and f_2 are complex-valued smooth functions on $\mathbb{H}_n$ such that*

$$\inf_{x\in\mathbb{H}_n, \lambda\geq\Lambda} \frac{|f_1(x) + f_2(x)\lambda|}{1+\lambda} > 0 \quad \textit{for some } \Lambda \geq 0,$$

and such that $X^{\alpha_1} f_1$, $X^{\alpha_2} f_2$ are bounded on $\mathbb{H}_n$ for each $\alpha_1, \alpha_2 \in \mathbb{N}_0^n$, then the differential operator $f_1(x) + f_2(x)\mathcal{R}$ is (locally) hypoelliptic. It is also globally hypoelliptic in the sense of Proposition 5.8.15. This operator also satisfies the following subelliptic estimates

$$\forall p \in (1,\infty) \quad \forall s \in \mathbb{R} \quad \forall N \in \mathbb{R} \quad \exists C > 0 \quad \forall \varphi \in \mathcal{S}(\mathbb{H}_n)$$
$$\|\varphi\|_{L^p_{s+4}} \leq C\Big(\|f_1\varphi + f_2\mathcal{R}\varphi\|_{L^p_s} + \|\varphi\|_{L^p_{-N}}\Big).$$

2. *Let $\psi \in C^\infty(\mathbb{R})$ be such that*

$$\psi_{|(-\infty,\Lambda_1]} = 0 \quad \textit{and} \quad \psi_{|[\Lambda_2,\infty)} = 1,$$

for some real numbers Λ_1, Λ_2 satisfying $0 < \Lambda_1 < \Lambda_2$. Let also f_1 be a continuous complex-valued function on $\mathbb{H}_n$ such that $\inf_{\mathbb{H}_n} |f_1| > 0$ and that $X^\alpha f_1$ is bounded on $\mathbb{H}_n$ for each $\alpha \in \mathbb{N}_0^n$. Then the operator $f_1(x)\psi(\mathcal{R})\mathcal{R} \in \Psi^4$ is (locally) hypoelliptic. It is also globally hypoelliptic in the sense of Proposition 5.8.15. This operator also satisfies the following subelliptic estimates

$$\forall p \in (1,\infty) \quad \forall s \in \mathbb{R} \quad \forall N \in \mathbb{R} \quad \exists C > 0 \quad \forall \varphi \in \mathcal{S}(\mathbb{H}_n)$$
$$\|\varphi\|_{L^p_{s+4}} \leq C\Big(\|f_1\psi(\mathcal{R})\mathcal{R}\varphi\|_{L^p_s} + \|\varphi\|_{L^p_{-N}}\Big).$$

Appendix A

Miscellaneous

In this chapter we collect a number of analytic tools that are used at some point in the monograph. These are all well-known, and we present them without proofs providing references to relevant sources when needed. Thus, here we make short expositions of topics including local hypoellipticity and solvability, operator semi-groups, fractional powers of operators, singular integrals, almost orthogonality, and the analytic interpolation.

A.1 General properties of hypoelliptic operators

In this section, we recall the definition and first properties of locally hypoelliptic operators. We will also point out the useful duality between local solvability and local hypoellipticity in Theorem A.1.3.

Roughly speaking, a differential operator L is (locally) *hypoelliptic* if whenever u and f are distributions satisfying $Lu = f$, u must be smooth where f is smooth. Usually, we omit the word 'local' and just speak of hypoellipticity. More precisely:

Definition A.1.1. Let Ω be an open subset of $\mathbb{R}^n$ and let L be a differential operator on Ω with smooth coefficients. Then L is said to be *hypoelliptic* if, for any distribution $u \in \mathcal{D}'(\Omega)$ and any open subset Ω' of Ω, the condition $Lu \in C^\infty(\Omega')$ implies that $u \in C^\infty(\Omega')$.

This definition extends to an open subset of a smooth manifold.

Of course elliptic operators such as Laplace operators are hypoelliptic. Less obvious examples are provided by the celebrated Hörmander's Theorem on sums of squares of vector fields [Hör67a] which we recall here even if we will not use it in this monograph:

Theorem A.1.2 (Hörmander sum of squares). *Let $X_o, X_1, \ldots, X_p$ be smooth real-valued vector fields on an open set $\Omega \subset \mathbb{R}^n$, and let $c_o \in C^\infty(\Omega)$. We assume* *that the vector fields $X_o, X_1, \ldots, X_p$ satisfy Hörmander's condition, that is, the Lie algebra generated by $\{X_o, X_1, \ldots, X_p\}$ is of dimension n at every point of Ω. Then the operator $X_1^2 + \ldots + X_p^2 + X_o + c$ is hypoelliptic on Ω.*

This extends to smooth manifolds.

Consequently any sub-Laplacian (see Definition 4.1.6) on a stratified Lie group is hypoelliptic on the whole group since any basis of the first stratum satisfies Hörmander's condition.

Hörmander's condition in Theorem A.1.2 is sufficient but not necessary for the hypoellipticity of sums of squares, thus allowing for sharper versions, see e.g. [BM95].

In the following sense, local hypoellipticity is dual to local solvability:

Theorem A.1.3. *Let L be hypoelliptic on Ω. Then L^t is locally solvable at every point of Ω.*

Let us briefly recall the definitions of the local solvability and of transpose:

Definition A.1.4. Let L be a linear differential operator with smooth coefficients on Ω. We say that L is *locally solvable* at $x \in \Omega$ if x has an open neighbourhood V in Ω such that, for every function $f \in \mathcal{D}(V)$ there is a distribution $u \in \mathcal{D}'(V)$ satisfying $Lu = f$ on V.

Definition A.1.5. The *transpose* of a differential operator L with smooth coefficients on an open subset Ω of $\mathbb{R}^n$ is the operator, denoted by L^t, given by

$$\forall \phi, \psi \in \mathcal{D}(\Omega) \qquad \langle L\phi, \psi \rangle = \langle \phi, L^t \psi \rangle.$$

This extends to manifolds.

Note that if

$$Lf(x) = \sum_{|\alpha| \leq m} a_\alpha(x) \partial^\alpha f(x),$$

then

$$L^t f(x) = \sum_{|\alpha| \leq m} \partial^\alpha \big(a_\alpha(x) f(x) \big) = \sum_{|\alpha| \leq m} b_\alpha(x) \partial^\alpha f(x),$$

where the b_α's are linear combinations of derivatives of the a_α's, in particular they are smooth functions.

We will need the following property:

Theorem A.1.6 (Schwartz-Trèves). *Let L be a differential operator with smooth coefficients on an open subset Ω of $\mathbb{R}^n$. We assume that L and L^t are hypoelliptic on $\Omega \subset \mathbb{R}^n$. Then the $\mathcal{D}'(\Omega)$ and $C^\infty(\Omega)$ topologies agree on*

$$N_L(\Omega) = \{ f \in \mathcal{D}'(\Omega) \ : \ Lf = 0 \}.$$

For its proof, we refer to [Tre67, Corollary 1 in Ch. 52].

A.2 Semi-groups of operators

In this section we discuss operator semi-groups and their infinitesimal generators.

Definition A.2.1. Suppose that for every $t \in (0, \infty)$, there is an associated bounded linear operator $Q(t)$ on a Banach space $\mathcal{X}$ in such a way that

$$\forall s, t > 0 \qquad Q(s+t) = Q(s)Q(t).$$

Then the family $\{Q(t)\}_{t>0}$ is called a *semi-group* of operators on $\mathcal{X}$.

If we have for every $x \in \mathcal{X}$, that

$$\|Q(t)x - x\|_{\mathcal{X}} \underset{t \to 0}{\longrightarrow} 0,$$

then the semi-group is said to be *strongly continuous*.

If the operator norm of each $Q(t)$ is less or equal to one, $\|Q(t)\|_{\mathscr{L}(\mathcal{X})} \leq 1$, then the semi-group is called a *contraction* semi-group.

Let $\{Q(t)\}_{t>0}$ be a semi-group of operators on $\mathcal{X}$. If $x \in \mathcal{X}$ is such that $\frac{1}{\epsilon}(Q(\epsilon)x - x)$ converges in the norm topology of $\mathcal{X}$ as $\epsilon \to 0$, then we denote its limit by Ax and we say that x is in the domain $\mathrm{Dom}(A)$ of A. Clearly $\mathrm{Dom}(A)$ is a linear subspace of $\mathcal{X}$ and A is a linear operator on $\mathrm{Dom}(A) \subset X$. This operator is essentially $A = Q'(0)$.

Definition A.2.2. The operator A defined just above is called the *infinitesimal generator* of the semi-group $\{Q(t)\}_{t>0}$.

We now collect some properties of semi-groups and their generators.

Proposition A.2.3. *Let $\{Q(t)\}_{t>0}$ be a strongly continuous semi-group with infinitesimal generator A. We also set $Q(0) := \mathrm{I}$, the identity operator. Then*

1. *there are constants C, γ such that for all $t \in [0, \infty)$,*

$$\|Q(t)\|_{\mathscr{L}(\mathcal{X})} \leq Ce^{\gamma t};$$

2. *for every $x \in \mathcal{X}$, the map $[0, \infty) \ni t \mapsto Q(t)x \in \mathcal{X}$ is continuous;*
3. *the operator A is closed with dense domain;*
4. *the differential equation*

$$\partial_t Q(t)x = Q(t)A\,x = AQ(t)x,$$

holds for every $x \in \mathrm{Dom}(A)$ and $t \geq 0$;

5. *for every $x \in \mathcal{X}$ and $t > 0$,*

$$Q(t)x = \lim_{\epsilon \to 0} \exp(tA_\epsilon)x,$$

where

$$A_\epsilon = \frac{1}{\epsilon}\left(Q(\epsilon) - \mathrm{I}\right) \quad \textit{and} \quad \exp(tA_\epsilon) = \sum_{k=0}^{\infty} \frac{1}{k!}(tA_\epsilon)^k;$$

furthermore the convergence is uniform on every compact subset of $[0, \infty)$*;*

6. *if* $\lambda \in \mathbb{C}$ *and* $\operatorname{Re}\lambda > \gamma$ *(where* γ *is any constant such that (1) holds), the integral*

$$R(\lambda)x = \int_0^\infty e^{-\lambda t} Q(t)x\, dt,$$

defines a bounded linear operator $R(\lambda)$ *on* $\mathcal{X}$ (often called the resolvent of the semi-group $\{Q(t)\}$) *whose range is* $\operatorname{Dom}(A)$ *and which inverts* $\lambda\mathrm{I} - A$*. In particular, the spectrum of* A *lies in the half plane* $\{\lambda : \operatorname{Re}\lambda \leq \gamma\}$*.*

For the proof, see e.g. Rudin [Rud91, §13.35].

Theorem A.2.4 (Hille-Yosida). *A densely defined operator* A *on a Banach space* $\mathcal{X}$ *is the infinitesimal generator of a strongly continuous semi-group* $\{Q(t)\}_{t>0}$ *if and only if there are constants* C, γ *such that*

$$\forall \lambda > \gamma,\ m \in \mathbb{N} \qquad \|(\lambda\mathrm{I} - A)^{-m}\| \leq C(\lambda - \gamma)^{-m}.$$

The constant γ *can be taken as in Proposition A.2.3.*

For the proof of the Hille-Yosida Theorem, see e.g [Rud91, §13.37].

In this case the operators of the semi-group $\{Q(t)\}_{t>0}$ generated by A are denoted by

$$Q(t) = e^{tA}.$$

Theorem A.2.5 (Lumer-Phillips). *A densely defined operator* A *on a Banach space* $\mathcal{X}$ *is the infinitesimal generator of a strongly continuous contraction semi-group* $\{Q(t)\}_{t>0}$ *if and only if*

- *A is dissipative, i.e.*

$$\forall \lambda > 0,\ x \in \operatorname{Dom}(A) \qquad \|(\lambda\mathrm{I} - A)x\| \geq \lambda\|x\|;$$

- *there is at least one* λ_o *such that* $A - \lambda_o\mathrm{I}$ *is surjective.*

For the proof of the Lumer-Phillips Theorem, see [LP61].

For this monograph, the facts given in this section will be enough. We refer for the general theory of semi-groups to the fundamental work of Hille and Phillips [HP57], or to later expositions e.g. by Davies [Dav80] or Pazy [Paz83].

A.3 Fractional powers of operators

Here we summarise the definition of fractional powers for certain operators. We refer the interested reader to the monograph of Martinez and Sanz [MCSA01] and all the explanations and historical discussions therein.

Let $A : \mathrm{Dom}(A) \subset \mathcal{X} \to \mathcal{X}$ be a linear operator on a Banach space $\mathcal{X}$. In order to present only the part of the theory that we use in this monograph, we make the following assumptions

(i) The operator A is closed and densely defined.

(ii) The operator A is injective, that is, A is one-to-one on its domain.

(iii) The operator A is Komatsu-non-negative, that is, $(-\infty, 0)$ is included in the resolvent $\rho(A)$ of A and

$$\exists M > 0 \quad \forall \lambda > 0 \quad \|(\lambda + A)^{-1}\| \leq M\lambda^{-1}.$$

Remark A.3.1. This implies (cf. [MCSA01, Proposition 1.1.3 (iii)]) that for all $n, m \in \mathbb{N}$, $\mathrm{Dom}(A^n)$ is dense in $\mathcal{X}$, and $\mathrm{Range}(A^m)$ as well as $\mathrm{Dom}(A^n) \cap \mathrm{Range}(A^m)$ are dense in the closure of $\mathrm{Range}(A)$.

The powers A^n, $n \in \mathbb{N}$, are defined using iteratively the following definition:

Definition A.3.2. The product of two (possibly) unbounded operators A and B acting on the same Banach space $\mathcal{X}$ is as follows. A vector x is in the domain of the operator AB whenever x is in the domain of B and Bx is in the domain of A. In this case $(AB)(x) = A(Bx)$.

Remark A.3.3. Note that if an operator A satisfies (i), (ii) and (iii), then it is also the case for $\mathrm{I} + A$.

Following Balakrishnan (cf. [MCSA01, Section 3.1]), the (Balakrishnan) operators J^α, $\alpha \in \mathbb{C}_+ := \{z \in \mathbb{C}, \mathrm{Re}\, z > 0\}$, are (densely) defined by the following:

- If $0 < \mathrm{Re}\,\alpha < 1$, $\mathrm{Dom}(J^\alpha) := \mathrm{Dom}(A)$ and for $\phi \in \mathrm{Dom}(A)$,

$$J^\alpha \phi := \frac{\sin \alpha\pi}{\pi} \int_0^\infty \lambda^{\alpha-1} (\lambda \mathrm{I} + A)^{-1} A\phi \, d\lambda.$$

- If $\mathrm{Re}\,\alpha = 1$, $\mathrm{Dom}(J^\alpha) := \mathrm{Dom}(A^2)$ and for $\phi \in \mathrm{Dom}(A^2)$,

$$J^\alpha \phi := \frac{\sin \alpha\pi}{\pi} \int_0^\infty \lambda^{\alpha-1} \left[(\lambda \mathrm{I} + A)^{-1} - \frac{\lambda}{\lambda^2 + 1} \right] A\phi \, d\lambda + \sin \frac{\alpha\pi}{2} A\phi.$$

- If $n < \mathrm{Re}\,\alpha < n + 1$, $n \in \mathbb{N}$, $\mathrm{Dom}(J^\alpha) := \mathrm{Dom}(A^{n+1})$ and for $\phi \in \mathrm{Dom}(A)$,

$$J^\alpha \phi := J^{\alpha - n} A^n \phi.$$

- If $\operatorname{Re}\alpha = n+1$, $n \in \mathbb{N}$, $\operatorname{Dom}(J^\alpha) := \operatorname{Dom}(A^{n+2})$ and for $\phi \in \operatorname{Dom}(A^{n+2})$,
$$J^\alpha\phi := J^{\alpha-n}A^n\phi.$$

We now define fractional powers distinguishing between three different cases:

Case 0: A is bounded.

Case I: A is unbounded and $0 \in \rho(A)$, that is, the resolvent of A contains zero; in other words, A^{-1} is bounded.

Case II: A is unbounded and $0 \in \sigma(A)$, that is, the spectrum of A contains zero.

The fractional powers A^α, $\alpha \in \mathbb{C}_+$, are defined in the following way (cf. [MCSA01, Section 5.1]):

Case 0: A being bounded, J^α is bounded and we define $A^\alpha := J^\alpha$, $\alpha \in \mathbb{C}_+$.

Case I: A^{-1} being bounded, we can use Case 0 to define $(A^{-1})^\alpha$ which is injective; then we define
$$A^\alpha := \left[(A^{-1})^\alpha\right]^{-1} \quad (\alpha \in \mathbb{C}_+).$$

Case II: Using Case I for $A + \epsilon \mathrm{I}$, $\epsilon > 0$, we define
$$A^\alpha := \lim_{\epsilon\to 0}(A+\epsilon\mathrm{I})^\alpha \quad (\alpha \in \mathbb{C}_+);$$
that is, the domain of A^α is composed of all the elements $\phi \in \operatorname{Dom}\left[(A+\epsilon\mathrm{I})^\alpha\right]$, $\epsilon > 0$ close to zero, and such that $(A+\epsilon\mathrm{I})\phi$ is convergent for the norm topology of $\mathcal{X}$ as $\epsilon \to 0$; the limit defines $A^\alpha\phi$.

In all cases, J^α is closable and we have (cf. [MCSA01, Theorem 5.2.1]):
$$A^\alpha = (A+\lambda\mathrm{I})^n\overline{J^\alpha}(A+\lambda\mathrm{I})^{-n} \qquad (\alpha \in \mathbb{C}_+,\ \lambda \in \rho(-A),\ n \in \mathbb{N}).$$

Hence A^α, $\alpha \in \mathbb{C}_+$, can be understood as the maximal domain operator which extends J^α and commutes with the resolvent of A (in other words *commutes strongly with* A).

We can now define the powers for complex numbers also with non-positive real parts (cf. [MCSA01, Section 7.1]):

- Given $\alpha \in \mathbb{C}_+$, the operators A^α, $\alpha \in \mathbb{C}_+$, are injective, and we can define
$$A^{-\alpha} := (A^\alpha)^{-1}.$$

- Given $\tau \in \mathbb{R}$, we define
$$A^{i\tau} := (A+\mathrm{I})^2A^{-1}A^{1+i\tau}(A+\mathrm{I})^{-2}.$$

We now collect properties of fractional powers.

Theorem A.3.4. *Let $A : \mathrm{Dom}(A) \subset \mathcal{X} \to \mathcal{X}$ be a linear operator on a Banach space $\mathcal{X}$. Assume that the operator A satisfies Properties (i), (ii) and (iii), and define its fractional powers A^α as above.*

1. *For every $\alpha \in \mathbb{C}$, the operator A^α is closed and injective with $(A^\alpha)^{-1} = A^{-\alpha}$. In particular, $A^0 = \mathrm{I}$.*
2. *For $\alpha \in \mathbb{C}_+$, the operator A^α coincides with the closure of J^α.*
3. *If A has dense range and for all $\tau \in \mathbb{R}$, $A^{i\tau}$ is bounded, then there exist $C > 0$ and $\theta \in (0, \pi)$ such that*
$$\forall \tau \in \mathbb{R} \qquad \|A^{i\tau}\|_{\mathscr{L}(\mathcal{X})} \leq Ce^{\theta\tau}.$$
Given $\tau \in \mathbb{R}\backslash\{0\}$, if $A^{i\tau}$ is bounded then $\mathrm{Dom}(A^\alpha) \subset \mathrm{Dom}(A^{\alpha+i\tau})$ for all $\alpha \in \mathbb{R}$. Conversely, if $\mathrm{Dom}(A^\alpha) \subset \mathrm{Dom}(A^{\alpha+i\tau})$ for all $\alpha \in \mathbb{R}\backslash\{0\}$, then $A^{i\tau}$ is bounded.
4. *For any $\alpha, \beta \in \mathbb{C}$, we have $A^\alpha A^\beta \subset A^{\alpha+\beta}$, and if $\mathrm{Range}(A)$ is dense in $\mathcal{X}$ then the closure of $A^\alpha A^\beta$ is $A^{\alpha+\beta}$.*
5. *Let $\alpha_o \in \mathbb{C}_+$.*
   - *If $\phi \in \mathrm{Range}(A^{\alpha_o})$ then $\phi \in \mathrm{Dom}(A^\alpha)$ for all $\alpha \in \mathbb{C}$ with $0 < -\mathrm{Re}\,\alpha < \mathrm{Re}\,\alpha_o$ and the function $\alpha \mapsto A^\alpha\phi$ is holomorphic in $\{\alpha \in \mathbb{C} : -\mathrm{Re}\,\alpha_o < \mathrm{Re}\,\alpha < 0\}$.*
   - *If $\phi \in \mathrm{Dom}(A^{\alpha_o})$ then $\phi \in \mathrm{Dom}(A^\alpha)$ for all $\alpha \in \mathbb{C}$ with $0 < \mathrm{Re}\,\alpha < \mathrm{Re}\,\alpha_o$ and the function $\alpha \mapsto A^\alpha\phi$ is holomorphic in $\{\alpha \in \mathbb{C} : 0 < \mathrm{Re}\,\alpha < \mathrm{Re}\,\alpha_o\}$.*
   - *If $\phi \in \mathrm{Dom}(A^{\alpha_o}) \cap \mathrm{Range}(A^{\alpha_o})$ then $\phi \in \mathrm{Dom}(A^\alpha)$ for all $\alpha \in \mathbb{C}$ with $|\mathrm{Re}\,\alpha| < \mathrm{Re}\,\alpha_o$ and the function $\alpha \mapsto A^\alpha\phi$ is holomorphic in $\{\alpha \in \mathbb{C} : -\mathrm{Re}\,\alpha_o < \mathrm{Re}\,\alpha < \mathrm{Re}\,\alpha_o\}$.*
6. *If $\alpha, \beta \in \mathbb{C}_+$ with $\mathrm{Re}\,\beta > \mathrm{Re}\,\alpha$, then*
$$\exists C = C_{A,\alpha,\beta} > 0 \quad \forall \phi \in \mathrm{Dom}(A^\beta) \quad \|A^\alpha\phi\|_{\mathcal{X}} \leq C\|\phi\|_{\mathcal{X}}^{1-\frac{\mathrm{Re}\,\alpha}{\mathrm{Re}\,\beta}} \|A^\beta\phi\|_{\mathcal{X}}^{\frac{\mathrm{Re}\,\alpha}{\mathrm{Re}\,\beta}}.$$
7. *If B^* denotes the dual of an operator B on $\mathcal{X}$, then $(A^\alpha)^* = (A^*)^\alpha$.*
8. *For $\alpha \in \mathbb{C}_+$ and $\epsilon > 0$, $\mathrm{Dom}\left[(A + \epsilon\mathrm{I})^\alpha\right] = \mathrm{Dom}(A^\alpha)$.*
9. *Let $\tau \in \mathbb{R}$. Let $S_{i\tau}$ be the strong limit of $(A + \epsilon\mathrm{I})^{i\tau}$ as $\epsilon \to 0^+$, with domain $\mathrm{Dom}(S_{i\tau}) = \{\phi \in \mathrm{Dom}\left[(A + \epsilon)^{i\tau}\right] : \exists \lim_{\epsilon\to 0^+}(A + \epsilon)^{i\tau}\phi\}$. Then $S_{i\tau}$ is closable and the closure of (the graph of) $J^{i\tau}$ is included in the closure of (the graph of) $S_{i\tau}$ which is included in (the graph of) $A^{i\tau}$.*

   In particular, if A has dense domain and range, then the closure of $S_{i\tau}$ is $A^{i\tau}$.

10. *Let us assume that A generates an equibounded semi-group $\{e^{-tA}\}_{t>0}$ on $\mathcal{X}$, that is,*

$$\exists M \qquad \forall t>0 \qquad \|e^{-tA}\|_{\mathcal{X}} \leq M. \tag{A.1}$$

If $0 < \operatorname{Re}\alpha < 1$ and $\phi \in \operatorname{Range}(A)$ then

$$A^{-\alpha}\phi = \frac{1}{\Gamma(\alpha)} \int_0^{\infty} t^{\alpha-1} e^{-tA} \phi \, dt, \tag{A.2}$$

in the sense that $\lim_{N\to\infty} \int_0^N$ converges in the $\mathcal{X}$-norm.

Moreover, if $\{e^{-tA}\}_{t>0}$ is exponentially stable, that is,

$$\exists M, \mu > 0 \quad \forall t > 0 \qquad \|e^{-tA}\|_{\mathscr{L}(\mathcal{X})} \leq M e^{-t\mu},$$

then Formula (A.2) holds for all $\alpha \in \mathbb{C}_+$ and $\phi \in \mathcal{X}$, and the integral converges absolutely: $\int_0^\infty \|t^{\alpha-1}e^{-tA}\phi\|_{\mathcal{X}} dt < \infty$.

References for these results are in [MCSA01] as follows:

(1) Corollary 5.2.4 and Section 7.1;

(2) Corollary 5.1.12;

(3) Proposition 8.1.1, Section 7.1 and Corollary 7.1.2;

(4) Theorem 7.1.1;

(5) Proposition 7.1.5 with its proof, and Corollary 5.1.13;

(6) Corollary 5.1.13;

(7) Corollary 5.2.4 for $\alpha \in \mathbb{C}_+$, consequently for any $\alpha \in \mathbb{C}$;

(8) Theorem 5.1.7;

(9) Theorem 7.4.6;

(10) Lemma 6.1.5.

In Theorem A.3.4 Part (10), Γ denotes the Gamma function. Let us recall briefly its definition. For each $\alpha \in \mathbb{C}_+$, it is defined by the convergent integral

$$\Gamma(\alpha) := \int_0^\infty t^{\alpha-1} e^{-t} dt.$$

A direct computation gives $\Gamma(1) = \int_0^\infty e^{-t} dt = 1$ and an integration by parts yields the functional equation $\alpha\Gamma(\alpha) = \Gamma(\alpha+1)$. Hence the Gamma function coincides with the factorial in the sense that if $\alpha \in \mathbb{N}$, then the equality $\Gamma(\alpha) = (\alpha-1)!$ holds. It is easy to see that Γ is analytic on the half plane $\{\operatorname{Re}\alpha > 0\}$. Because of the functional equation, it admits a unique analytic continuation to the whole complex plane except for non-positive integers where it has simple pole. We keep the same notation Γ for its analytic continuation.

For $\mathrm{Re}\, z > 0$, we have the Sterling estimate

$$\Gamma(z) = \sqrt{\frac{2\pi}{z}} \left(\frac{z}{e}\right)^z (1 + O(\frac{1}{z})). \tag{A.3}$$

Also, the following known relation will be of use to us,

$$\int_{t=0}^{1} t^{x-1}(1-t)^{y-1} dt = \frac{\Gamma(x)\Gamma(y)}{\Gamma(x+y)}, \quad \mathrm{Re}\, x > 0, \ \mathrm{Re}\, y > 0. \tag{A.4}$$

We will use Part (6) also in the following form: let $\alpha, \beta, \gamma \in \mathbb{C}$ with $\mathrm{Re}\,\alpha < \mathrm{Re}\,\beta$ and $\mathrm{Re}\,\alpha \leq \mathrm{Re}\,\gamma \leq \mathrm{Re}\,\beta$; then there exists $C = C_{\alpha,\beta,\gamma,A} > 0$ such that for any $f \in \mathrm{Dom}(A^\alpha)$ with $A^\alpha f \in \mathrm{Dom}(A^{\beta-\alpha})$, we have

$$\|A^\gamma f\|_{\mathcal{X}} \leq C \|A^\alpha f\|_{\mathcal{X}}^{1-\theta} \|A^\beta f\|_{\mathcal{X}}^{\theta} \qquad \text{where} \quad \theta := \frac{\mathrm{Re}\,(\gamma - \alpha)}{\mathrm{Re}\,(\beta - \alpha)}.$$

A.4 Singular integrals (according to Coifman-Weiss)

The operators appearing 'in practice' in the theory of partial differential equations on $\mathbb{R}^n$ often have kernels κ satisfying the following properties:

1. the restriction of $\kappa(x,y)$ to $(\mathbb{R}^n_x \times \mathbb{R}^n_y) \backslash \{x = y\}$ coincides with a smooth function $\kappa_o = \kappa_o(x,y) \in C^\infty((\mathbb{R}^n_x \times \mathbb{R}^n_y) \backslash \{x = y\})$;
2. away from the diagonal $x = y$, the function κ_o decays rapidly;
3. at the diagonal, κ_o is singular but not completely wild: κ_o and some of its first derivatives admit a control of the form $|\kappa_o(x,y)| \leq C_x |x-y|^k$ for some power $k \in (-\infty, \infty)$ with C_x varying slowly in x.

These types of operators include all the (Hörmander, Shubin, semi-classical, ...) pseudo-differential operators, and these types of operators appear when looking for fundamental solutions or parametrices of differential operators.

In general, we want our operator T to map continuously some well-known functional space to another. For example, we are looking for conditions to ensure that our operator extends to a bounded operator from L^p to L^q. This is the subject of the theory of singular integrals on $\mathbb{R}^n$, especially when the power k above equals $-n$. In the classical Euclidean case, we refer to the monograph [Ste93] by Stein for a detailed presentation of this theory.

Here, let us present the main lines of the generalisation of the theory of singular integrals to the setting of 'spaces of homogeneous type' where there is no (apparent) trace of a group structure. This generalisation is relevant for us since examples of such spaces are compact manifolds and homogeneous nilpotent Lie groups. We omit the proofs, referring to [CW71a, Chapitre III] for details.

Definition A.4.1. A *quasi-distance* on a set X is a function $d : X \times X \to [0, \infty)$ such that

1. $d(x, y) > 0$ if and only if $x \neq y$;
2. $d(x, y) = d(y, x)$;
3. there exists a constant $K > 0$ such that

$$\forall x, y, z \in X \qquad d(x, z) \leq K \left(d(x, y) + d(y, z)\right).$$

We call

$$B(x, r) := \{y \in G \ : \ d(x, y) < r\},$$

the *quasi-ball of radius* r *around* x.

Definition A.4.2. A *space of homogeneous type* is a topological space X endowed with a quasi-distance d such that

1. The quasi-balls $B(x, r)$ form a basis of open neighbourhood at x;
2. homogeneity property

   there exists $N \in \mathbb{N}$ such that for every $x \in X$ and every $r > 0$ the ball $B(x, r)$ contains at most N points x_i such that $d(x_i, x_j) > r/2$.

The constants K in Definition A.4.1 and N in Definition A.4.2 are called the constants of the space of homogeneous type X.

Some authors (like in the original text of [CW71a]) prefer using the vocabulary pseudo-norms, pseudo-distance, etc. instead of quasi-norms, quasi-distance, etc. In this monograph, following e.g. both Stein [Ste93] and Wikipedia, we choose the perhaps more widely adapted convention of the term quasi-norm.

Examples of spaces of homogeneous type:

1. A homogeneous Lie group endowed with the quasi-distance associated to any homogeneous quasi-norm (see Lemma 3.2.12).
2. The unit sphere $\mathbb{S}^{n-1}$ in $\mathbb{R}^n$ with the quasi-distance

$$d(x, y) = |1 - x \cdot y|^\alpha,$$

   where $\alpha > 0$ and $x \cdot y = \sum_{j=1}^n x_j y_j$ is the real scalar product of $x, y \in \mathbb{R}^n$.
3. The unit sphere $\mathbb{S}^{2n-1}$ embedded in $\mathbb{C}^n$ with the quasi-distance

$$d(z, w) = |1 - (z, w)|^\alpha,$$

   where $\alpha > 0$ and $(z, w) = \sum_{j=1}^n z_j \bar{w}_j$.
4. Any compact Riemannian manifold.

The proof that these spaces are effectively of homogeneous type comes easily from the following lemma:

Lemma A.4.3. *Let X be a topological set endowed with a quasi-distance d satisfying (1) of Definition A.4.2.*

Assume that there exist a Borel measure μ on X satisfying

$$0 < \mu\left(B(x,r)\right) \leq C\mu\left(B(x,\frac{r}{2})\right) < \infty. \tag{A.5}$$

Then X is a space of homogeneous type.

The condition (A.5) is called the *doubling condition.* For instance, the Riemannian measure of a Riemannian compact manifold or the Haar measure of a homogeneous Lie group satisfy the doubling condition; we omit the proof of these facts, as well as the proof of Lemma A.4.3.

Let (X,d) be a space of homogeneous type. The hypotheses are 'just right' to obtain a covering lemma. We assume now that X is also equipped with a measure μ satisfying the doubling condition (A.5). A maximal function with respect to the quasi-balls may be defined. Then given a level, any function f can be decomposed 'in the usual way' into good and bad functions $f = g+\sum_j b_j$. The Euclidean proof of the Singular Integral Theorem can be adapted to obtain

Theorem A.4.4 (Singular integrals). *Let (X,d) be a space of homogeneous type equipped with a measure μ satisfying the doubling condition given in (A.5).*

Let T be an operator which is bounded on $L^2(X)$:

$$\exists C_o \qquad \forall f \in L^2 \quad \|Tf\|_2 \leq C_o\|f\|_2. \tag{A.6}$$

We assume that there exists a locally integrable function κ on $(X\times X)\setminus \{(x,y)\in X\times X : x=y\}$ such that for any compactly supported function $f\in L^2(X)$, we have

$$\forall x \notin \text{supp} f \qquad Tf(x) = \int_X \kappa(x,y)f(y)d\mu(y).$$

We also assume that there exist $C_1, C_2 > 0$ such that

$$\forall y, y_o \in X \qquad \int_{d(x,y_o)>C_1 d(y,y_o)} |\kappa(x,y) - \kappa(x,y_o)|d\mu(x) \leq C_2. \tag{A.7}$$

Then for all p, $1<p\leq 2$, T extends to a bounded operator on L^p because

$$\exists A_p \qquad \forall f \in L^2\cap L^p \quad \|Tf\|_p \leq A_p\|f\|_p;$$

for $p=1$, the operator T extends to a weak-type (1,1) operator since

$$\exists A_1 \qquad \forall f \in L^2\cap L^1 \quad \mu\{x \,:\, |Tf(x)| > \alpha\} \leq A_1\frac{\|f\|_1}{\alpha};$$

the constants A_p, $1\leq p\leq 2$, depend only on C_o, C_1 and C_2.

Remark A.4.5. 1. In the statement of the fundamental theorem of singular integrals on spaces of homogeneous types, cf. [CW71a, Théorème 2.4 Chapitre III], the kernel κ is assumed to be square integrable in $L^2(X \times X)$. However, the proof requires only that the kernel κ is locally integrable away from the diagonal, beside the L^2-boundedness of the operator T. We have therefore chosen to state it in the form given above.

2. Following the constants in the proof of [CW71a, Théorème 2.4 Chapitre III], we find
$$A_2 = C_1 \qquad \text{and} \qquad A_1 = C(C_1^2 + C_3),$$
where C is a constant which depends only on the constants of the space of homogeneous type. The constants A_p for $p \in (1, 2)$ are obtained via the constants appearing in the Marcinkiewicz interpolation theorem (see e.g. [DiB02, Theorem 9.1]):
$$A_p = \frac{2p}{(2-p)(1-p)} A_1^\delta A_2^{1-\delta} \quad \text{with } \delta = 2(\frac{1}{p} - \frac{1}{2}).$$

Let us discuss the two main hypotheses of Theorem A.4.4.

About Condition (A.7) in the Euclidean case. As explained at the beginning of this section, we are interested in 'nice' kernels $\kappa_o(x, y)$ with a control of the form $|\kappa_o(x, y)| \leq C_x |x - y|^k$ with a particular interest for $k = -n$, and similar estimates for their derivatives with power $-n-1$. Hence they should satisfy Condition (A.7). They are called Calderón-Zygmund kernels, which we now briefly recall:

Calderón-Zygmund kernels on $\mathbb{R}^n$

A *Calderón-Zygmund kernel* on $\mathbb{R}^n$ is a measurable function κ_o defined on $(\mathbb{R}^n_x \times \mathbb{R}^n_y) \backslash \{x = y\}$ satisfying for some γ, $0 < \gamma \leq 1$, the inequalities
$$\begin{aligned} |\kappa_o(x, y)| &\leq A|x - y|^{-n}, \\ |\kappa_o(x, y) - \kappa_o(x', y)| &\leq A\frac{|x - x'|^\gamma}{|x - y|^{n+\gamma}} \qquad \text{if } |x - x'| \leq \frac{|x - y|}{2}, \\ |\kappa_o(x, y) - \kappa_o(x, y')| &\leq A\frac{|y - y'|^\gamma}{|x - y|^{n+\gamma}} \qquad \text{if } |y - y'| \leq \frac{|x - y|}{2}. \end{aligned}$$

Sometimes the condition of Calderón-Zygmund kernels refers to a smooth function κ_o defined on $(\mathbb{R}^n_x \times \mathbb{R}^n_y) \backslash \{x = y\}$ satisfying
$$\forall \alpha, \beta \quad \exists C_{\alpha,\beta} \qquad \left|\partial_x^\alpha \partial_y^\beta \kappa_o(x, y)\right| \leq C_{\alpha,\beta}|x - y|^{-n-\alpha-\beta}.$$

For a detailed discussion, the reader is directed to [Ste93, ch.VII].

A *Calderón-Zygmund operator* on $\mathbb{R}^n$ is an operator $T : \mathcal{S}(\mathbb{R}^n) \to \mathcal{S}'(\mathbb{R}^n)$ such that the restriction of its kernel κ to $(\mathbb{R}^n_x \times \mathbb{R}^n_y)\backslash\{x = y\}$ is a Calderón-Zygmund kernel κ_o. In other words, $T : \mathcal{S}(\mathbb{R}^n) \to \mathcal{S}'(\mathbb{R}^n)$ is a Calderón-Zygmund operator if there exists a Calderón-Zygmund kernel κ_o satisfying

$$Tf(x) = \int_{\mathbb{R}^n} \kappa_o(x,y)f(y)dy,$$

for $f \in \mathcal{S}(\mathbb{R}^n)$ with compact support and $x \in \mathbb{R}^n$ outside the support of f.

The Calderón-Zygmund conditions imply Condition (A.7) for the operator T and its formal adjoint T^* but they are not sufficient to imply the L^2-boundedness for which some additional 'cancellation' conditions are needed.

About Condition (A.6). The difficulty with applying the main theorem of singular integrals (i.e. Theorem A.4.4) is often to know that the operator is L^2-bounded. The next section explains the Cotlar-Stein lemma which may help to prove the L^2-boundedness in many cases.

A.5 Almost orthogonality

On $\mathbb{R}^n$, a convolution operator (for the usual convolution) is bounded on $L^2(\mathbb{R}^n)$ if and only if the Fourier transform of its kernel is bounded. Similar result is valid on compact Lie groups, see (2.23), and more generally on any Hausdorff locally compact separable group, see the decomposition of group von Neumann algebras in the abstract Plancherel theorem in Theorem B.2.32. For operators on spaces without readily available Fourier transform or with no control on the Fourier transform of its kernel, or for non-convolution operators this becomes more complicated (however, see Theorem 2.2.5 for the case of non-invariant operators on compact Lie groups).

Fortunately, the space L^2 is a Hilbert space and to prove that an operator is bounded on L^2, it suffices to do the same for TT^* (or T^*T). The reason that this observation is useful in practice is that if T is formally representable by a kernel κ (see Schwartz kernel theorem, Theorem 1.4.1), then T^*T is representable by the kernel

$$\int \overline{\kappa(z,x)}\kappa(z,y)\, dz;$$

the latter kernel is often better than κ because the integration can have a smoothing effect and/or can take into account the cancellation properties of κ. This remark alone does not always suffice to prove the L^2-boundedness. Sometimes some 'smart' decomposition $T = \sum_k T_k$ of the operator is needed and again the properties of a Hilbert space may help.

The next statement is an easy case of 'exact' orthogonality:

Proposition A.5.1. *Let $\mathcal{H}$ be a Hilbert space and let $\{T_k,\ k \in \mathbb{Z}\}$ be a sequence of linear operators on $\mathcal{H}$. We assume that the operators $\{T_k\}$ are uniformly bounded:*

$$\exists C > 0 \qquad \forall k \in \mathbb{Z} \qquad \|T_k\|_{\mathscr{L}(\mathcal{H})} \leq C,$$

and that

$$\forall j \neq k \qquad T_j^* T_k = 0 \quad \text{and} \quad T_j T_k^* = 0. \tag{A.8}$$

Then the series $\sum_{k \in \mathbb{Z}} T_k$ converges in the strong operator norm topology to an operator S satisfying $\|S\|_{\mathscr{L}(\mathcal{H})} \leq C$.

Note that (A.8) is equivalent to

$$\forall j \neq k \qquad (\ker T_j)^\perp \perp (\ker T_k)^\perp \quad \text{and} \quad \operatorname{Im} T_j \perp \operatorname{Im} T_k.$$

Proof. Let $v \in \mathcal{H}$ and $N \in \mathbb{N}$. Since the images of the T_j's are orthogonal, the Pythagoras equality implies

$$\|\sum_{|j| \leq N} T_j v\|^2 = \sum_{|j| \leq N} \|T_j v\|^2.$$

Denoting by P_j the orthogonal projection onto $(\ker T_j)^\perp$, we have

$$\|T_j v\| = \|T_j P_j v\| \leq C \|P_j v\|,$$

since $\|T_j\|_{\mathscr{L}(\mathcal{H})} \leq C$. Thus

$$\|\sum_{|j| \leq N} T_j v\|^2 \leq C^2 \sum_{|j| \leq N} \|P_j v\|^2.$$

As the kernels of the T_j's are mutually orthogonal, we have

$$\sum_{|j| \leq N} \|P_j v\|^2 \leq \|v\|^2.$$

We have obtained that

$$\|\sum_{|j| \leq N} T_j v\|^2 \leq C^2 \|v\|^2,$$

for any $N \in \mathbb{N}$ and $v \in \mathcal{H}$. The constant C here is the uniform bound of the operator norms of the T_j's and is independent of v or N. The same proof shows that the sequence $(\sum_{|j| \leq N} T_j v)_{N \in \mathbb{N}}$ is Cauchy when v is in a finite number of $(\ker T_j)^\perp$. This allows us to define the operator S on the dense subspace $\sum_j (\ker T_j)^\perp$. The conclusion follows. □

In practice, the orthogonality assumption above is rather demanding, and is often substituted by a condition of 'almost' orthogonality:

Theorem A.5.2 (Cotlar-Stein lemma). *Let $\mathcal{H}$ be a Hilbert space and $\{T_k,\ k \in \mathbb{Z}\}$ be a sequence of linear operators on $\mathcal{H}$. We assume that we are given a sequence of positive constants $\{\gamma_j\}_{j=-\infty}^{\infty}$ with*

$$A = \sum_{j=-\infty}^{\infty} \gamma_j < \infty.$$

If for any $i, j \in \mathbb{Z}$,

$$\max\left(\|T_i^* T_j\|_{\mathscr{L}(\mathcal{H})}, \|T_i T_j^*\|_{\mathscr{L}(\mathcal{H})}\right) \leq \gamma_{i-j}^2,$$

then the series $\sum_{k\in\mathbb{Z}} T_k$ converges in the strong operator topology to an operator S satisfying $\|S\|_{\mathscr{L}(\mathcal{H})} \leq A$.

For the proof of the Cotlar-Stein lemma, see e.g. [Ste93, Ch. VII §2], and for its history see Knapp and Stein [KS69].

When working on groups, one sometimes has to deal with operators mapping the L^2-space on the group to the L^2-space on its unitary dual. This requires one to use the version of Cotlar's lemma for operators mapping between two different Hilbert spaces. In this case, the statement of Theorem A.5.2 still holds, for an operator $T : \mathcal{H} \to \mathcal{G}$, provided we take the operator norms $T_i^* T_j$ and $T_i T_j^*$ in appropriate spaces. For details, we refer to [RT10a, Theorem 4.14.1].

The following crude version of the Cotlar lemma will be also useful to us:

Proposition A.5.3 (Cotlar-Stein lemma; crude version). *Let $\mathcal{H}$ be a Hilbert space and $\{T_k,\ k \in \mathbb{Z}\}$ be a sequence of linear operators on $\mathcal{H}$. We assume that*

$$T_i T_j^* = 0 \qquad \text{if } i \neq j. \tag{A.9}$$

We also assume that the operators T_k, $k \in \mathbb{Z}$, are uniformly bounded,

$$\text{i.e.} \quad \sup_{k\in\mathbb{Z}} \|T_k\|_{\mathscr{L}(\mathcal{H})} < \infty, \tag{A.10}$$

and that the following sum is finite

$$\sum_{i\neq j} \|T_i^* T_j\|_{\mathscr{L}(\mathcal{H})} < \infty. \tag{A.11}$$

Then the series $\sum_{k\in\mathbb{Z}} T_k$ converges in the strong operator topology to an operator S satisfying

$$\|S\|_{\mathscr{L}(\mathcal{H})}^2 \leq 2\max\left(\sup_{k\in\mathbb{Z}} \|T_k\|_{\mathscr{L}(\mathcal{H})}^2, \sum_{i\neq j} \|T_i^* T_j\|_{\mathscr{L}(\mathcal{H})}\right).$$

For the proof of this statement, see [Ste93, Ch. VII §2.3].

Remark A.5.4. The condition (A.9) can can be relaxed slightly with the following modifications.

For instance, (A.9) can be replaced with

$$T_i^* T_j = 0 \quad \text{if} \quad i \neq j \text{ have the same parity.}$$

(This condition appears often when considering dyadic decomposition.) Indeed, applying Proposition A.5.3 to $\{T_{2k+1}\}_{k\in\mathbb{Z}}$ and to $\{T_{2k}\}_{k\in\mathbb{Z}}$, we obtain that the series $\sum_k T_k = \sum_k T_{2k} + \sum_k T_{2k+1}$ converges in the strong operator norm topology to an operator S satisfying

$$\|S\|_{\mathscr{L}(\mathcal{H})} \leq 2^{1/2} \times 2 \times \max\left(\sup_{k\in\mathbb{Z}} \|T_k\|_{\mathscr{L}(\mathcal{H})}, \big(2 \sum_{i-j\in 2\mathbb{N}} \|T_i^* T_j\|_{\mathscr{L}(\mathcal{H})}\big)^{1/2}\right).$$

More generally, (A.9) can be replaced with

$$T_i^* T_j = 0 \quad \text{for} \quad |i-j| > a,$$

where $a \in \mathbb{N}$ is a fixed positive integer. It suffices to apply Proposition A.5.3 to each $\{T_{ak+b}\}_{k\in\mathbb{Z}}$ for $b = 0, \dots, a-1$. Then the series $\sum T_k = \sum_{0\leq b<a} T_{ak+b}$ converges in the strong operator norm topology to an operator S satisfying

$$\|S\|_{\mathscr{L}(\mathcal{H})} \leq 2^{1/2} \times a \times \max\left(\sup_{k} \|T_k\|_{\mathscr{L}(\mathcal{H})}, \big(2 \sum_{i-j>a} \|T_i^* T_j\|_{\mathscr{L}(\mathcal{H})}\big)^{1/2}\right).$$

A.6 Interpolation of analytic families of operators

Let $(M, \mathcal{M}, \mu)$ and $(N, \mathcal{N}, \nu)$ be measure spaces. We suppose that to each $z \in \mathbb{C}$ in the strip

$$S := \{z \in \mathbb{C} \ : \ 0 \leq \operatorname{Re} z \leq 1\},$$

there corresponds a linear operator T_z from the space of simple functions in $L^1(M)$ to measurable functions on N, in such a way that $(T_z f)g$ is integrable on N whenever f is a simple function in $L^1(M)$ and g is a simple function in $L^1(N)$. (Recall that a simple function is a measurable function which takes only a finite number of values.)

We assume that the family $\{T_z\}_{z\in S}$ is admissible in the sense that the mapping

$$z \mapsto \int_N (T_z f) g \, d\nu$$

is analytic in the interior of S, continuous on S, and there exists a constant $a < \pi$ such that

$$e^{-a|\operatorname{Im} z|} \ln \left| \int_N (T_z f) g \, d\nu \right|,$$

is uniformly bounded from above in the strip S.

Theorem A.6.1. *Let $\{T_z\}_{z\in S}$ be an admissible family as above. We assume that*

$$\|T_{iy}f\|_{q_0} \leq M_0(y)\|f\|_{p_0} \quad \text{and} \quad \|T_{1+iy}f\|_{q_1} \leq M_1(y)\|f\|_{p_1},$$

for all simple functions in $L^1(M)$ where $1 \leq p_j, q_j \leq \infty$, and functions $M_j(y)$, $j = 1, 2$ are independent of f and satisfy

$$\sup_{y\in\mathbb{R}} e^{-b|y|} \ln M_j(y) < \infty,$$

for some $b < \pi$. Then if $0 \leq t \leq 1$, there exists a constant M_t such that

$$\|T_t f\|_{q_t} \leq M_t \|f\|_{p_t},$$

for all simple functions f in $L^1(M)$, provided that

$$\frac{1}{p_t} = (1-t)\frac{1}{p_0} + t\frac{1}{p_1} \quad \text{and} \quad \frac{1}{q_t} = (1-t)\frac{1}{q_0} + t\frac{1}{q_1}.$$

For the proof of this theorem, we refer e.g. to [SW71, ch. V §4].

Remark A.6.2. The following remarks are useful.

- The constant M_t depends only on t and on $a, b, M_0(y), M_1(y)$, but not on T.
- From the proof, it appears that, if $N = M = \mathbb{R}^n$ is endowed with the usual Borel structure and the Lebesgue measures, one can require the assumptions and the conclusion to be on simple functions f with compact support.

We also refer to Definition 6.4.17 for the notion of the complex interpolation (which requires stronger estimates).

Appendix B
Group C^* and von Neumann algebras

In this chapter we make a short review of the machinery related to group von Neumann algebras that will be useful for setting up the Fourier analysis in other parts of book, in particular in Section 1.8.2. We try to make a short and concise presentation of notions and ideas without proofs trying to make the presentation as informal as possible. All the material presented in this chapter is well known but is often scattered over the literature in different languages and with different notation. Here we collect what is necessary for us giving references along the exposition. The final aim of this chapter is to introduce the notion of the von Neumann algebra of the group (or the group von Neumann algebra) and describe its main properties.

B.1 Direct integral of Hilbert spaces

We start by describing direct integrals of Hilbert spaces. For more details and overall proofs we can refer to more classical literature such as Bruhat [Bru68] or to more modern exposition of Folland [Fol95, p. 219].

B.1.1 Convention: Hilbert spaces are assumed separable

All the Hilbert spaces considered in this chapter are separable, unless stated otherwise. Let us recall the definition and some properties of separable spaces.

Definition B.1.1. A topological space is *separable* if its topology admits a countable basis of neighbourhoods.

When a topological space is metrisable, being separable is equivalent to having a (countable) sequence which is dense in the space.

Moreover, a separable Hilbert space of infinite dimension is unitarily equivalent to the Hilbert space of square integrable complex sequences: that is, to

$$\ell^2(\mathbb{N}_0) = \{(x_j)_{j\in\mathbb{N}_0}, \sum_{j=0}^{\infty} |x_j|^2 < \infty\}.$$

Naturally a separable Hilbert space of finite dimension n is unitarily equivalent to $\mathbb{C}^n$.

We can refer e.g. to Rudin [Rud91] for different topological implications of the separability.

B.1.2 Measurable fields of vectors

Here we recall the definitions of measurable fields of Hilbert spaces, of vectors and of operators.

Definition B.1.2. Let Z be a set and let $(\mathcal{H}_\zeta)_{\zeta\in Z}$ is a family of vector spaces (on the same field) indexed by Z. Then $\prod_{\zeta\in Z}\mathcal{H}_\zeta$ denotes the *direct product* of $(\mathcal{H}_\zeta)_{\zeta\in Z}$, that is, the set of all tuples $v = (v(\zeta))_{\zeta\in Z}$ with $v(\zeta) \in \mathcal{H}_\zeta$ for each $\zeta \in Z$. It is naturally endowed with a structure of a vector space with addition and scalar multiplication being performed componentwise.

An element of $\prod_{\zeta\in Z}\mathcal{H}_\zeta$, that is, a tuple $v = (v(\zeta))_{\zeta\in Z}$, may be called a field of vectors parametrised by Z, or, when no confusion is possible, a *vector field.*

We will use this definition for a measurable space Z. In practice, for the set Γ in the following definition, we may also choose $\Gamma \subset \prod_{\zeta\in Z}\mathcal{H}_\zeta^\infty$ in view of Gårding's theorem (see Proposition 1.7.7).

Definition B.1.3. Let Z be a measurable space and μ a positive sigma-finite measure on Z. A *μ-measurable field of Hilbert spaces* over Z is a pair $\mathcal{E} = \big((\mathcal{H}_\zeta)_{\zeta\in Z}, \Gamma\big)$ where $(\mathcal{H}_\zeta)_{\zeta\in Z}$ is a family of (separable) Hilbert spaces indexed by Z and where $\Gamma \subset \prod_{\zeta\in Z}\mathcal{H}_\zeta$ satisfies the following conditions:

(i) Γ is a vector subspace of $\prod_{\zeta\in Z}\mathcal{H}_\zeta$;

(ii) there exists a sequence $(x_\ell)_{\ell\in\mathbb{N}}$ of elements of Γ such that for every $\zeta \in Z$, the sequence $(x_\ell(\zeta))_{\ell\in\mathbb{N}}$ spans $\mathcal{H}_\zeta$ (in the sense that the subspace formed by the finite linear combination of the $x_\ell(\zeta), \ell \in \mathbb{N}$, is dense in $\mathcal{H}_\zeta$);

(iii) for every $x \in \Gamma$, the function $\zeta \mapsto \|x(\zeta)\|_{\mathcal{H}_\zeta}$ is μ-measurable;

(iv) if $x \in \prod_{\zeta\in Z}\mathcal{H}_\zeta$ is such that for every $y \in \Gamma$, the function

$$Z \ni \zeta \mapsto (x(\zeta), y(\zeta))_{\mathcal{H}_\zeta}$$

is measurable, then $x \in \Gamma$.

Under these conditions, the elements of Γ are called the *measurable vector fields* of $\mathcal{E}$. We always identify two vector fields which are equal almost everywhere. This means that we identify two elements x and x' of Γ when, for every $y \in \Gamma$, the two mappings

$$Z \ni \zeta \mapsto (x(\zeta), y(\zeta))_{\mathcal{H}_\zeta} \quad \text{and} \quad Z \ni \zeta \mapsto (x'(\zeta), y(\zeta))_{\mathcal{H}_\zeta},$$

can be identified as measurable functions.

A vector field x is *square integrable* if $x \in \Gamma$ and $\int_Z \|x(\zeta)\|^2_{\mathcal{H}_\zeta} d\mu(\zeta) < \infty$. One may write then

$$x = \int_Z^{\oplus} x(\zeta) d\mu(\zeta).$$

The set of square integrable vector fields form a (possibly non-separable) Hilbert space denoted by

$$\mathcal{H} := \int_Z^{\oplus} \mathcal{H}_\zeta d\mu(\zeta),$$

and called the *direct integral* of the $\mathcal{H}_\zeta$. The inner product is given via

$$(x|y)_{\mathcal{H}} = \int_Z^{\oplus} (x(\zeta)|y(\zeta)_{\mathcal{H}_\zeta} d\mu(\zeta), \quad x, y \in \mathcal{H}.$$

B.1.3 Direct integral of tensor products of Hilbert spaces

After a brief recollection of the definitions of tensor products, we will be able to analyse the direct integral of tensor products of Hilbert spaces, as well as their decomposable operators.

Definition of tensor products

Here we define firstly the algebraic tensor product of two vector spaces, and secondly the tensor products of Hilbert spaces.

Definition B.1.4. Let V and W be two complex vector spaces.

The free space generated by V and W is the vector space $\mathbb{F}(V \times W)$ linearly spanned by $V \times W$, that is, the space of finite $\mathbb{C}$-linear combinations of elements of $V \times W$.

The *algebraic tensor product* of V and W is the quotient of $\mathbb{F}(V \times W)$ by its subspace generated by the following elements

$$\begin{aligned} (v_1, w) + (v_2, w) - (v_1 + v_2, w), &\qquad (v, w_1) + (v, w_2) - (v, w_1 + w_2), \\ c(v, w) - (cv, w), &\qquad c(v, w) - (v, cw), \end{aligned}$$

where v, v_1, v_2 are arbitrary elements of V, w, w_1, w_2 are arbitrary elements of W, and c is an arbitrary complex number.

The equivalence class of an element $(v, w) \in V \times W \subset \mathbb{F}(V \times W)$ is denoted $v \otimes w$.

The algebraic tensor product of V and W is naturally a complex vector space which we will denote in this monograph by

$$V \overset{alg}{\otimes} W.$$

The algebraic tensor product has the following universal property (which may be given as an alternate definition):

Proposition B.1.5 (Universal property). *Let V, W and X be (complex) vector spaces and let $\Psi : V \times W \to X$ be a bilinear mapping. Then there exists a unique map $\tilde{\Psi} : V \overset{alg}{\otimes} W \to X$ such that*

$$\Psi = \tilde{\Psi} \circ \pi$$

where $\pi : V \times W \to V \overset{alg}{\otimes} W$ is the map defined by $\pi(v, w) = v \otimes w$.

More can be said when the complex vector spaces are also Hilbert spaces. Indeed one checks easily:

Lemma B.1.6. *Let $\mathcal{H}_1$ and $\mathcal{H}_2$ be Hilbert spaces. Then the mapping defined on $\mathcal{H}_1 \overset{alg}{\otimes} \mathcal{H}_2$ via*

$$(u_1 \otimes v_1, u_2 \otimes v_2) := (u_1, u_2)(v_1, v_2), \quad u_1, u_2 \in \mathcal{H}_1,\ v_1, v_2 \in \mathcal{H}_2,$$

is a complex inner product on $\mathcal{H}_1 \overset{alg}{\otimes} \mathcal{H}_2$.

This shows that $\mathcal{H}_1 \overset{alg}{\otimes} \mathcal{H}_2$ is a pre-Hilbert space.

Definition B.1.7. The *tensor product of the Hilbert spaces* $\mathcal{H}_1$ and $\mathcal{H}_2$ is the completion of $\mathcal{H}_1 \overset{alg}{\otimes} \mathcal{H}_2$ for the natural sesquilinear form from Lemma B.1.6. It is denoted by $\mathcal{H}_1 \otimes \mathcal{H}_2$.

Naturally we have the universal property of tensor products of Hilbert spaces:

Proposition B.1.8 (Universal property). *Let $\mathcal{H}_1$, $\mathcal{H}_2$ and $\mathcal{H}$ be Hilbert spaces and let $\Psi : \mathcal{H}_1 \times \mathcal{H}_2 \to \mathcal{H}$ be a continuous bilinear mapping. Then there exists a unique continuous map $\tilde{\Psi} : \mathcal{H}_1 \otimes \mathcal{H}_2 \to \mathcal{H}$ such that*

$$\Psi = \tilde{\Psi} \circ \pi$$

where $\pi : \mathcal{H}_1 \times \mathcal{H}_2 \to \mathcal{H}_1 \otimes \mathcal{H}_2$ is the map defined by $\pi(v, w) = v \otimes w$.

Tensor products of Hilbert spaces as Hilbert-Schmidt spaces

The tensor product of two Hilbert spaces may be identified with a space of Hilbert Schmidt operators in the following way. To any vector $w \in \mathcal{H}_2$, we associate the continuous linear form on $\mathcal{H}_2$

$$w^* : v \longmapsto (v, w)_{\mathcal{H}_2}.$$

Conversely any element of $\mathcal{H}_2^*$, that is, any continuous linear form on $\mathcal{H}_2$, is of this form. To any $u \in \mathcal{H}_1$ and $v \in \mathcal{H}_2$, we associate the rank-one operator

$$\Psi_{u,v} : \begin{cases} \mathcal{H}_2^* & \longrightarrow & \mathcal{H}_1 \\ w^* & \longmapsto & w^*(v)u \end{cases}$$

Lemma B.1.9. *With the notation above, the continuous bilinear mapping*

$$\Psi : \mathcal{H}_1 \times \mathcal{H}_2 \to \mathtt{HS}(\mathcal{H}_2^*, \mathcal{H}_1)$$

extends to an isometric isomorphism of Hilbert spaces

$$\tilde{\Psi} : \mathcal{H}_1 \otimes \mathcal{H}_2 \to \mathtt{HS}(\mathcal{H}_2^*, \mathcal{H}_1).$$

Moreover, if $T_1 \in \mathscr{L}(\mathcal{H}_1)$ *and* $T_2 \in \mathscr{L}(\mathcal{H}_2)$, *then the operator* $T_1 \otimes T_2$ *defined via*

$$(T_1 \otimes T_2)(v_1 \otimes v_2) := (T_1 v_1) \otimes (T_2 v_2), \quad v_1 \in \mathcal{H}_1, v_2 \in \mathcal{H}_2,$$

is in $\mathscr{L}(\mathcal{H}_1 \otimes \mathcal{H}_2)$ *and corresponds to the bounded operator*

$$\tilde{\Psi}(T_1 \otimes T_2)\tilde{\Psi}^{-1} : \begin{cases} \mathtt{HS}(\mathcal{H}_2^*, \mathcal{H}_1) & \longrightarrow & \mathtt{HS}(\mathcal{H}_2^*, \mathcal{H}_1) \\ A & \longmapsto & T_1 A T_2 \end{cases} .$$

Recall that the scalar product of $\mathtt{HS}(\mathcal{H}_2^*, \mathcal{H}_1)$ is given by

$$(T_1, T_2)_{\mathtt{HS}(\mathcal{H}_2^*, \mathcal{H}_1)} = \sum_j (T_1 f_j^*, T_2 f_j^*)_{\mathcal{H}_1}.$$

where $(f_j^*)_{j \in \mathbb{N}}$ is any orthonormal basis of $\mathcal{H}_2^*$.

Proof. By Proposition B.1.8, Ψ leads to a continuous linear mapping $\tilde{\Psi} : \mathcal{H}_1 \otimes \mathcal{H}_2 \to \mathtt{HS}(\mathcal{H}_2^*, \mathcal{H}_1)$. The image of $\tilde{\Psi}$ contains the rank-one operators, thus all the finite ranked operators which form a dense subset of $\mathtt{HS}(\mathcal{H}_2^*, \mathcal{H}_1)$. Thus $\tilde{\Psi}$ is surjective.

If $(f_j^*)_{j \in \mathbb{N}}$ is an orthonormal basis of $\mathcal{H}_2^*$, we can compute easily the scalar product between Ψ_{u_1,v_1} and Ψ_{u_2,v_2}:

$$\begin{aligned}
(\Psi_{u_1,v_1}, \Psi_{u_2,v_2})_{\mathtt{HS}(\mathcal{H}_2^*, \mathcal{H}_1)} &= \sum_j (\Psi_{u_1,v_1} f_j^*, \Psi_{u_2,v_2} f_j^*)_{\mathcal{H}_1} \\
&= \sum_j (f_j^*(v_1)u_1, f_j^*(v_2)u_2)_{\mathcal{H}_1} = (u_1, u_2)_{\mathcal{H}_1} \sum_j f_j^*(v_1)\overline{f_j^*(v_2)} \\
&= (u_1, u_2)_{\mathcal{H}_1} \sum_j (v_1, f_j)\overline{(v_2, f_j)} = (u_1, u_2)_{\mathcal{H}_1}(v_1, v_2)_{\mathcal{H}_2}.
\end{aligned}$$

This implies that the mapping $\tilde{\Psi} : \mathcal{H}_1 \otimes \mathcal{H}_2 \to \mathtt{HS}(\mathcal{H}_2^*, \mathcal{H}_1)$ is an isometry.

For the last part of the statement, one checks easily that

$$(T_1 \Psi_{u,v} T_2)(w^*) = w^*(T_2 v)\, T_1 u,$$

concluding the proof. □

Let us apply this to $\mathcal{H}_1 = \mathcal{H}$ and $\mathcal{H}_2 = \mathcal{H}^*$.

Corollary B.1.10. *Let $\mathcal{H}$ be a Hilbert space. The Hilbert space given by the tensor product $\mathcal{H} \otimes \mathcal{H}^*$ of Hilbert spaces is isomorphic to* $\mathtt{HS}(\mathcal{H})$ *via*

$$u \otimes v^* \longleftrightarrow \Psi_{u,v}, \quad \Psi_{u,v}(w) = (w, v)_{\mathcal{H}} u.$$

Via this isomorphism, the bounded operator $T_1 \otimes T_2^*$ where $T_1, T_2 \in \mathscr{L}(\mathcal{H})$, corresponds to the bounded operator

$$\tilde{\Psi}(T_1 \otimes T_2)\tilde{\Psi}^{-1} : \left\{ \begin{array}{rcl} \mathtt{HS}(\mathcal{H}) & \longrightarrow & \mathtt{HS}(\mathcal{H}) \\ A & \longmapsto & T_1 A T_2^* \end{array} \right. .$$

Direct integral of tensor products of Hilbert spaces

Let μ be a positive sigma-finite measure on a measurable space Z and $\mathcal{E} = \big((\mathcal{H}_\zeta)_{\zeta \in Z}, \Gamma\big)$ a *μ-measurable field of Hilbert spaces* over Z. Then

$$\mathcal{E}^{\otimes} := \big((\mathcal{H}_\zeta \otimes \mathcal{H}_\zeta^*)_{\zeta \in Z}, \Gamma \otimes \Gamma^*\big)$$

is a *μ-measurable field of Hilbert spaces* over Z.

Identifying each tensor product $\mathcal{H}_\zeta \otimes \mathcal{H}_\zeta^*$ with $\mathtt{HS}(\mathcal{H}_\zeta)$, see Corollary B.1.10, we may write

$$\int_Z^{\oplus} \mathcal{H}_\zeta \otimes \mathcal{H}_\zeta^* d\mu(\zeta) \equiv \int_Z^{\oplus} \mathtt{HS}(\mathcal{H}_\zeta) d\mu(\zeta).$$

Furthermore if $x \in \int_Z^{\oplus} \mathcal{H}_\zeta \otimes \mathcal{H}_\zeta^* d\mu(\zeta)$ then

$$\|x\|^2 = \int_Z \|x(\zeta)\|^2_{\mathtt{HS}(\mathcal{H}_\zeta)} d\mu(\zeta).$$

B.1.4 Separability of a direct integral of Hilbert spaces

In this chapter, we are always concerned with separable Hilbert spaces (see Section B.1.1). A sufficient condition to ensure the separability of a direct integral is that the measured space is standard (the definition of this notion is recalled below):

Proposition B.1.11. *Keeping the setting of Definition B.1.3, if (Z, μ) is a standard space, then $\int_Z^{\oplus} \mathcal{H}_\zeta d\mu(\zeta)$ is a separable Hilbert space.*

For the proof we refer to Dixmier [Dix96, §II.1.6].

Definition B.1.12. A measurable space Z is a *standard Borel space* if Z is a Polish space (i.e. a separable complete metrisable topological space) and the considered sigma-algebra is the Borel sigma-algebra of Z (i.e. the smallest sigma-algebra containing the open sets of Z).

These Borel spaces have a simple classification: they are isomorphic (as Borel spaces) either to a (finite or infinite) countable set, or to $[0,1]$. For these and other details see, for instance, Kechris [Kec95, Chapter II, Theorem 15.6] and its proof.

Definition B.1.13. A positive measure μ on a measure space Z is a *standard measure* if μ is sigma-finite, (i.e. there exists a sequence of mutually disjoint measurable sets $Y_1, Y_2, \ldots$ such that $\mu(Y_j) < \infty$ and $Z = Y_1 \cup Y_2 \cup \ldots$) and there exists a null set E such that $Z \backslash E$ is a standard Borel space.

In this monograph, we consider only the setting described in Proposition B.2.24 which is standard.

B.1.5 Measurable fields of operators

Let Z be a measurable space and μ a positive sigma-finite measure on Z. The main application for our analysis of these constructions will be in Section 1.8.3 dealing with measurable fields of operators over $\widehat{G}$.

Definition B.1.14. Let $\mathcal{E} = \big((\mathcal{H}_\zeta)_{\zeta\in Z}, \Gamma\big)$ be a μ-measurable field of Hilbert spaces over Z. A *μ-measurable field of operators* over Z is a collection of operators $(T(\zeta))_{\zeta\in Z}$ such that $T(\zeta) \in \mathscr{L}(\mathcal{H}_\zeta)$ and for any $x \in \Gamma$, the field $(T(\zeta)x(\zeta))_{\zeta\in Z}$ is measurable. If furthermore the function $\zeta \mapsto \|T(\zeta)\|_{\mathscr{L}(\mathcal{H}_\zeta)}$ is μ-essentially bounded, then the field of operators $(T(\zeta))_{\zeta\in Z}$ is *essentially bounded.*

Let us continue with the notation of Definition B.1.14. Let $(T(\zeta))_{\zeta\in Z}$ be an essentially bounded field of operators. Then we can define the operator T on the Hilbert space $\mathcal{H} = \int_Z^\oplus \mathcal{H}_\zeta d\mu(\zeta)$ via $(Tx)(\zeta) := T(\zeta)x(\zeta)$. Clearly the operator T is linear and bounded. It is often denoted by

$$T := \int_Z^\oplus T(\zeta) d\mu(\zeta).$$

Naturally two fields of operators which are equal up to a μ-negligible set yield the same operator on $\mathcal{H}$ and may be identified. Furthermore the operator norm of $T \in \mathscr{L}(\mathcal{H})$ is

$$\|T\|_{\mathscr{L}(\mathcal{H})} = \sup_{\zeta\in Z} \|T(\zeta)\|_{\mathscr{L}(\mathcal{H}_\zeta)},$$

where sup denotes here the essential supremum with respect to μ.

Definition B.1.15. An operator on $\mathcal{H}$ as above, that is, obtained via

$$T := \int_Z^\oplus T(\zeta) d\mu(\zeta)$$

where $(T(\zeta))_{\zeta\in Z}$ is an essentially bounded field of operators, is said to be *decomposable.*

The set of decomposable operators form a subspace of $\mathscr{L}(\mathcal{H})$ stable by composition and taking the adjoint.

B.1.6 Integral of representations

In the following definition, μ is a positive sigma-finite measure on a measurable space Z, $\mathcal{A}$ is a separable C^*-algebra, and G is a (Hausdorff) locally compact separable group. For further details on the constructions of this section we refer to Dixmier [Dix77, §8]. For the definition of representations of C^*-algebras see Definition B.2.16.

Definition B.1.16. Let $\mathcal{E} = \big((\mathcal{H}_\zeta)_{\zeta\in Z}, \Gamma\big)$ be a μ-measurable field of Hilbert spaces over Z. A *μ-measurable field of representations* of $\mathcal{A}$, resp. G, is a μ-measurable field of operator $(T(\zeta))_{\zeta\in Z}$ (see Definition B.1.14) such that for each $\zeta \in Z$, $T(\zeta) = \pi_\zeta$ is a representation of $\mathcal{A}$, resp. a unitary continuous representation of G, in $\mathcal{H}_\zeta$.

In this case, for each $x \in G$, we can define the operator

$$\pi(x) := \int_Z^\oplus \pi_\zeta(x) d\mu(\zeta) \quad \text{acting on } \mathcal{H} := \int_Z^\oplus \mathcal{H}_\zeta d\mu(\zeta).$$

One checks easily that this yields a representation π of $\mathcal{A}$, resp. a unitary continuous representation of G, on $\mathcal{H}$ denoted by

$$\pi := \int_Z^\oplus \pi_\zeta d\mu(\zeta),$$

often called *the integral of the representations* $(\pi_\zeta)_{\zeta\in Z}$.

The following technical properties give sufficient conditions for two integrals of representations to yield equivalent representations. Again $\mathcal{A}$ is a separable C^*-algebra and G a (Hausdorff) locally compact separable group.

Proposition B.1.17. *Let μ_1 and μ_2 be two positive sigma-finite measures on measurable spaces Z_1 and Z_2 respectively. For $j = 1, 2$, let $\mathcal{E}_j = \big((\mathcal{H}^{(j)}_{\zeta_j})_{\zeta_j\in Z_j}, \Gamma_j\big)$ be a μ_j-measurable field of Hilbert spaces over Z_j and let $(\pi^{(j)}_{\zeta_j})$ be a measurable field of representations of $\mathcal{A}$, resp. of unitary continuous representations of G.*

We assume that μ_1 and μ_2 are standard. We also assume that there exist a Borel μ_1-negligible part $E_1 \subset Z_1$, a Borel μ_2-negligible part $E_2 \subset Z_2$ and a Borel isomorphism $\eta : Z_1\backslash E_1 \to Z_2\backslash E_2$ which transforms μ_1 to μ_2 and such that $\pi^{(1)}_{\zeta_1}$ and $\pi^{(2)}_{\eta(\zeta_1)}$ are equivalent for any $\zeta_1 \in Z_1\backslash E_1$. Then there exists a unitary mapping from $\mathcal{H}^{(1)} := \int^\oplus_{Z_1} \mathcal{H}^{(1)}_{\zeta_1} d\mu_1(\zeta_1)$ onto $\mathcal{H}^{(2)} := \int^\oplus_{Z_2} \mathcal{H}^{(2)}_{\zeta_2} d\mu_2(\zeta_2)$ which intertwines the representations of $\mathcal{A}$, resp. the unitary continuous representations of G,

$$\pi^{(1)} := \int_{Z_1}^\oplus \pi^{(1)}_{\zeta_1} d\mu_1(\zeta_1) \quad \textit{and} \quad \pi^{(2)} := \int_{Z_2}^\oplus \pi^{(2)}_{\zeta_2} d\mu_2(\zeta_2).$$

B.2 C^*- and von Neumann algebras

The main reference for this section are Dixmier's books [Dix81, Dix77], Arveson [Arv76] or Blackadar [Bla06]. For a more basic introduction to C^*-algebras and elements of the Gelfand theory see also Ruzhansky and Turunen [RT10a, Chapter D].

B.2.1 Generalities on algebras

Here we recall the definitions of an algebra, together with its possible additional structures (involution, norm) and sets usually associated with it (spectrum, bi-commutant).

Algebra

Let us start with the definition of an algebra over a field.

Let $\mathcal{A}$ be a vector space over a field $\mathbb{K}$ equipped with an additional binary operation

$$\begin{array}{rcl} \mathcal{A}\times\mathcal{A} & \longrightarrow & \mathcal{A}, \\ (x,y) & \longmapsto & x\cdot y. \end{array}$$

It is an *algebra* over $\mathbb{K}$ when the binary operation (then often called the product) satisfies:

- left distributivity: $(x+y)\cdot z = x\cdot z + y\cdot z$ for any $x,y,z\in\mathcal{A}$,
- right distributivity: $z\cdot(x+y) = z\cdot x + z\cdot y$ for any $x,y,z\in\mathcal{A}$,
- compatibility with scalars: $(ax)\cdot(by) = (ab)(x\cdot y)$ for any $x,y\in\mathcal{A}$ and $a,b\in\mathbb{K}$.

The algebra $\mathcal{A}$ is said to be *unital* when there exists a unit, that is, an element $1\in\mathcal{A}$ such that $x\cdot 1 = 1\cdot x = x$ for every $x\in\mathcal{A}$.

A subspace $\mathcal{Y}\subset\mathcal{A}$ is a *sub-algebra* of $\mathcal{A}$ whenever $y_1\cdot y_2\in\mathcal{Y}$ for any $y_1,y_2\in\mathcal{Y}$.

Commutant and bi-commutant

We will need the notion of commutant:

Definition B.2.1. Let $\mathcal{M}$ be a subset of the algebra $\mathcal{A}$. The *commutant* of $\mathcal{M}$ is the set denoted by $\mathcal{M}'$ of the elements which commute with all the elements of $\mathcal{M}$, that is,

$$\mathcal{M}' := \{x\in\mathcal{A}\ :\ xm = mx \ \text{ forall } \ m\in\mathcal{M}\}.$$

The *bi-commutant* of $\mathcal{M}$ is the commutant of the commutant of $\mathcal{M}$, that is,

$$\mathcal{M}'' := (\mathcal{M}')'.$$

Keeping the notation of Definition B.2.1, one checks easily that a commutant $\mathcal{M}'$ is a sub-algebra of $\mathcal{A}$. It contains the unit if $\mathcal{A}$ is unital. Furthermore, in any case, $\mathcal{M} \subset \mathcal{M}''$.

Involution and norms

We consider now algebras endowed with an involution:

Definition B.2.2. Let $\mathcal{A}$ be an algebra over the complex numbers $\mathbb{C}$. It is called an *involutive algebra* or a $*$*-algebra* when there exists a map $* : \mathcal{A} \to \mathcal{A}$ which is

- sesquilinear (that is, $(ax + by)^* = \bar{a}x^* + \bar{b}y^*$ for every $x, y \in \mathcal{A}$ and $a, b \in \mathbb{C}$),
- involutive (that is, $(x^*)^* = x$ for every $x \in \mathcal{A}$).

In this case, x^* may be called the *adjoint* of $x \in \mathcal{A}$. An element $x \in \mathcal{A}$ is *hermitian* if $x^* = x$. An element $x \in \mathcal{A}$ is *unitary* if $xx^* = x^*x = 1$.

Example B.2.3. Let $\mathcal{A}$ be a $*$-algebra. If $\mathcal{M}$ is a subset of $\mathcal{A}$ stable under the involution (that is, $m^* \in \mathcal{M}$ for every $m \in \mathcal{M}$), then its commutant $\mathcal{M}'$ is a $*$-subalgebra of $\mathcal{A}$.

Definition B.2.4. A *normed involutive algebra* is an involutive algebra $\mathcal{A}$ endowed with a norm $\|\cdot\|$ such that

$$\|x^*\| = \|x\|$$

for each $x \in \mathcal{A}$. If, in addition, $\mathcal{A}$ is $\|\cdot\|$-complete, then $\mathcal{A}$ is called an *involutive Banach algebra.*

The notions of (involutive, normed involutive / involutive Banach) sub-algebra and morphism between (involutive / normed involutive / involutive Banach) algebras follow naturally. Furthermore if $\mathcal{A}$ is a (involutive / normed involutive / involutive Banach) non unital algebra, then there exists a unique (involutive / normed involutive / involutive Banach) unital algebra $\tilde{\mathcal{A}} = \mathcal{A} \oplus \mathbb{C}1$, up to isomorphism, which contains $\mathcal{A}$ as a (involutive, normed involutive / involutive Banach) sub-algebra.

Examples

Example B.2.5. The complex field $\mathcal{A} = \mathbb{C}$ is naturally a unital commutative involutive Banach algebra.

Example B.2.6. Let X be a locally compact space and let $\mathcal{A} = C_o(X)$ be the space of continuous functions $f : X \to \mathbb{C}$ vanishing at infinity, that is, for every $\epsilon > 0$, there exists a compact neighbourhood out of which $|f| < \epsilon$. Then $\mathcal{A}$ is a commutative involutive Banach algebra when endowed with pointwise multiplication and involution $f \mapsto \bar{f}$. When X is a singleton, this reduces to Example B.2.5.

Example B.2.7. If η is a positive measure on a measurable space X and if $\mathcal{A}$ is the space of η-essentially bounded functions $f : X \to \mathbb{C}$, that is, $\mathcal{A} = L^\infty(X, \eta)$, then $\mathcal{A}$ is a unital commutative involutive Banach algebra when endowed with pointwise multiplication and involution $f \mapsto \bar{f}$. When X is a singleton, this reduces to Example B.2.5.

Recall that all the Hilbert spaces we consider are separable.

Example B.2.8. The space $\mathscr{L}(\mathcal{H})$ of continuous linear operators on a Hilbert space $\mathcal{H}$ is naturally a unital involutive Banach algebra for the usual structure. This means that the product is given by the composition of operators $(A, B) \mapsto AB$, the involution by the adjoint and the norm by the operator norm. The unit is the identity mapping $\mathrm{I}_{\mathcal{H}} = \mathrm{I} : v \mapsto v$.

Example B.2.9. If G is a locally compact (Hausdorff) group which is unimodular, then $L^1(G)$ is naturally an involutive Banach algebra where the product is given by the convolution and the involution $f \mapsto f^*$ by $f^*(x) = \bar{f}(x^{-1})$. If G is separable then $L^1(G)$ is separable.

Example B.2.9 can be generalised to locally compact groups which are not necessarily unimodular. First, let us recall the following definitions:

Definition B.2.10. Let G be a locally compact (Hausdorff) group. Let us fix a left Haar measure dx. We also denote by $|E|$ the volume of a Borel set for this measure. Then there exists a unique function Δ such that

$$|Ex| = \Delta(x)|E|$$

for any Borel set E and $x \in G$. It is called the *modular* function of G and is independent of the chosen left Haar measure. It is a group homomorphism $G \to (\mathbb{R}^+, \times)$.

If the modular function is constant then $\Delta \equiv 1$ and G is said to be *unimodular.*

Remark B.2.11. Any Lie group is a separable locally compact (Hausdorff) group. Any compact (Hausdorff) group is necessarily a locally compact (Hausdorff) group and it is also unimodular. Any abelian locally compact (Hausdorff) group is unimodular. Any nilpotent or semi-simple Lie group is unimodular.

Example B.2.12. If G is a locally compact (Hausdorff) group then $L^1(G)$ is naturally an involutive Banach algebra often called the *group algebra.* The product is given by the convolution and the involution $f \mapsto f^*$ by

$$f^*(x) = \bar{f}(x^{-1})\Delta(x)^{-1},$$

where Δ is the modular function (see Definition B.2.10).

The space $M(G)$ of complex measures on G is also naturally an involutive Banach algebra and $L^1(G)$ may be viewed as a closed involutive sub-algebra. The algebra $M(G)$ always admits the Dirac measure δ_e at the neutral element of the group as unit.

Note that $L^1(G)$ is unital if and only if G is discrete and in this case $L^1(G) = M(G)$.

B.2.2 C^*-algebras

In this subsection we briefly review the notion of C^*-algebra and its main properties. We can refer to Ruzhansky and Turunen [RT10a, Chapter D] for a longer exposition.

Definition B.2.13. A *C^*-algebra* is an involutive Banach algebra $\mathcal{A}$ such that

$$\|x\|^2 = \|x^* x\|$$

for every $x \in \mathcal{A}$.

Example B.2.14. Examples B.2.5, B.2.6, B.2.7, and B.2.8 are C^*-algebras.

Remark B.2.15.

1. If we choose a Hilbert space $\mathcal{H}$ of finite dimension n in Example B.2.8, the Banach algebra $\mathscr{L}(\mathcal{H}) \sim \mathscr{L}(\mathbb{C}^n) \sim \mathbb{C}^{n\times n}$ is a C^*-algebra if endowed with the operator norm, but is not a C^*-algebra when equipped with the Euclidean norm of $\mathbb{C}^{n^2}$ for instance.

2. Example B.2.6 is fundamental in the sense that one can show that any commutative C^*-algebra $\mathcal{A}$ is isomorphic to $C_o(X)$, where X is the spectrum of $\mathcal{A}$, that is, the set of non-zero complex homomorphisms with its usual topology. Moreover the isomorphism often called the Gelfand-Fourier transform is $*$-isometric. For further details see e.g. Rudin [Rud91] but with a different vocabulary.

3. In the non-commutative setting, the previous point may be generalised via the Gelfand-Naimark theorem: this theorem states that any C^*-algebra is $*$-isometric to a closed sub-$*$-algebra of $\mathscr{L}(\mathcal{H})$ for a suitable Hilbert space $\mathcal{H}$. Note that Example B.2.8 give the precise structure of $\mathscr{L}(\mathcal{H})$ and shows that a closed sub-$*$-algebra of $\mathscr{L}(\mathcal{H})$ is indeed a C^*-algebra. The proof is based on the Gelfand-Naimark-Segal construction, see e.g. Arveson [Arv76] for more precise statements.

The general definition of the spectrum of a (not necessarily commutative) C^* algebra is more involved than in the commutative case (Remark B.2.15 (2)):

Definition B.2.16 (Representations of C^*-algebras). Let $\mathcal{A}$ be a C^*-algebra.

A *representation* of $\mathcal{A}$ is a continuous mapping $\mathcal{A} \to \mathscr{L}(\mathcal{H})$ for some Hilbert space $\mathcal{H}$, this mapping being a homomorphism of involutive algebras. Two representations $\pi_j : \mathcal{A} \to \mathscr{L}(\mathcal{H}_j)$, $j = 1, 2$, of $\mathcal{A}$, are *unitarily equivalent* if there exists a unitary operator $U : \mathcal{H}_1 \to \mathcal{H}_2$ such that $U\pi_1(x) = \pi_2(x)U$ for every $x \in \mathcal{A}$. A representation $\pi : \mathcal{A} \to \mathscr{L}(\mathcal{H})$ is *irreducible* if the only subspaces of $\mathcal{H}$ which are invariant under π, that is, under every $\pi(x)$, $x \in \mathcal{A}$, are trivial: $\{0\}$ and $\mathcal{H}$.

The *dual* (or spectrum) of $\mathcal{A}$ is the set of unitary irreducible representations of $\mathcal{A}$ modulo unitary equivalence. It is denoted by $\widehat{\mathcal{A}}$.

Remark B.2.17. The dual of a C^*-algebra is equipped with the hull-kernel topology due to Jacobson, and, if it is separable, with a structure of measurable space due to Mackey, see Dixmier [Dix77, §3].

B.2.3 Group C^*-algebras

In general, the group algebra of a locally compact (Hausdorff) group G, that is, the involutive Banach algebra $L^1(G)$ in Example B.2.12, is not a C^* algebra (see Remark B.2.26 below). The group C^* algebra is the C^*-enveloping algebra of $L^1(G)$, meaning that it is a 'small' C^* algebra containing $L^1(G)$ and built in the following way.

First, let us mention that many authors, for instance Jacques Dixmier, prefer to use for the Fourier transform

$$\pi_{\mathscr{D}}(f) := \int_G \pi(x) f(x) dx, \quad f \in L^1(G), \tag{B.1}$$

instead of $\pi(f)$ defined via

$$\pi(f) = \int_G \pi(x)^* f(x) dx, \quad f \in L^1(G), \tag{B.2}$$

which we adopt in this monograph, starting from (1.2), see Remark 1.1.4 for the explanation of this choice.

An advantage of using $\pi_{\mathscr{D}}$ would be that it yields a morphism of involutive Banach algebras from $L^1(G)$ to $\mathscr{L}(\mathcal{H}_\pi)$ as one checks readily:

Lemma B.2.18. *Let π be a unitary continuous representation of G. Then $\pi_{\mathscr{D}}$ is a (non-degenerate) representation of the involutive Banach algebra $L^1(G)$:*

$$\forall f, g \in L^1(G) \qquad \pi_{\mathscr{D}}(f * g) = \pi_{\mathscr{D}}(f)\pi_{\mathscr{D}}(g), \quad \pi_{\mathscr{D}}(f)^* = \pi_{\mathscr{D}}(f^*),$$

and

$$\|\pi_{\mathscr{D}}(f)\|_{\mathscr{L}(\mathcal{H}_\pi)} \leq \|f\|_{L^1(G)}.$$

For the proof, see Dixmier [Dix77, Proposition 13.3.1].

The choice of the Fourier transform in (B.2) made throughout this monograph, yields in contrast

$$\forall f, g \in L^1(G) \qquad \pi(f * g) = \pi(g)\pi(f)$$

and still

$$\pi(f)^* = \pi(f^*), \quad \|\pi(f)\|_{\mathscr{L}(\mathcal{H}_\pi)} \leq \|f\|_{L^1(G)}.$$

The main advantage of our choice of Fourier transform is the fact that the Fourier transform of left-invariant operators will act on the left, as is customary in harmonic analysis, see our presentation of the abstract Plancherel theorem in Section 1.8.2.

Definition B.2.19. On $L^1(G)$, we can define $\|\cdot\|_*$ via

$$\|f\|_* := \sup_\pi \|\pi_{\mathcal{D}}(f)\|_{\mathscr{L}(\mathcal{H}_\pi)}, \quad f \in L^1(G),$$

where the supremum runs over all continuous unitary irreducible representations π of the group G.

One checks easily that $\|\cdot\|_*$ is a seminorm on $L^1(G)$ which satisfies

$$\|f\|_* \leq \|f\|_{L^1} < \infty.$$

One can show that it is in fact also a norm on $L^1(G)$, see Dixmier [Dix77, §13.9.1].

Definition B.2.20. The *group C^*-algebra* is the Banach space obtained by completion of $L^1(G)$ for the norm $\|\cdot\|_*$. It is often denoted by $C^*(G)$.

Remark B.2.21. Choosing the definition of $\|\cdot\|_*$ using $\pi_{\mathcal{D}}$ as above or using our usual Fourier transform leads to the same C^*-algebra of the group. Indeed one checks easily that the adjoint of the operator $\pi(f)$ acting on $\mathcal{H}_\pi$ is $\pi_{\mathscr{D}}(\bar{f})$:

$$\pi(f) = \pi_{\mathscr{D}}(\bar{f})^* = \pi_{\mathscr{D}}(\bar{f}^*) \quad \text{and} \quad \|\pi(f)\|_{\mathscr{L}(\mathcal{H}_\pi)} = \|\pi_{\mathscr{D}}(\bar{f})\|_{\mathscr{L}(\mathcal{H}_\pi)}, \tag{B.3}$$

for all $f \in L^1(G)$.

Naturally $C^*(G)$ is a C^*-algebra and there are natural one-to-one correspondences between the representation theories of the group G, of the involutive Banach algebra $L^1(G)$, and of the C^*-algebra $C^*(G)$ in the following sense:

Lemma B.2.22. *If π is a continuous unitary representation of G, then $f \mapsto \pi_{\mathscr{D}}(f)$ defined via (B.1) is a non-degenerate $*$-representation of $L^1(G)$ which extends naturally to $C^*(G)$. Conversely any non-degenerate $*$-representation of $L^1(G)$ or $C^*(G)$ arise in this way.*

Hence

$$\|f\|_* = \sup_\pi \|\pi(f)\|_{\mathscr{L}(\mathcal{H}_\pi)}, \quad f \in L^1(G),$$

where the supremum runs over all representations π of the involutive Banach algebra $L^1(G)$ or over all representations π of the C^*-algebra $C^*(G)$.

For the proof see Dixmier [Dix77, §13.3.5 and §13.9.1].

Definition B.2.23. The dual of the group G is the set $\widehat{G}$ of (continuous) irreducible unitary representations of G modulo equivalence, see (1.1).

Given the correspondence explained in Lemma B.2.22, $\widehat{G}$ can be identified with the dual of $C^*(G)$ and inherit the structure that may occur on $\widehat{C^*(G)}$, see Remark B.2.17.

In particular, $\widehat{G}$ inherits a topology, called the *Fell topology*, corresponding to the hull-kernel (Jacobson) topology on $C^*(G)$, see e.g. Folland [Fol95, §7.2], Dixmier [Dix77, §18.1 and §3]. If G is separable, then $C^*(G)$ is separable, see [Dix77, §13.9.2], and $\widehat{G}$ also inherits the Mackey structure of measurable space.

Proposition B.2.24. *Let G be a separable locally compact group of type I. Then its dual $\widehat{G}$ is a standard Borel space. Moreover the Mackey structure coincides with the sigma-algebra associated with the Fell topology.*

For the definition of groups of type I, see Dixmier [Dix77, §13.9.4] or Folland [Fol95, §7.2]. See also hypothesis (H) in Section 1.8.2 for a relevant discussion. For the definition of the Plancherel measure, see (1.28), as well as Dixmier [Dix77, Definition 8.8.3] or Folland [Fol95, §7.5].

References for the proof of Proposition B.2.24. As G is of type I and separable, its group C^*-algebra $C^*(G)$ is of type I, postliminar and separable, see Dixmier [Dix77, §13.9]. Hence the Mackey Borel structure on the spectrum of this C^*-algebra (cf. [Dix77, §3.8]) is a standard Borel space by Dixmier [Dix77, Proposition 4.6.1]. □

Reduced group C^*-algebra

Although we do not use the following in this monograph, let us mention that one can also define another 'small' C^* algebra which contains $L^1(G)$.

Let us recall that the left regular representation π_L is defined on the group via

$$\pi_L(x)\phi(y) := \phi(x^{-1}y), \quad x, y \in G, \ \phi \in L^2(G). \tag{B.4}$$

This leads to the representation of $L^1(G)$ given by

$$(\pi_L)_{\mathscr{D}}(f)\phi = \int_G f(x)\pi_L(x)\phi \, dx = \int_G f(x)\phi(x^{-1}\cdot) \, dx = f * \phi, \tag{B.5}$$

which may be extended onto the closure $\overline{(\pi_L)_{\mathcal{D}}(L^1(G))}$ of $(\pi_L)_{\mathcal{D}}(L^1(G))$ for the operator norm, see Lemma B.2.22. This closure is naturally a C^*-algebra, often called the *reduced* C^*-algebra of the group and denoted by $C_r^*(G)$. Equivalently, $C_r^*(G)$ may be realised as the closure of $L^1(G)$ for the norm given by

$$\|f\|_{C_r^*} = \|(\pi_L)_{\mathscr{D}}(f)\|_{\mathscr{L}(L^2(G))} = \{\|f * \phi\|_{L^2}, \phi \in L^2(G) \text{ with } \|\phi\|_{L^2} = 1\}.$$

The 'full' and reduced C^* algebras of a group may be different. When they are equal, that is, $C_r^*(G) = C^*(G)$, then the group G is said to be *amenable*. Amenability can be described in many other ways. The advantage of considering the 'full'

C^*-algebra of a group is the one-to-one correspondence between the representations theories of G, $L^1(G)$, and $C^*(G)$.

The groups considered in this monograph, that is, compact groups and nilpotent Lie groups, are amenable.

Pontryagin duality

Although we do not use it in this monograph, let us recall briefly the Pontryagin duality, as this may be viewed as one of the historical motivation to develop the theory of (noncommutative) C^*-algebras.

The case of a locally compact (Hausdorff) abelian ($\equiv$ commutative) group G is described by the Pontryagin duality, see Section 1.1. In this case, the group algebra $L^1(G)$ (see Example B.2.9) is an abelian involutive Banach algebra. Its spectrum $\widehat{G}$ may be identified with the set of the continuous characters of G and is naturally equipped with the structure of a locally compact (Hausdorff) abelian group. The group G is amenable, that is, the full and reduced group C^*-algebras coincide: $C^*(G) = C^*_r(G)$. Moreover, the Fourier-Gelfand transform (see Remark B.2.15 (2)) extends into an isometry of C^*-algebra from $C^*(G)$ onto $C_o(\widehat{G})$.

Example B.2.25. In the particular example of the abelian group $G = \mathbb{R}^n$, the dual $\widehat{G}$ may also be identified with $\mathbb{R}^n$ and the Fourier-Gelfand transform in this case is the (usual) Euclidean Fourier transform $\mathcal{F}_{\mathbb{R}^n}$.

The group C^*-algebra $C^*(\mathbb{R}^n) = C^*_r(\mathbb{R}^n)$ may be viewed as a subspace of $\mathcal{S}'(\mathbb{R}^n)$ which contains $L^1(\mathbb{R}^n)$. Recall that, by the Riemann-Lebesgue Theorem (see e.g. [RT10a, Theorem 1.1.8]), the Euclidean Fourier transform $\mathcal{F}_{\mathbb{R}^n}$ maps $L^1(\mathbb{R}^n)$ to $C_o(\mathbb{R}^n)$, and one can show that

$$C^*(\mathbb{R}^n) = \mathcal{F}^{-1}_{\mathbb{R}^n} C_o(\mathbb{R}^n).$$

Remark B.2.26. Note that the inclusion $\mathcal{F}_{\mathbb{R}^n}(L^1(\mathbb{R}^n)) \subset C_o(\mathbb{R}^n)$ is strict. Indeed for $n > 1$, the kernel of the Bochner Riesz means $\mathcal{F}^{-1}_{\mathbb{R}^n}\{\sqrt{1-|\xi|^2} 1_{|\xi|\leq 1}\}$ is not in $L^1(\mathbb{R}^n)$ but its Fourier transform is in $C_o(\mathbb{R}^n)$. For $n = 1$, see e.g. Stein and Weiss [SW71, Ch 1, §4.1].

B.2.4 Von Neumann algebras

Let us recall the von Neumann bi-commutant theorem:

Theorem B.2.27. *Let $\mathscr{L}(\mathcal{H})$ be the space of continuous linear operators on a Hilbert space $\mathcal{H}$ with its natural structure (see Example B.2.8). Let $\mathcal{M}$ be a $*$-subalgebra of $\mathscr{L}(\mathcal{H})$ containing the identity mapping* I. *Then the following are equivalent:*

(i) $\mathcal{M}$ is equal to its bi-commutant (in the sense of Definition B.2.1):

$$\mathcal{M} = \mathcal{M}''.$$

(ii) *$\mathcal{M}$ is closed in the weak-operator topology, i.e. the topology given by the family of seminorms $\{T \mapsto (Tv, w)_{\mathcal{H}},\ v, w \in \mathcal{H}\}$.*

(iii) *$\mathcal{M}$ is closed in the strong-operator topology, i.e. the topology on $\mathscr{L}(\mathcal{H})$ given by the family of seminorms $\{T \mapsto \|Tv\|_{\mathcal{H}},\ v \in \mathcal{H}\}$.*

This leads to the notion of a von Neumann algebra where we take the above equivalent properties as its definition:

Definition B.2.28. We keep the notation of Theorem B.2.27. A *von Neumann algebra in $\mathcal{H}$ is a $*$-subalgebra $\mathcal{M}$ of $\mathscr{L}(\mathcal{H})$ which satisfies any of the equivalent properties (i), (ii), or (iii) in Theorem B.2.27.*

Note that the operator-norm topology on $\mathscr{L}(\mathcal{H})$ is stronger than the strong-operator topology, which in turn is stronger than the weak-operator topology. Thus a von Neumann algebra in $\mathcal{H}$ is a $*$-subalgebra of $\mathscr{L}(\mathcal{H})$ closed for the operator-norm topology, hence is a C^*-subalgebra of $\mathscr{L}(\mathcal{H})$ and a C^*-algebra itself. Among C^*-algebras, the von Neumann algebras are the C^*-algebras which are realised as a closed $*$-subalgebra of $\mathscr{L}(\mathcal{H})$ and furthermore satisfy any of the equivalent properties (*i*), (*ii*), or (*iii*) in Theorem B.2.27.

It is also possible to define the von Neumann algebras abstractly as the C^*-algebras having a predual, see e.g. Sakai [Sak98].

Example B.2.29. Naturally $\mathscr{L}(\mathcal{H})$ and $\mathbb{C}\mathrm{I}_{\mathcal{H}}$ are von Neumann algebras in $\mathcal{H}$.

Example B.2.30. If η is a positive and sigma-finite measure on a locally compact space X, then $\mathcal{A} = L^\infty(X, \eta)$ is a commutative unital C^*-algebra (see Example B.2.7). The operator of pointwise multiplication

$$L^\infty(X,\eta) \ni f \mapsto T_f \in \mathscr{L}(L^2(X,\mu)), \quad T_f(\phi) = f\phi,$$

is an isometric ($*$-algebra) morphism. This yields a C^*-algebra isomorphism from $\mathcal{A} = L^\infty(X, \eta)$ onto an abelian von Neumann algebra acting on the separable Hilbert space $L^2(X, \mu)$.

Conversely any abelian von Neumann algebra on a separable Hilbert space may be realised in the way described in Example B.2.30, see Dixmier [Dix96, §I.7.3].

The main example of von Neumann algebras of interest for us is the one associated with a group. This is explained in the next subsection.

B.2.5 Group von Neumann algebra

In this section we follow Dixmier [Dix77, §13]. The main application of these constructions are in Section 1.8.2, see Definition 1.8.7 and the subsequent discussion.

Now, first let us define the (isomorphic) left and right von Neumann algebras of a (Hausdorff) locally compact group G.

The left, resp. right, von Neumann algebra of G is the von Neumann algebra $\mathrm{VN}_L(G)$, resp. $\mathrm{VN}_R(G)$, in $L^2(G)$ generated by the left, resp. right, regular representation. This means that $\mathrm{VN}_L(G)$ *is the smallest von Neumann algebra containing all the operators* $\pi_L(x)$, $x \in G$, where π_L is defined in (B.4), i.e.

$$\pi_L(x)\phi(y) := \phi(x^{-1}y), \quad x, y \in G, \ \phi \in L^2(G).$$

Let us recall that the right regular representation π_R is given by

$$\pi_R(x)\phi(y) = \Delta(x)^{\frac{1}{2}}\phi(yx).$$

Here Δ denotes the modular function (see Definition B.2.10).

One checks easily that the isomorphism U of $L^2(G)$ given by

$$U\phi(y) = \Delta(y)^{\frac{1}{2}}\phi(y^{-1}), \quad \phi \in L^2(G), \ y \in G,$$

intertwines π_L and π_R:

$$\forall x \in G \qquad U\pi_L(x) = \pi_R(x)U.$$

Thus one is sometimes allowed to speak of 'the regular representation' and 'the group von Neumann algebra'. However, in this subsection, we will keep making the distinction between left and right regular representations.

Let us assume that the group G is also separable. In this case, the group von Neumann algebra can be described further.

Clearly $\mathrm{VN}_L(G)$, resp. $\mathrm{VN}_R(G)$, is the smallest von Neumann algebra containing all the operators $(\pi_L)_{\mathscr{D}}(f)$, $f \in C_c(G)$, resp. $(\pi_R)_{\mathscr{D}}(f)$, $f \in C_c(G)$, see [Dix77, §13.10.2]. Here $C_c(G)$ denotes the space of continuous functions with compact support on G. For the definitions of $(\pi_L)_{\mathscr{D}}(f)$ and $(\pi_R)_{\mathscr{D}}$, see (B.5) and (B.1). This easily implies that $\mathrm{VN}_L(G)$, resp. $\mathrm{VN}_R(G)$, is the smallest von Neumann algebra containing all the operators $(\pi_L)_{\mathscr{D}}(f)$, resp. $(\pi_R)_{\mathscr{D}}(f)$, where f runs over $L^1(G)$ or $C^*(G)$.

Applying the commutation theorem (cf. Dixmier [Dix96, Ch 1, §5.2]) to the quasi-Hilbertian algebra $C_c(G)$ ([Eym72, p. 210]) we see that

$$\mathrm{VN}_L(G) = (\mathrm{VN}_R(G))' \quad \text{and} \quad \mathrm{VN}_R(G) = (\mathrm{VN}_L(G))'.$$

See Definition B.2.1 for the definition of the commutant. This implies

Proposition B.2.31. *The group von Neumann algebra coincides with the invariant bounded operators in the following sense:*

- $\mathrm{VN}_L(G)$ *is the space* $\mathscr{L}_R(L^2(G))$ *of operators in* $\mathscr{L}(L^2(G))$ *which commute with* $\pi_R(x)$*, for all* $x \in G$*,*
- $\mathrm{VN}_R(G)$ *is the space* $\mathscr{L}_L(L^2(G))$ *of operators in* $\mathscr{L}(L^2(G))$ *which commute with* $\pi_L(x)$*, for all* $x \in G$*:*

$$\mathrm{VN}_L(G) = \mathscr{L}_R(L^2(G)) \quad \textit{and} \quad \mathrm{VN}_R(G) = \mathscr{L}_L(L^2(G)).$$

Denoting by J the involutive anti-automorphism on $L^2(G)$ given by

$$J(\phi)(x) := \bar{\phi}(x^{-1})\Delta(x)^{-\frac{1}{2}}, \quad \phi \in L^2(G),\ x \in G,$$

we also have

$$J\ \mathrm{VN}_L(G)\ J = \mathrm{VN}_R(G) \quad \textit{and} \quad J\ \mathrm{VN}_R(G)\ J = \mathrm{VN}_L(G).$$

Under our hypotheses, it is possible to describe the group von Neumann algebra as a space of convolution operators, see Eymard [Eym72, Theorem 3.10 and Proposition 3.27]. In the special case of Lie groups, this is a consequence of the Schwartz kernel theorem, see Corollary 3.2.1 and its right-invariant version.

B.2.6 Decomposition of group von Neumann algebras and abstract Plancherel theorem

The full abstract version of the Plancherel theorem allows us to decompose not only the Hilbert space $L^2(G)$ (thus obtaining the Plancherel formula) but also the operators in $\mathrm{VN}_R(G)$ and $\mathrm{VN}_L(G)$:

Theorem B.2.32 (Plancherel theorem). *We assume that the (Hausdorff locally compact separable) group G is also unimodular and of type I and that a (left) Haar measure has been fixed.*

Then there exist

- *a positive sigma-finite measure μ on $\widehat{G}$,*
- *a μ-measurable field of unitary continuous representations $(\pi_\zeta)_{\zeta\in\widehat{G}}$ of G on the μ-measurable field of Hilbert spaces $(\mathcal{H}_\zeta)_{\zeta\in\widehat{G}}$,*
- *and a unitary map W from $L^2(G)$ onto*

$$\int_{\widehat{G}}^{\oplus} (\mathcal{H}_\zeta \otimes \mathcal{H}_\zeta^*)\, d\mu(\zeta) \equiv \int_{\widehat{G}}^{\oplus} \mathtt{HS}(\mathcal{H}_\zeta)\, d\mu(\zeta),$$

(see Subsection B.1.3)

such that W satisfies the following properties:

1. *If $\phi \in L^2(G)$, then $W\phi = \int_{\widehat{G}}^{\oplus} v_\zeta d\mu(\zeta)$ where each v_ζ is a Hilbert-Schmidt operator on $\mathcal{H}_\zeta$ and we have*

$$WJ\phi = \int_{\widehat{G}}^{\oplus} v_\zeta^*\, d\mu(\zeta), \quad \textit{where} \quad (J\phi)(x) = \bar{\phi}(x^{-1}).$$

2. *For any $f \in L^1(G)$ (or $C^*(G)$), the operators $(\pi_R)_{\mathcal{D}}(f)$ and $(\pi_L)_{\mathcal{D}}(f)$ acting on $L^2(G)$ are transformed via W into the decomposable operators (in the sense of Definition B.1.15) on $\int_{\widehat{G}}^{\oplus}(\mathcal{H}_\zeta \otimes \mathcal{H}_\zeta^*)d\mu(\zeta)$,*

$$W\left\{(\pi_L)_{\mathcal{D}}(f)\right\}W^{-1} = \int_{\widehat{G}}^{\oplus} (\pi_\zeta)_{\mathcal{D}}(f) \otimes \mathrm{I}_{\mathcal{H}_\zeta^*}\, d\mu(\zeta),$$

and

$$W\left\{(\pi_R)_{\mathcal{D}}(f)\right\}W^{-1} = \int_{\widehat{G}}^{\oplus} \mathrm{I}_{\mathcal{H}_\zeta} \otimes (\pi_\zeta^{dual})_{\mathcal{D}}(f)\, d\mu(\zeta).$$

See (B.1) for the notation $(\pi)_{\mathcal{D}}$, and here π_ζ^{dual} denotes the dual representation to π_ζ which acts on $\mathcal{H}_\zeta^*$ via

$$(\pi_\zeta^{dual}(x))v^* : w \mapsto (\pi_\zeta(x^{-1})w, v)_{\mathcal{H}_\zeta}.$$

3. *If T is a bounded operator on $L^2(G)$ which commutes with $\pi_L(x)$, for all $x \in G$, that is, $T \in \mathrm{VN}_R(G) = \mathscr{L}_L(L^2(G))$, then T is transformed via W into a decomposable operator (in the sense of Definition B.1.15) on the Hilbert space $\int_{\widehat{G}}^{\oplus}(\mathcal{H}_\zeta \otimes \mathcal{H}_\zeta^*)d\mu(\zeta)$ of the form*

$$WTW^{-1} = \int_{\widehat{G}}^{\oplus} T_\zeta \otimes \mathrm{I}_{\mathcal{H}_\zeta^*}\, d\mu(\zeta).$$

Conversely any decomposable operator of this type yields an operator in $\mathscr{L}_L(L^2(G))$. Hence we may summarise this by writing

$$\mathrm{VN}_R(G) = \mathscr{L}_L(L^2(G)) = W^{-1}\int_{\widehat{G}}^{\oplus} \mathscr{L}(\mathcal{H}_\zeta) \otimes \mathbb{C}\, d\mu(\zeta)\, W.$$

Similarly

$$\mathrm{VN}_L(G) = \mathscr{L}_R(L^2(G)) = W^{-1}\int_{\widehat{G}}^{\oplus} \mathbb{C} \otimes \mathscr{L}(\mathcal{H}_\zeta^*)\, d\mu(\zeta)\, W.$$

A consequence of Points 1. and 2. is that if $f \in L^1(G) \cap L^2(G)$, then $(\pi_\zeta)_{\mathcal{D}}(f) \in \mathtt{HS}(\mathcal{H}_\zeta)$ for almost every $\zeta \in \widehat{G}$ and

$$Wf = \int_{\widehat{G}}^{\oplus} (\pi_\zeta)_{\mathcal{D}}(f) d\mu(\zeta) \quad \text{thus} \quad \|f\|_{L^2(G)}^2 = \int_{\widehat{G}} \|(\pi_\zeta)_{\mathcal{D}}(f)\|_{\mathtt{HS}(\mathcal{H}_\zeta)}^2 d\mu(\zeta).$$

The measure μ is standard (in the sense of Definition B.1.13, see also Proposition B.2.24) and unique modulo equivalence (see Proposition B.1.17).

Reference for the proof of Theorem B.2.32. For the Plancherel measure being standard, see Dixmier [Dix77, Proposition 18.7.7 and Theorems 8.8.1 and 8.8.2]. For the Plancherel theorem expressed in terms of the canonical fields, see [Dix77, 18.8.1 and 18.8.2]. □

The main application of the above theorem for us is Theorem 1.8.11.

Definition B.2.33. The measure μ is called the *Plancherel measure* (associated to the fixed Haar measure).

A different choice of the Haar measure would lead to a different Plancherel measure. Up to this choice, the Plancherel measure is unique. Proposition B.1.17 then implies that we do not need to specify the choice of a measurable field of continuous representations.

In our monograph, our group Fourier transform and Dixmier's defined in (B.2) and (B.1) respectively, are related via (B.3). This implies that the statement of Theorem B.2.32 remains valid if we replace firstly $(\pi)_{\mathcal{D}}$ with our definition of the group Fourier transform and, secondly, W with the isometric isomorphism

$$\tilde{W} : L^2(G) \to \int_{\widehat{G}}^{\oplus} \mathtt{HS}(\mathcal{H}_\zeta) d\mu(\zeta)$$

given by

$$\tilde{W}\phi := W(\phi \circ \text{inv}) \quad \text{where} \quad \text{inv}(x) = x^{-1}.$$

In particular, if $\phi \in L^2(G)$ then

$$\tilde{W}\phi = \int_{\widehat{G}}^{\oplus} \phi_\zeta d\mu(\zeta), \tag{B.6}$$

and we understand $(\phi_\zeta)_{\zeta \in \widehat{G}}$ as the group Fourier transform of ϕ. If $T \in \mathscr{L}_L(L^2(G))$ then it may be decomposed by

$$\tilde{W}T\tilde{W}^{-1} = \int_{\widehat{G}}^{\oplus} T_\zeta \otimes \mathrm{I}_{\mathcal{H}_\zeta^*} \, d\mu(\zeta),$$

which means that if $\phi \in L^2(G)$ with (B.6), then

$$\tilde{W}(T\phi) = \int_{\widehat{G}}^{\oplus} T_\zeta \phi_\zeta d\mu(\zeta).$$

Theorem B.2.32 is reformulated in Theorem 1.8.11 with our choice of group Fourier transform.

We end this appendix with the following observation. Comparing closely the contents of Chapter 1 and Chapter B, there is a small discrepancy about the separability of Hilbert spaces. Indeed, in Chapter B, all the Hilbert spaces on which the representations act are assumed separable, see Section B.1.1, whereas the separability of the Hilbert spaces is not mentioned in Chapter 1. This leeds however to no contradiction when considering a continuous irreducible unitary representation π of a Hausdorff locally compact separable group G on a Hilbert space $\mathcal{H}_\pi$. Indeed, in this case, this yields a continuous non-degenerate representation of $L^1(G)$ on $\mathcal{H}_\pi$ as in Lemma B.2.18. As $L^1(G)$ is separable [Dix77, §13.2.4] and π is irreducible, one can easily adapt the arguments in [Dix77, §2.3.3] to show that $\mathcal{H}_\pi$ is separable. Consequently, the dual $\widehat{G}$ of a Hausdorff locally compact separable group G may be defined as in Section 1.1 as the equivalence classes of the continuous unitary representations, without stating the hypothesis of separability on the representation spaces.

Schrödinger representations and Weyl quantization

Here we summarise the choices of normalisations and give some relations between the Schrödinger representations π_λ, $\lambda \in \mathbb{R}\backslash 0$, of the Heisenberg group $\mathbb{H}_n$ and the Weyl quantization on $L^2(\mathbb{R}^n)$. Detailed justifications and some proofs are given in Section 6.2.

Euclidean Fourier transform (for $f \in \mathcal{S}(\mathbb{R}^N)$ and $\xi \in \mathbb{R}^N$)

$$\boxed{\mathcal{F}_{\mathbb{R}^N} f(\xi) = (2\pi)^{-\frac{N}{2}} \int_{\mathbb{R}^N} f(x) e^{-ix\xi} dx}$$

Weyl quantization (for $f \in \mathcal{S}(\mathbb{R}^N)$ and $u \in \mathbb{R}^N$)

$$\boxed{\mathrm{Op}^W(a) f(u) = (2\pi)^{-N} \int_{\mathbb{R}^N} \int_{\mathbb{R}^N} e^{i(u-v)\xi} a(\xi, \frac{u+v}{2}) f(v) dv d\xi}$$

The useful convention for abbreviating the expressions below is

$$\sqrt{\lambda} := \mathrm{sgn}(\lambda)\sqrt{|\lambda|} = \begin{cases} \sqrt{\lambda} & \text{if } \lambda > 0, \\ -\sqrt{|\lambda|} & \text{if } \lambda < 0. \end{cases} \tag{B.7}$$

Schrödinger representations (for $(x, y, t) \in \mathbb{H}_n$, $h \in L^2(\mathbb{R}^n)$, and $u \in \mathbb{R}^n$)

$$\boxed{\pi_\lambda(x, y, t) h(u) = e^{i\lambda(t + \frac{1}{2}xy)} e^{i\sqrt{\lambda} y u} h(u + \sqrt{|\lambda|} x)}$$

Notation for the group Fourier transform

$$\pi_\lambda(\kappa) \equiv \widehat{\kappa}(\pi_\lambda) = \int_{\mathbb{H}_n} \kappa(x,y,t)\ \pi_\lambda(x,y,t)^*\ dxdydt$$

Relation between Schrödinger representation and Weyl quantization

$$\pi_\lambda(\kappa) = (2\pi)^{\frac{2n+1}{2}} \mathrm{Op}^W \left[\mathcal{F}_{\mathbb{R}^{2n+1}}(\kappa)(\sqrt{|\lambda|}\,\cdot, \sqrt{\lambda}\,\cdot, \lambda)\right]$$

or, with more details,

$$\begin{aligned} \pi_\lambda(\kappa)h(u) &= \int_{\mathbb{H}_n} \kappa(x,y,t)\ \pi_\lambda(x,y,t)^* h(u)\ dxdydt \\ &= \int_{\mathbb{R}^{2n+1}} \kappa(x,y,t) e^{i\lambda(-t+\frac{1}{2}xy)} e^{-i\sqrt{\lambda}yu} h(u-\sqrt{|\lambda|}x) dxdydt \\ &= (2\pi)^{\frac{2n+1}{2}} \int_{\mathbb{R}^n}\int_{\mathbb{R}^n} e^{i(u-v)\xi} \mathcal{F}_{\mathbb{R}^{2n+1}}(\kappa)(\sqrt{|\lambda|}\xi, \sqrt{\lambda}\frac{u+v}{2}, \lambda) h(v) dv d\xi. \end{aligned}$$

Plancherel formula

$$\int_{\mathbb{H}_n} |f(x,y,t)|^2 dxdydt = c_n \int_{\lambda\in\mathbb{R}\backslash\{0\}} \|\widehat{f}(\pi_\lambda)\|^2_{\mathrm{HS}(L^2(\mathbb{R}^n))} |\lambda|^n d\lambda$$

Explicit symbolic calculus on the Heisenberg group

Here we give a summary of some explicit formulae for symbolic analysis of concrete operators on the Heisenberg group $\mathbb{H}_n$. We refer to Section 6.3.3 for more details. We always employ the convention in (B.7) for $\sqrt{\lambda}$.

Symbols of left-invariant vector fields and the sub-Laplacian

$$\begin{aligned} \pi_\lambda(X_j) &= \sqrt{|\lambda|}\partial_{u_j} &&= \mathrm{Op}^W\left(i\sqrt{|\lambda|}\xi_j\right) \\ \pi_\lambda(Y_j) &= i\sqrt{\lambda}u_j &&= \mathrm{Op}^W\left(i\sqrt{\lambda}u_j\right) \\ \pi_\lambda(T) &= i\lambda \mathrm{I} &&= \mathrm{Op}^W(i\lambda) \\ \pi_\lambda(\mathcal{L}) &= |\lambda|\textstyle\sum_j(\partial_{u_j}^2 - u_j^2) &&= \mathrm{Op}^W\left(|\lambda|\textstyle\sum_j(-\xi_j^2 - u_j^2)\right) \end{aligned}$$

Difference operators

$$\begin{aligned} \Delta_{x_j}|_{\pi_\lambda} &= \tfrac{1}{i\lambda}\mathrm{ad}\,(\pi_\lambda(Y_j)) &&= \tfrac{1}{\sqrt{|\lambda|}}\mathrm{ad}u_j \\ \Delta_{y_j}|_{\pi_\lambda} &= -\tfrac{1}{i\lambda}\mathrm{ad}\,(\pi_\lambda(X_j)) &&= -\tfrac{1}{i\sqrt{\lambda}}\mathrm{ad}\partial_{u_j} \\ \Delta_t|_{\pi_\lambda} &= i\partial_\lambda + \tfrac{1}{2}\textstyle\sum_{j=1}^n \Delta_{x_j}\Delta_{y_j}|_{\pi_\lambda} + \tfrac{i}{2\lambda}\textstyle\sum_{j=1}^n \{\pi_\lambda(Y_j)|_{\pi_\lambda}\Delta_{y_j} + \Delta_{x_j}|_{\pi_\lambda}\pi_\lambda(X_j)\} \end{aligned}$$

Difference operators acting on symbols of left-invariant vector fields

	$\pi_\lambda(X_k)$	$\pi_\lambda(Y_k)$	$\pi_\lambda(T)$	$\pi_\lambda(\mathcal{L})$
Δ_{x_j}	$-\delta_{j=k}$	0	0	$-2\pi_\lambda(X_j)$
Δ_{y_j}	0	$-\delta_{j=k}$	0	$-2\pi_\lambda(Y_j)$
Δ_t	0	0	$-\mathrm{I}$	0

Relation between the group Fourier transform and the λ-symbols

$$\widehat{\kappa}(\pi_\lambda) \equiv \pi_\lambda(\kappa) = \mathrm{Op}^W(a_\lambda) = \mathrm{Op}^W(\tilde{a}_\lambda(\sqrt{|\lambda|}\cdot, \sqrt{\lambda}\cdot))$$

$$\text{with} \quad \begin{array}{rcl} a_\lambda & = & \{a_\lambda(\xi, u) = \sqrt{2\pi}\mathcal{F}_{\mathbb{R}^{2n+1}}(\kappa)(\sqrt{|\lambda|}\xi, \sqrt{\lambda}u, \lambda)\} \\ \tilde{a}_\lambda & = & \{\tilde{a}_\lambda(\xi, u) = \sqrt{2\pi}\mathcal{F}_{\mathbb{R}^{2n+1}}(\kappa)(\xi, u, \lambda)\} \end{array}$$

Difference operators in terms of the Weyl quantization of λ-symbols

$$\begin{array}{rclcl} \Delta_{x_j}\pi_\lambda(\kappa) & = & i\mathrm{Op}^W\left(\frac{1}{\sqrt{|\lambda|}}\partial_{\xi_j} a_\lambda\right) & = & i\mathrm{Op}^W\left(\partial_{\xi_j}\tilde{a}_\lambda\right) \\ \Delta_{y_j}\pi_\lambda(\kappa) & = & i\mathrm{Op}^W\left(\frac{1}{\sqrt{\lambda}}\partial_{u_j} a_\lambda\right) & = & i\mathrm{Op}^W\left(\partial_{u_j}\tilde{a}_\lambda\right) \\ \Delta_t\pi_\lambda(\kappa) & = & i\mathrm{Op}^W\left(\tilde{\partial}_{\lambda,\xi,u} a_\lambda\right) & = & i\mathrm{Op}^W\left(\partial_\lambda\tilde{a}_\lambda\right) \end{array}$$

$$\left(\text{with } \tilde{\partial}_{\lambda,\xi,u} = \partial_\lambda - \frac{1}{2\lambda}\sum_{j=1}^{n}\{u_j\partial_{u_j} + \xi_j\partial_{\xi_j}\}\right)$$

List of quantizations

We refer to Sections 2.2, 5.1.3 and 6.5.1 for the cases of compact, graded, and Heisenberg groups, respectively.

Quantization on compact Lie groups (for $\varphi \in C^\infty(G)$ and $x \in G$)

$$A\varphi(x) = \sum_{\pi \in \widehat{G}} d_\pi \operatorname{Tr}\left(\pi(x)\ \sigma_A(x,\pi)\ \widehat{\varphi}(\pi)\right)$$

with the formula for the symbol

$$\sigma_A(x,\pi) = \pi(x)^*(A\pi)(x)$$

Quantization on general graded Lie groups (for $\varphi \in \mathcal{S}(G)$ and $x \in G$)

$$A\varphi(x) = \int_{\widehat{G}} \operatorname{Tr}\left(\pi(x)\ \sigma_A(x,\pi)\ \widehat{\varphi}(\pi)\right) d\mu(\pi)$$

Symbols of vector fields $\sigma_X(\pi) \equiv d\pi(X) = X\pi(e)$, see (1.22)

In the compact and graded cases, relation with the right-convolution kernel

$$\boxed{A\varphi(x) = \varphi * \kappa_x(x) = \int_G \varphi(y)\kappa_x(y^{-1}x)dy \quad \text{with} \quad \widehat{\kappa_x}(\pi) = \sigma_A(x,\pi)}$$

Quantization on the Heisenberg group (for $\varphi \in \mathcal{S}(\mathbb{H}_n)$ and $g = (x,y,t) \in \mathbb{H}_n$)

$$\boxed{A\varphi(g) = c_n \int_{\mathbb{R}\backslash\{0\}} \mathrm{Tr}\left(\pi_\lambda(g)\ \sigma_A(g,\lambda)\ \widehat{\varphi}(\pi_\lambda)\right) |\lambda|^n d\lambda}$$

and in terms of λ-symbols $a_{g,\lambda} : \mathbb{R}^n \times \mathbb{R}^n \to \mathbb{C}$,

$$\boxed{\sigma_A(g,\lambda) = \mathrm{Op}^W(a_{g,\lambda})} \quad (g \in \mathbb{H}_n,\ \lambda \in \mathbb{R}\backslash\{0\})$$

$$\begin{aligned}
&A\varphi(g) \\
&= c_n' \int_{\mathbb{R}\backslash\{0\}} \mathrm{Tr}\left(\pi_\lambda(g)\mathrm{Op}^W(a_{g,\lambda})\ \mathrm{Op}^W\left[\mathcal{F}_{\mathbb{R}^{2n+1}}(\varphi)(\sqrt{|\lambda|}\,\cdot, \sqrt{\lambda}\,\cdot, \lambda)\right]\right) |\lambda|^n d\lambda \\
&= c_n' \int_{\mathbb{R}\backslash\{0\}} \mathrm{Tr}\left(\mathrm{Op}^W(a_{g,\lambda})\ \mathrm{Op}^W\left[\mathcal{F}_{\mathbb{R}^{2n+1}}(\varphi(g\,\cdot))(\sqrt{|\lambda|}\,\cdot, \sqrt{\lambda}\,\cdot, \lambda)\right]\right) |\lambda|^n d\lambda
\end{aligned}$$

Bibliography

[ACC05] G. Arena, A. O. Caruso, and A. Causa. Taylor formula for Carnot groups and applications. *Matematiche (Catania)*, 60(2):375–383 (2006), 2005.

[ADBR09] F. Astengo, B. Di Blasio, and F. Ricci. Gelfand pairs on the Heisenberg group and Schwartz functions. *J. Funct. Anal.*, 256(5):1565–1587, 2009.

[ADBR13] F. Astengo, B. Di Blasio, and F. Ricci. Fourier transform of Schwartz functions on the Heisenberg group. *Studia Math.*, 214(3):201–222, 2013.

[Ale94] G. Alexopoulos. Spectral multipliers on Lie groups of polynomial growth. *Proc. Amer. Math. Soc.*, 120(3):973–979, 1994.

[AR15] R. Akylzhanov and M. Ruzhansky. Hausdorff-Young-Paley inequalities and L^p-L^q Fourier multipliers on locally compact groups. *arXiv:1510.06321*, 2015.

[Arv76] W. Arveson. *An invitation to C^*-algebras.* Springer-Verlag, New York-Heidelberg, 1976. Graduate Texts in Mathematics, No. 39.

[AtER94] P. Auscher, A. F. M. ter Elst, and D. W. Robinson. On positive Rockland operators. *Colloq. Math.*, 67(2):197–216, 1994.

[Bea77a] R. Beals. Characterization of pseudodifferential operators and applications. *Duke Math. J.*, 44(1):45–57, 1977.

[Bea77b] R. Beals. Opérateurs invariants hypoelliptiques sur un groupe de Lie nilpotent. In *Séminaire Goulaouic-Schwartz 1976/1977: Équations aux dérivées partielles et analyse fonctionnelle, Exp. No. 19*, page 8. Centre Math., École Polytech., Palaiseau, 1977.

[BFKG12a] H. Bahouri, C. Fermanian-Kammerer, and I. Gallagher. Phase-space analysis and pseudodifferential calculus on the Heisenberg group. *Astérisque*, (342):vi+127, 2012.

[BFKG12b] H. Bahouri, C. Fermanian-Kammerer, and I. Gallagher. Refined inequalities on graded Lie groups. *C. R. Math. Acad. Sci. Paris*, 350(7-8):393–397, 2012.

[BG88] R. Beals and P. Greiner. *Calculus on Heisenberg manifolds*, volume 119 of *Annals of Mathematics Studies*. Princeton University Press, Princeton, NJ, 1988.

[BGGV86] R. W. Beals, B. Gaveau, P. C. Greiner, and J. Vauthier. The Laguerre calculus on the Heisenberg group. II. *Bull. Sci. Math. (2)*, 110(3):225–288, 1986.

[BGJR89] O. Bratteli, F. Goodman, P. Jorgensen, and D. W. Robinson. Unitary representations of Lie groups and Gårding's inequality. *Proc. Amer. Math. Soc.*, 107(3):627–632, 1989.

[BGX00] H. Bahouri, P. Gérard, and C.-J. Xu. Espaces de Besov et estimations de Strichartz généralisées sur le groupe de Heisenberg. *J. Anal. Math.*, 82:93–118, 2000.

[BJR98] C. Benson, J. Jenkins, and G. Ratcliff. The spherical transform of a Schwartz function on the Heisenberg group. *J. Funct. Anal.*, 154(2):379–423, 1998.

[BL76] J. Bergh and J. Löfström. *Interpolation spaces. An introduction.* Springer-Verlag, Berlin, 1976. Grundlehren der Mathematischen Wissenschaften, No. 223.

[Bla06] B. Blackadar. *Operator algebras*, volume 122 of *Encyclopaedia of Mathematical Sciences*. Springer-Verlag, Berlin, 2006. Theory of C^*-algebras and von Neumann algebras, Operator Algebras and Non-commutative Geometry, III.

[BLU07] A. Bonfiglioli, E. Lanconelli, and F. Uguzzoni. *Stratified Lie groups and potential theory for their sub-Laplacians*. Springer Monographs in Mathematics. Springer, Berlin, 2007.

[BM95] D. R. Bell and S. E. A. Mohammed. An extension of Hörmander's theorem for infinitely degenerate second-order operators. *Duke Math. J.*, 78(3):453–475, 1995.

[Bon09] A. Bonfiglioli. Taylor formula for homogeneous groups and applications. *Math. Z.*, 262(2):255–279, 2009.

[Bou98] N. Bourbaki. *Lie groups and Lie algebras. Chapters 1–3*. Elements of Mathematics (Berlin). Springer-Verlag, Berlin, 1998. Translated from the French, Reprint of the 1989 English translation.

[BP08] E. Binz and S. Pods. *The geometry of Heisenberg groups*, volume 151 of *Mathematical Surveys and Monographs*. American Mathematical

Society, Providence, RI, 2008. With applications in signal theory, optics, quantization, and field quantization, With an appendix by Serge Preston.

[Bru68] F. Bruhat. *Lectures on Lie groups and representations of locally compact groups.* Tata Institute of Fundamental Research, Bombay, 1968. Notes by S. Ramanan, Tata Institute of Fundamental Research Lectures on Mathematics, No. 14.

[BV08] I. Birindelli and E. Valdinoci. The Ginzburg-Landau equation in the Heisenberg group. *Commun. Contemp. Math.*, 10(5):671–719, 2008.

[Cap99] L. Capogna. Regularity for quasilinear equations and 1-quasiconformal maps in Carnot groups. *Math. Ann.*, 313(2):263–295, 1999.

[CCG07] O. Calin, D.-C. Chang, and P. Greiner. *Geometric analysis on the Heisenberg group and its generalizations*, volume 40 of *AMS/IP Studies in Advanced Mathematics.* American Mathematical Society, Providence, RI, 2007.

[CdG71] R. R. Coifman and M. de Guzmán. Singular integrals and multipliers on homogeneous spaces. *Rev. Un. Mat. Argentina*, 25:137–143, 1970/71. Collection of articles dedicated to Alberto González Domínguez on his sixty-fifth birthday.

[CDPT07] L. Capogna, D. Danielli, S. D. Pauls, and J. T. Tyson. *An introduction to the Heisenberg group and the sub-Riemannian isoperimetric problem*, volume 259 of *Progress in Mathematics.* Birkhäuser Verlag, Basel, 2007.

[CG84] M. Christ and D. Geller. Singular integral characterizations of Hardy spaces on homogeneous groups. *Duke Math. J.*, 51(3):547–598, 1984.

[CG90] L. J. Corwin and F. P. Greenleaf. *Representations of nilpotent Lie groups and their applications. Part I*, volume 18 of *Cambridge Studies in Advanced Mathematics.* Cambridge University Press, Cambridge, 1990. Basic theory and examples.

[CGGP92] M. Christ, D. Geller, P. Głowacki, and L. Polin. Pseudodifferential operators on groups with dilations. *Duke Math. J.*, 68(1):31–65, 1992.

[Che99] C. Chevalley. *Theory of Lie groups. I*, volume 8 of *Princeton Mathematical Series.* Princeton University Press, Princeton, NJ, 1999. Fifteenth printing, Princeton Landmarks in Mathematics.

[Chr84] M. Christ. Characterization of H^1 by singular integrals: necessary conditions. *Duke Math. J.*, 51(3):599–609, 1984.

[Cow83] M. G. Cowling. Harmonic analysis on semigroups. *Ann. of Math. (2)*, 117(2):267–283, 1983.

[CR81] L. Corwin and L. P. Rothschild. Necessary conditions for local solvability of homogeneous left invariant differential operators on nilpotent Lie groups. *Acta Math.*, 147(3-4):265–288, 1981.

[CS01] M. Cowling and A. Sikora. A spectral multiplier theorem for a sublaplacian on SU(2). *Math. Z.*, 238(1):1–36, 2001.

[CW71a] R. R. Coifman and G. Weiss. *Analyse harmonique non-commutative sur certains espaces homogènes.* Lecture Notes in Mathematics, Vol. 242. Springer-Verlag, Berlin, 1971. Étude de certaines intégrales singulières.

[CW71b] R. R. Coifman and G. Weiss. Multiplier transformations of functions on SU(2) and $\sum_2$. *Rev. Un. Mat. Argentina*, 25:145–166, 1971. Collection of articles dedicated to Alberto González Domínguez on his sixty-fifth birthday.

[CW74] R. R. Coifman and G. Weiss. Central multiplier theorems for compact Lie groups. *Bull. Amer. Math. Soc.*, 80:124–126, 1974.

[Cyg81] J. Cygan. Subadditivity of homogeneous norms on certain nilpotent Lie groups. *Proc. Amer. Math. Soc.*, 83(1):69–70, 1981.

[Dav80] E. B. Davies. *One-parameter semigroups*, volume 15 of *London Mathematical Society Monographs.* Academic Press Inc. [Harcourt Brace Jovanovich Publishers], London, 1980.

[DHZ94] J. Dziubański, W. Hebisch, and J. Zienkiewicz. Note on semigroups generated by positive Rockland operators on graded homogeneous groups. *Studia Math.*, 110(2):115–126, 1994.

[DiB02] E. DiBenedetto. *Real analysis.* Birkhäuser Advanced Texts: Basler Lehrbücher. [Birkhäuser Advanced Texts: Basel Textbooks]. Birkhäuser Boston, Inc., Boston, MA, 2002.

[Dix53] J. Dixmier. Formes linéaires sur un anneau d'opérateurs. *Bull. Soc. Math. France*, 81:9–39, 1953.

[Dix77] J. Dixmier. *C^*-algebras.* North-Holland Publishing Co., Amsterdam, 1977. Translated from the French by Francis Jellett, North-Holland Mathematical Library, Vol. 15.

[Dix81] J. Dixmier. *von Neumann algebras*, volume 27 of *North-Holland Mathematical Library.* North-Holland Publishing Co., Amsterdam, 1981. With a preface by E. C. Lance, Translated from the second French edition by F. Jellett.

[Dix96] J. Dixmier. *Les algèbres d'opérateurs dans l'espace hilbertien (algèbres de von Neumann).* Les Grands Classiques Gauthier-Villars. [Gauthier-Villars Great Classics]. Éditions Jacques Gabay, Paris, 1996. Reprint of the second (1969) edition.

[DM78] J. Dixmier and P. Malliavin. Factorisations de fonctions et de vecteurs indéfiniment différentiables. *Bull. Sci. Math. (2)*, 102(4):307–330, 1978.

[DR14a] A. Dasgupta and M. Ruzhansky. Gevrey functions and ultradistributions on compact Lie groups and homogeneous spaces. *Bull. Sci. Math.*, 138(6):756–782, 2014.

[DR14b] J. Delgado and M. Ruzhansky. L^p-nuclearity, traces, and Grothendieck-Lidskii formula on compact Lie groups. *J. Math. Pures Appl. (9)*, 102(1):153–172, 2014.

[DR16] A. Dasgupta and M. Ruzhansky. Eigenfunction expansions of ultradifferentiable functions and ultradistributions. *arXiv:1410.2637. To appear in Trans. Amer. Math. Soc.*, 2016.

[DtER03] N. Dungey, A. F. M. ter Elst, and D. W. Robinson. *Analysis on Lie groups with polynomial growth*, volume 214 of *Progress in Mathematics*. Birkhäuser Boston Inc., Boston, MA, 2003.

[Dye70] J. L. Dyer. A nilpotent Lie algebra with nilpotent automorphism group. *Bull. Amer. Math. Soc.*, 76:52–56, 1970.

[Dyn76] A. S. Dynin. An algebra of pseudodifferential operators on the Heisenberg groups. Symbolic calculus. *Dokl. Akad. Nauk SSSR*, 227(4):792–795, 1976.

[Dyn78] A. Dynin. Pseudodifferential operators on Heisenberg groups. In *Pseudodifferential operator with applications (Bressanone, 1977)*, pages 5–18. Liguori, Naples, 1978.

[Dzi93] J. Dziubański. On semigroups generated by subelliptic operators on homogeneous groups. *Colloq. Math.*, 64(2):215–231, 1993.

[Edw72] R. E. Edwards. *Integration and harmonic analysis on compact groups.* Cambridge Univ. Press, London, 1972. London Mathematical Society Lecture Note Series, No. 8.

[Eym72] P. Eymard. *Moyennes invariantes et représentations unitaires.* Lecture Notes in Mathematics, Vol. 300. Springer-Verlag, Berlin-New York, 1972.

[Feg91] H. D. Fegan. *Introduction to compact Lie groups*, volume 13 of *Series in Pure Mathematics.* World Scientific Publishing Co. Inc., River Edge, NJ, 1991.

[Fis15] V. Fischer. Intrinsic pseudo-differential calculi on any compact Lie group. *J. Funct. Anal.*, 268(11):3404–3477, 2015.

[FMV07] G. Furioli, C. Melzi, and A. Veneruso. Strichartz inequalities for the wave equation with the full Laplacian on the Heisenberg group. *Canad. J. Math.*, 59(6):1301–1322, 2007.

[Fol75] G. B. Folland. Subelliptic estimates and function spaces on nilpotent Lie groups. *Ark. Mat.*, 13(2):161–207, 1975.

[Fol77a] G. B. Folland. Applications of analysis on nilpotent groups to partial differential equations. *Bull. Amer. Math. Soc.*, 83(5):912–930, 1977.

[Fol77b] G. B. Folland. On the Rothschild-Stein lifting theorem. *Comm. Partial Differential Equations*, 2(2):165–191, 1977.

[Fol89] G. B. Folland. *Harmonic analysis in phase space*, volume 122 of *Annals of Mathematics Studies*. Princeton University Press, Princeton, NJ, 1989.

[Fol94] G. B. Folland. Meta-heisenberg groups. In *Fourier analysis (Orono, ME, 1992)*, volume 157 of *Lecture Notes in Pure and Appl. Math.*, pages 121–147. Dekker, New York, 1994.

[Fol95] G. B. Folland. *A course in abstract harmonic analysis.* Studies in Advanced Mathematics. CRC Press, Boca Raton, FL, 1995.

[Fol99] G. B. Folland. *Real analysis.* Pure and Applied Mathematics (New York). John Wiley & Sons Inc., New York, second edition, 1999. Modern techniques and their applications, A Wiley-Interscience Publication.

[FP78] C. Fefferman and D. H. Phong. On positivity of pseudo-differential operators. *Proc. Nat. Acad. Sci. U.S.A.*, 75(10):4673–4674, 1978.

[FR66] E. B. Fabes and N. M. Rivière. Singular integrals with mixed homogeneity. *Studia Math.*, 27:19–38, 1966.

[FR13] V. Fischer and M. Ruzhansky. Lower bounds for operators on graded Lie groups. *C. R. Math. Acad. Sci. Paris*, 351(1-2):13–18, 2013.

[FR14a] V. Fischer and M. Ruzhansky. A pseudo-differential calculus on graded nilpotent Lie groups. In *Fourier analysis*, Trends Math., pages 107–132. Birkhäuser/Springer, Cham, 2014.

[FR14b] V. Fischer and M. Ruzhansky. A pseudo-differential calculus on the Heisenberg group. *C. R. Math. Acad. Sci. Paris*, 352(3):197–204, 2014.

[Fri70] K. O. Friedrichs. *Pseudo-diferential operators. An introduction.* Notes prepared with the assistance of R. Vaillancourt. Revised edition. Courant Institute of Mathematical Sciences New York University, New York, 1970.

[FS74] G. B. Folland and E. M. Stein. Estimates for the $\bar{\partial}_b$ complex and analysis on the Heisenberg group. *Comm. Pure Appl. Math.*, 27:429–522, 1974.

[FS82] G. B. Folland and E. M. Stein. *Hardy spaces on homogeneous groups*, volume 28 of *Mathematical Notes*. Princeton University Press, Princeton, N.J., 1982.

[Går53] L. Gårding. Dirichlet's problem for linear elliptic partial differential equations. *Math. Scand.*, 1:55–72, 1953.

[Gel80] D. Geller. Fourier analysis on the Heisenberg group. I. Schwartz space. *J. Funct. Anal.*, 36(2):205–254, 1980.

[Gel83] D. Geller. Liouville's theorem for homogeneous groups. *Comm. Partial Differential Equations*, 8(15):1665–1677, 1983.

[Gel90] D. Geller. *Analytic pseudodifferential operators for the Heisenberg group and local solvability*, volume 37 of *Mathematical Notes*. Princeton University Press, Princeton, NJ, 1990.

[GGV86] B. Gaveau, P. Greiner, and J. Vauthier. Intégrales de Fourier quadratiques et calcul symbolique exact sur le groupe d'Heisenberg. *J. Funct. Anal.*, 68(2):248–272, 1986.

[GKS75] P. C. Greiner, J. J. Kohn, and E. M. Stein. Necessary and sufficient conditions for solvability of the Lewy equation. *Proc. Nat. Acad. Sci. U.S.A.*, 72(9):3287–3289, 1975.

[Gło89] P. Głowacki. The Rockland condition for nondifferential convolution operators. *Duke Math. J.*, 58(2):371–395, 1989.

[Gło91] P. Głowacki. The Rockland condition for nondifferential convolution operators. II. *Studia Math.*, 98(2):99–114, 1991.

[Gło04] P. Głowacki. A symbolic calculus and L^2-boundedness on nilpotent Lie groups. *J. Funct. Anal.*, 206(1):233–251, 2004.

[Gło07] P. Głowacki. The Melin calculus for general homogeneous groups. *Ark. Mat.*, 45(1):31–48, 2007.

[Gło12] P. Głowacki. Invertibility of convolution operators on homogeneous groups. *Rev. Mat. Iberoam.*, 28(1):141–156, 2012.

[Goo76] R. W. Goodman. *Nilpotent Lie groups: structure and applications to analysis*. Lecture Notes in Mathematics, Vol. 562. Springer-Verlag, Berlin, 1976.

[Goo80] R. Goodman. Singular integral operators on nilpotent Lie groups. *Ark. Mat.*, 18(1):1–11, 1980.

[Gro96] M. Gromov. Carnot-Carathéodory spaces seen from within. In *Sub-Riemannian geometry*, volume 144 of *Progr. Math.*, pages 79–323. Birkhäuser, Basel, 1996.

[GS85] A. Grigis and J. Sjöstrand. Front d'onde analytique et sommes de carrés de champs de vecteurs. *Duke Math. J.*, 52(1):35–51, 1985.

[Hel82] B. Helffer. Conditions nécessaires d'hypoanalyticité pour des opérateurs invariants à gauche homogènes sur un groupe nilpotent gradué. *J. Differential Equations*, 44(3):460–481, 1982.

[Hel84a] B. Helffer. *Théorie spectrale pour des opérateurs globalement elliptiques*, volume 112 of *Astérisque*. Société Mathématique de France, Paris, 1984. With an English summary.

[Hel84b] S. Helgason. *Groups and geometric analysis*, volume 113 of *Pure and Applied Mathematics*. Academic Press Inc., Orlando, FL, 1984. Integral geometry, invariant differential operators, and spherical functions.

[Hel01] S. Helgason. *Differential geometry, Lie groups, and symmetric spaces*, volume 34 of *Graduate Studies in Mathematics*. American Mathematical Society, Providence, RI, 2001. Corrected reprint of the 1978 original.

[HJL85] A. Hulanicki, J. W. Jenkins, and J. Ludwig. Minimum eigenvalues for positive, Rockland operators. *Proc. Amer. Math. Soc.*, 94(4):718–720, 1985.

[HN79] B. Helffer and J. Nourrigat. Caracterisation des opérateurs hypoelliptiques homogènes invariants à gauche sur un groupe de Lie nilpotent gradué. *Comm. Partial Differential Equations*, 4(8):899–958, 1979.

[HN05] B. Helffer and F. Nier. *Hypoelliptic estimates and spectral theory for Fokker-Planck operators and Witten Laplacians*, volume 1862 of *Lecture Notes in Mathematics*. Springer-Verlag, Berlin, 2005.

[Hör60] L. Hörmander. Estimates for translation invariant operators in L^p spaces. *Acta Math.*, 104:93–140, 1960.

[Hör66] L. Hörmander. Pseudo-differential operators and non-elliptic boundary problems. *Ann. of Math. (2)*, 83:129–209, 1966.

[Hör67a] L. Hörmander. Hypoelliptic second order differential equations. *Acta Math.*, 119:147–171, 1967.

[Hör67b] L. Hörmander. Pseudo-differential operators and hypoelliptic equations. In *Singular integrals (Proc. Sympos. Pure Math., Vol. X, Chicago, Ill., 1966)*, pages 138–183. Amer. Math. Soc., Providence, R.I., 1967.

[Hör77] L. Hörmander. The Cauchy problem for differential equations with double characteristics. *J. Analyse Math.*, 32:118–196, 1977.

[Hör03] L. Hörmander. *The analysis of linear partial differential operators. I.* Classics in Mathematics. Springer-Verlag, Berlin, 2003. Distribution theory and Fourier analysis, Reprint of the second (1990) edition [Springer, Berlin; MR1065993 (91m:35001a)].

[How80] R. Howe. On the role of the Heisenberg group in harmonic analysis. *Bull. Amer. Math. Soc. (N.S.)*, 3(2):821–843, 1980.

[How84] R. Howe. A symbolic calculus for nilpotent groups. In *Operator algebras and group representations, Vol. I (Neptun, 1980)*, volume 17 of *Monogr. Stud. Math.*, pages 254–277. Pitman, Boston, MA, 1984.

[HP57] E. Hille and R. S. Phillips. *Functional analysis and semi-groups.* American Mathematical Society Colloquium Publications, vol. 31. American Mathematical Society, Providence, R. I., 1957. rev. ed.

[HR70] E. Hewitt and K. A. Ross. *Abstract harmonic analysis. Vol. II: Structure and analysis for compact groups. Analysis on locally compact Abelian groups.* Die Grundlehren der mathematischen Wissenschaften, Band 152. Springer-Verlag, New York, 1970.

[HS90] W. Hebisch and A. Sikora. A smooth subadditive homogeneous norm on a homogeneous group. *Studia Math.*, 96(3):231–236, 1990.

[Hul84] A. Hulanicki. A functional calculus for Rockland operators on nilpotent Lie groups. *Studia Math.*, 78(3):253–266, 1984.

[Hun56] G. A. Hunt. Semi-groups of measures on Lie groups. *Trans. Amer. Math. Soc.*, 81:264–293, 1956.

[Kec95] A. S. Kechris. *Classical descriptive set theory*, volume 156 of *Graduate Texts in Mathematics.* Springer-Verlag, New York, 1995.

[Kg81] H. Kumano-go. *Pseudodifferential operators.* MIT Press, Cambridge, Mass., 1981. Translated from the Japanese by the author, Rémi Vaillancourt and Michihiro Nagase.

[Kir04] A. A. Kirillov. *Lectures on the orbit method*, volume 64 of *Graduate Studies in Mathematics.* American Mathematical Society, Providence, RI, 2004.

[Kna01] A. W. Knapp. *Representation theory of semisimple groups.* Princeton Landmarks in Mathematics. Princeton University Press, Princeton, NJ, 2001. An overview based on examples, Reprint of the 1986 original.

[Koh73] J. J. Kohn. Pseudo-differential operators and hypoellipticity. In *Partial differential equations (Proc. Sympos. Pure Math., Vol. XXIII, Univ. California, Berkeley, Calif., 1971)*, pages 61–69. Amer. Math. Soc., Providence, R.I., 1973.

[Kol34] A. Kolmogoroff. Zufällige Bewegungen (zur Theorie der Brownschen Bewegung). *Ann. of Math. (2)*, 35(1):116–117, 1934.

[Kor72] A. Koranyi. Harmonic functions on symmetric spaces. In *Symmetric spaces (Short Courses, Washington Univ., St. Louis, Mo., 1969–1970)*, pages 379–412. Pure and Appl. Math., Vol. 8. Dekker, New York, 1972.

[Kra09] S. G. Krantz. *Explorations in harmonic analysis*. Applied and Numerical Harmonic Analysis. Birkhäuser Boston Inc., Boston, MA, 2009. With applications to complex function theory and the Heisenberg group, With the assistance of Lina Lee.

[KS69] A. W. Knapp and E. M. Stein. Singular integrals and the principal series. I, II. *Proc. Nat. Acad. Sci. U.S.A. 63 (1969), 281–284; ibid.*, 66:13–17, 1969.

[Kun58] R. A. Kunze. L_p Fourier transforms on locally compact unimodular groups. *Trans. Amer. Math. Soc.*, 89:519–540, 1958.

[KV71] A. Korányi and S. Vági. Singular integrals on homogeneous spaces and some problems of classical analysis. *Ann. Scuola Norm. Sup. Pisa (3)*, 25:575–648 (1972), 1971.

[Lan60] R. P. Langlands. Some holomorphic semi-groups. *Proc. Nat. Acad. Sci. U.S.A.*, 46:361–363, 1960.

[Ler10] N. Lerner. *Metrics on the phase space and non-selfadjoint pseudo-differential operators*, volume 3 of *Pseudo-Differential Operators. Theory and Applications*. Birkhäuser Verlag, Basel, 2010.

[Liz75] P. I. Lizorkin. Interpolation of weighted L_p spaces. *Dokl. Akad. Nauk SSSR*, 222(1):32–35, 1975.

[LN66] P. D. Lax and L. Nirenberg. On stability for difference schemes: A sharp form of Gårding's inequality. *Comm. Pure Appl. Math.*, 19:473–492, 1966.

[LP61] G. Lumer and R. S. Phillips. Dissipative operators in a Banach space. *Pacific J. Math.*, 11:679–698, 1961.

[LP94] E. Lanconelli and S. Polidoro. On a class of hypoelliptic evolution operators. *Rend. Sem. Mat. Univ. Politec. Torino*, 52(1):29–63, 1994. Partial differential equations, II (Turin, 1993).

[Man91] D. Manchon. Formule de Weyl pour les groupes de Lie nilpotents. *J. Reine Angew. Math.*, 418:77–129, 1991.

[MCSA01] C. Martínez Carracedo and M. Sanz Alix. *The theory of fractional powers of operators*, volume 187 of *North-Holland Mathematics Studies*. North-Holland Publishing Co., Amsterdam, 2001.

[Mel71] A. Melin. Lower bounds for pseudo-differential operators. *Ark. Mat.*, 9:117–140, 1971.

[Mel81] A. Melin. Parametrix constructions for some classes of right-invariant differential operators on the Heisenberg group. *Comm. Partial Differential Equations*, 6(12):1363–1405, 1981.

[Mel83] A. Melin. Parametrix constructions for right invariant differential operators on nilpotent groups. *Ann. Global Anal. Geom.*, 1(1):79–130, 1983.

[Mét80] G. Métivier. Hypoellipticité analytique sur des groupes nilpotents de rang 2. *Duke Math. J.*, 47(1):195–221, 1980.

[Mih56] S. G. Mihlin. On the theory of multidimensional singular integral equations. *Vestnik Leningrad. Univ.*, 11(1):3–24, 1956.

[Mih57] S. G. Mihlin. Singular integrals in L_p spaces. *Dokl. Akad. Nauk SSSR (N.S.)*, 117:28–31, 1957.

[Mil80] K. G. Miller. Parametrices for hypoelliptic operators on step two nilpotent Lie groups. *Comm. Partial Differential Equations*, 5(11):1153–1184, 1980.

[MPP07] M. Mughetti, C. Parenti, and A. Parmeggiani. Lower bound estimates without transversal ellipticity. *Comm. Partial Differential Equations*, 32(7-9):1399–1438, 2007.

[MPR99] D. Müller, M. M. Peloso, and F. Ricci. On local solvability for complex coefficient differential operators on the Heisenberg group. *J. Reine Angew. Math.*, 513:181–234, 1999.

[MR15] M. Mantoiu and M. Ruzhansky. Pseudo-differential operators, Wigner transform and Weyl systems on type I locally compact groups. *arXiv:1506.05854*, 2015.

[MS87] G. A. Meladze and M. A. Shubin. A functional calculus of pseudodifferential operators on unimodular Lie groups. *Trudy Sem. Petrovsk.*, (12):164–200, 245, 1987.

[MS99] D. Müller and E. M. Stein. L^p-estimates for the wave equation on the Heisenberg group. *Rev. Mat. Iberoamericana*, 15(2):297–334, 1999.

[MS11] P. McKeag and Y. Safarov. Pseudodifferential operators on manifolds: a coordinate-free approach. In *Partial differential equations and spectral theory*, volume 211 of *Oper. Theory Adv. Appl.*, pages 321–341. Birkhäuser Basel AG, Basel, 2011.

[Nac82] A. I. Nachman. The wave equation on the Heisenberg group. *Comm. Partial Differential Equations*, 7(6):675–714, 1982.

[Nag77] M. Nagase. A new proof of sharp Gårding inequality. *Funkcial. Ekvac.*, 20(3):259–271, 1977.

[Nik77] S. M. Nikolskii. *Priblizhenie funktsii mnogikh peremennykh i teoremy vlozheniya (Russian) [Approximation of functions of several variables and imbedding theorems].* "Nauka", Moscow, 1977. Second edition, revised and supplemented.

[Nom56] K. Nomizu. *Lie groups and differential geometry.* The Mathematical Society of Japan, 1956.

[NS59] E. Nelson and W. F. Stinespring. Representation of elliptic operators in an enveloping algebra. *Amer. J. Math.*, 81:547–560, 1959.

[NSW85] A. Nagel, E. M. Stein, and S. Wainger. Balls and metrics defined by vector fields. I. Basic properties. *Acta Math.*, 155(1-2):103–147, 1985.

[OR73] O. A. Oleĭnik and E. V. Radkevič. *Second order equations with nonnegative characteristic form.* Plenum Press, New York, 1973. Translated from the Russian by Paul C. Fife.

[Pan89] P. Pansu. Métriques de Carnot-Carathéodory et quasiisométries des espaces symétriques de rang un. *Ann. of Math. (2)*, 129(1):1–60, 1989.

[Paz83] A. Pazy. *Semigroups of linear operators and applications to partial differential equations*, volume 44 of *Applied Mathematical Sciences.* Springer-Verlag, New York, 1983.

[Pon66] L. S. Pontryagin. *Topological groups.* Translated from the second Russian edition by Arlen Brown. Gordon and Breach Science Publishers, Inc., New York, 1966.

[Pon08] R. S. Ponge. Heisenberg calculus and spectral theory of hypoelliptic operators on Heisenberg manifolds. *Mem. Amer. Math. Soc.*, 194(906):viii+ 134, 2008.

[Puk67] L. Pukánszky. *Leçons sur les représentations des groupes.* Monographies de la Société Mathématique de France, No. 2. Dunod, Paris, 1967.

[PW27] F. Peter and H. Weyl. Die Vollständigkeit der primitiven Darstellungen einer geschlossenen kontinuierlichen Gruppe. *Math. Ann.*, 97(1):737–755, 1927.

[Ric] F. Ricci. Sub-Laplacians on nilpotent Lie groups. Unpublished lecture notes accessible on webpage http://homepage.sns.it/fricci/corsi.html.

[Roc78] C. Rockland. Hypoellipticity on the Heisenberg group-representation-theoretic criteria. *Trans. Amer. Math. Soc.*, 240:1–52, 1978.

[Rot83] L. P. Rothschild. A remark on hypoellipticity of homogeneous invariant differential operators on nilpotent Lie groups. *Comm. Partial Differential Equations*, 8(15):1679–1682, 1983.

[RS75] M. Reed and B. Simon. *Methods of modern mathematical physics. II. Fourier analysis, self-adjointness.* Academic Press [Harcourt Brace Jovanovich Publishers], New York, 1975.

[RS76] L. P. Rothschild and E. M. Stein. Hypoelliptic differential operators and nilpotent groups. *Acta Math.*, 137(3-4):247–320, 1976.

[RS80] M. Reed and B. Simon. *Methods of modern mathematical physics. I.* Academic Press Inc. [Harcourt Brace Jovanovich Publishers], New York, second edition, 1980. Functional analysis.

[RT10a] M. Ruzhansky and V. Turunen. *Pseudo-differential operators and symmetries. Background analysis and advanced topics*, volume 2 of *Pseudo-Differential Operators. Theory and Applications.* Birkhäuser Verlag, Basel, 2010.

[RT10b] M. Ruzhansky and V. Turunen. Quantization of pseudo-differential operators on the torus. *J. Fourier Anal. Appl.*, 16(6):943–982, 2010.

[RT11] M. Ruzhansky and V. Turunen. Sharp Gårding inequality on compact Lie groups. *J. Funct. Anal.*, 260(10):2881–2901, 2011.

[RT13] M. Ruzhansky and V. Turunen. Global quantization of pseudo-differential operators on compact Lie groups, SU(2), 3-sphere, and homogeneous spaces. *Int. Math. Res. Not. IMRN*, (11):2439–2496, 2013.

[RTW14] M. Ruzhansky, V. Turunen, and J. Wirth. Hörmander class of pseudo-differential operators on compact Lie groups and global hypoellipticity. *J. Fourier Anal. Appl.*, 20(3):476–499, 2014.

[Rud87] W. Rudin. *Real and complex analysis.* McGraw-Hill Book Co., New York, third edition, 1987.

[Rud91] W. Rudin. *Functional analysis.* International Series in Pure and Applied Mathematics. McGraw-Hill Inc., New York, second edition, 1991.

[RW13] M. Ruzhansky and J. Wirth. On multipliers on compact Lie groups. *Funct. Anal. Appl.*, 47(1):87–91, 2013.

[RW14] M. Ruzhansky and J. Wirth. Global functional calculus for operators on compact Lie groups. *J. Funct. Anal.*, 267(1):144–172, 2014.

[RW15] M. Ruzhansky and J. Wirth. L^p Fourier multipliers on compact Lie groups. *Math. Z.*, 280(3-4):621–642, 2015.

[Saf97] Y. Safarov. Pseudodifferential operators and linear connections. *Proc. London Math. Soc. (3)*, 74(2):379–416, 1997.

[Sak79] K. Saka. Besov spaces and Sobolev spaces on a nilpotent Lie group. *Tôhoku Math. J. (2)*, 31(4):383–437, 1979.

[Sak98] S. Sakai. *C^*-algebras and W^*-algebras.* Classics in Mathematics. Springer-Verlag, Berlin, 1998. Reprint of the 1971 edition.

[See69] R. T. Seeley. Eigenfunction expansions of analytic functions. *Proc. Amer. Math. Soc.*, 21:734–738, 1969.

[Seg50] I. E. Segal. An extension of Plancherel's formula to separable unimodular groups. *Ann. of Math. (2)*, 52:272–292, 1950.

[Seg53] I. E. Segal. A non-commutative extension of abstract integration. *Ann. of Math. (2)*, 57:401–457, 1953.

[Sem03] S. Semmes. An introduction to Heisenberg groups in analysis and geometry. *Notices Amer. Math. Soc.*, 50(6):640–646, 2003.

[Sha05] V. A. Sharafutdinov. Geometric symbol calculus for pseudodifferential operators. I [Translation of Mat. Tr. **7** (2004), no. 2, 159–206]. *Siberian Adv. Math.*, 15(3):81–125, 2005.

[Shu87] M. A. Shubin. *Pseudodifferential operators and spectral theory.* Springer Series in Soviet Mathematics. Springer-Verlag, Berlin, 1987. Translated from the Russian by Stig I. Andersson.

[Sjö82] J. Sjöstrand. Singularités analytiques microlocales. In *Astérisque, 95*, volume 95 of *Astérisque*, pages 1–166. Soc. Math. France, Paris, 1982.

[Ste70a] E. M. Stein. *Singular integrals and differentiability properties of functions.* Princeton Mathematical Series, No. 30. Princeton University Press, Princeton, N.J., 1970.

[Ste70b] E. M. Stein. *Topics in harmonic analysis related to the Littlewood-Paley theory.* Annals of Mathematics Studies, No. 63. Princeton University Press, Princeton, N.J., 1970.

[Ste93] E. M. Stein. *Harmonic analysis: real-variable methods, orthogonality, and oscillatory integrals*, volume 43 of *Princeton Mathematical Series.* Princeton University Press, Princeton, NJ, 1993. With the assistance of Timothy S. Murphy, Monographs in Harmonic Analysis, III.

[Str72] R. S. Strichartz. Invariant pseudo-differential operators on a Lie group. *Ann. Scuola Norm. Sup. Pisa (3)*, 26:587–611, 1972.

[SW71] E. M. Stein and G. Weiss. *Introduction to Fourier analysis on Euclidean spaces.* Princeton University Press, Princeton, N.J., 1971. Princeton Mathematical Series, No. 32.

[SZ02] A. Sikora and J. Zienkiewicz. A note on the heat kernel on the Heisenberg group. *Bull. Austral. Math. Soc.*, 65(1):115–120, 2002.

[Tar78] D. S. Tartakoff. Local analytic hypoellipticity for $\Box_b$ on nondegenerate Cauchy-Riemann manifolds. *Proc. Nat. Acad. Sci. U.S.A.*, 75(7):3027–3028, 1978.

[Tar80] D. S. Tartakoff. The local real analyticity of solutions to $\Box_b$ and the $\bar{\partial}$-Neumann problem. *Acta Math.*, 145(3-4):177–204, 1980.

[Tay81] M. E. Taylor. *Pseudodifferential operators*, volume 34 of *Princeton Mathematical Series.* Princeton University Press, Princeton, N.J., 1981.

[Tay84] M. E. Taylor. Noncommutative microlocal analysis. I. *Mem. Amer. Math. Soc.*, 52(313):iv+182, (Revised version accessible at http://math.unc.edu/Faculty/met/ncmlms.pdf) 1984.

[Tay86] M. E. Taylor. *Noncommutative harmonic analysis*, volume 22 of *Mathematical Surveys and Monographs.* American Mathematical Society, Providence, RI, 1986.

[tER97] A. F. M. ter Elst and D. W. Robinson. Spectral estimates for positive Rockland operators. In *Algebraic groups and Lie groups*, volume 9 of *Austral. Math. Soc. Lect. Ser.*, pages 195–213. Cambridge Univ. Press, Cambridge, 1997.

[Tha98] S. Thangavelu. *Harmonic analysis on the Heisenberg group*, volume 159 of *Progress in Mathematics.* Birkhäuser Boston Inc., Boston, MA, 1998.

[Tre67] F. Treves. *Topological vector spaces, distributions and kernels.* Academic Press, New York, 1967.

[Trè78] F. Trèves. Analytic hypo-ellipticity of a class of pseudodifferential operators with double characteristics and applications to the $\bar{\partial}$-Neumann problem. *Comm. Partial Differential Equations*, 3(6-7):475–642, 1978.

[Vai70] R. Vaillancourt. A simple proof of Lax-Nirenberg theorems. *Comm. Pure Appl. Math.*, 23:151–163, 1970.

[vE10a] E. van Erp. The Atiyah-Singer index formula for subelliptic operators on contact manifolds. Part I. *Ann. of Math. (2)*, 171(3):1647–1681, 2010.

[vE10b] E. van Erp. The Atiyah-Singer index formula for subelliptic operators on contact manifolds. Part II. *Ann. of Math. (2)*, 171(3):1683–1706, 2010.

[Vil68] N. J. Vilenkin. *Special functions and the theory of group representations.* Translated from the Russian by V. N. Singh. Translations of Mathematical Monographs, Vol. 22. American Mathematical Society, Providence, R. I., 1968.

[VK91] N. J. Vilenkin and A. U. Klimyk. *Representation of Lie groups and special functions. Vol. 1*, volume 72 of *Mathematics and its Applications (Soviet Series).* Kluwer Academic Publishers Group, Dordrecht, 1991. Simplest Lie groups, special functions and integral transforms, Translated from the Russian by V. A. Groza and A. A. Groza.

[VK93] N. J. Vilenkin and A. U. Klimyk. *Representation of Lie groups and special functions. Vol. 2*, volume 74 of *Mathematics and its Applications (Soviet Series).* Kluwer Academic Publishers Group, Dordrecht, 1993. Class I representations, special functions, and integral transforms, Translated from the Russian by V. A. Groza and A. A. Groza.

[VSCC92] N. T. Varopoulos, L. Saloff-Coste, and T. Coulhon. *Analysis and geometry on groups*, volume 100 of *Cambridge Tracts in Mathematics.* Cambridge University Press, Cambridge, 1992.

[Wal73] N. R. Wallach. *Harmonic analysis on homogeneous spaces.* Marcel Dekker Inc., New York, 1973. Pure and Applied Mathematics, No. 19.

[Wal92] N. R. Wallach. *Real reductive groups. II*, volume 132 of *Pure and Applied Mathematics.* Academic Press Inc., Boston, MA, 1992.

[Wei72] N. J. Weiss. L^p estimates for bi-invariant operators on compact Lie groups. *Amer. J. Math.*, 94:103–118, 1972.

[Wid80] H. Widom. A complete symbolic calculus for pseudodifferential operators. *Bull. Sci. Math. (2)*, 104(1):19–63, 1980.

[Žel73] D. P. Želobenko. *Compact Lie groups and their representations.* American Mathematical Society, Providence, R.I., 1973. Translated from the Russian by Israel Program for Scientific Translations, Translations of Mathematical Monographs, Vol. 40.

[Zie04] J. Zienkiewicz. Schrödinger equation on the Heisenberg group. *Studia Math.*, 161(2):99–111, 2004.

[Zui93] C. Zuily. Existence globale de solutions régulières pour l'équation des ondes non linéaire amortie sur le groupe de Heisenberg. *Indiana Univ. Math. J.*, 42(2):323–360, 1993.

Permissions

All chapters in this book were first published in QNLG, by Springer by Veronique Fischer and Michael Ruzhansky; hereby published with permission under the Creative Commons Attribution License or equivalent. Every chapter published in this book has been scrutinized by our experts. Their significance has been extensively debated. The topics covered herein carry significant information for a comprehensive understanding. They may even be implemented as practical applications or may be referred to as a beginning point for further studies.

We would like to thank the editorial team for lending their expertise to make the book truly unique. They have played a crucial role in the development of this book. Without their invaluable contributions this book wouldn't have been possible. They have made vital efforts to compile up to date information on the varied aspects of this subject to make this book a valuable addition to the collection of many professionals and students.

This book was conceptualized with the vision of imparting up-to-date and integrated information in this field. To ensure the same, a matchless editorial board was set up. Every individual on the board went through rigorous rounds of assessment to prove their worth. After which they invested a large part of their time researching and compiling the most relevant data for our readers.

The editorial board has been involved in producing this book since its inception. They have spent rigorous hours researching and exploring the diverse topics which have resulted in the successful publishing of this book. They have passed on their knowledge of decades through this book. To expedite this challenging task, the publisher supported the team at every step. A small team of assistant editors was also appointed to further simplify the editing procedure and attain best results for the readers.

Apart from the editorial board, the designing team has also invested a significant amount of their time in understanding the subject and creating the most relevant covers. They scrutinized every image to scout for the most suitable representation of the subject and create an appropriate cover for the book.

The publishing team has been an ardent support to the editorial, designing and production team. Their endless efforts to recruit the best for this project, has resulted in the accomplishment of this book. They are a veteran in the field of academics and their pool of knowledge is as vast as their experience in printing. Their expertise and guidance has proved useful at every step. Their uncompromising quality standards have made this book an exceptional effort. Their encouragement from time to time has been an inspiration for everyone.

The publisher and the editorial board hope that this book will prove to be a valuable piece of knowledge for students, practitioners and scholars across the globe.

Index

www.ingramcontent.com/pod-product-compliance
Lightning Source LLC
Chambersburg PA
CBHW070738190326
41458CB00004B/1220
* 9 7 8 1 6 3 9 8 9 3 2 9 4 *